Advanced Topics in Global Information Management

Volume 4

M. Gordon Hunter
University of Lethbridge, Canada

Felix B. Tan
Auckland University of Technology, New Zealand

IDEA GROUP PUBLISHING
Hershey • London • Melbourne • Singapore

Acquisitions Editor:	Renée Davies
Development Editor:	Kristin Roth
Senior Managing Editor:	Amanda Appicello
Managing Editor:	Jennifer Neidig
Copy Editor:	Sue VanderHook
Typesetter:	Jennifer Neidig
Cover Design:	Integrated Book Technology
Printed at:	Integrated Book Technology

Published in the United States of America by
Idea Group Publishing (an imprint of Idea Group Inc.)
701 E. Chocolate Avenue, Suite 200
Hershey PA 17033
Tel: 717-533-8845
Fax: 717-533-8661
E-mail: cust@idea-group.com
Web site: http://www.idea-group.com

and in the United Kingdom by
Idea Group Publishing (an imprint of Idea Group Inc.)
3 Henrietta Street
Covent Garden
London WC2E 8LU
Tel: 44 20 7240 0856
Fax: 44 20 7379 3313
Web site: http://www.eurospan.co.uk

Advanced Topics in Global Information Management, Volume 4 is part of the Idea Group Publishing series named *Advanced Topics in Global Information Management* Series (ISSN: 1537-9302)

ISBN: 1-59140-468-1
Paperback ISBN: 1-59140-469-X
eISBN: 1-59140-470-3

British Cataloguing in Publication Data
A Cataloguing in Publication record for this book is available from the British Library.

Advanced Topics in Global Information Management

Volume 4

Table of Contents

Preface

The strategic use of information technology for global organizations represents a major component of their business plan. The strategic role involves using information technology to develop products, services, and capabilities to support strategic advantage in a global marketplace. Indeed, this role may support consideration of new business ventures with an international perspective.

Information technology supports current operations. Businesses employ intranets to improve information sharing and collaboration. Customer relationship management systems are used to maintain important contact with individuals or organizations that desire to acquire products and services. The Internet may be employed to communicate with customers, suppliers, and business partners. Web sites may be used to provide generic information, while e-mail may be the vehicle to support entity specific interaction.

Information technology facilitates business innovation. Cross-functional enterprise databases provide access to data to support innovative decision-making. Specific market niches may be identified in order to exploit global market penetration. The Internet may be used to optimize customer interaction. Online customer support may be provided via interactive Web sites. Innovative products and services may be developed. The Internet may be employed for online transaction processing for sales of products and provision of services.

In today's business environment, organizations must employ information technology strategically. Companies need employees with knowledge of and access to the technology. They also need customers and other business partners with access to the technology and the attitude that use of the technology is beneficial to all parties involved. Information technology should also be employed to link people, assets, and ideas. It is important that organizations be as agile as possible in order to respond to their ever-changing environment. Companies need to quickly respond to customers' needs in rapidly changing global markets. Information technology provides the information processing capability to treat masses of customers as individuals.

In order to sustain any competitive advantage gained through the application of information technology, companies must continually seek improvement. They must employ environmental scanning techniques to determine what technology is available. They must be willing to adopt new information technology. They need to know what the competition is planning or doing. The following chapters address some of these important issues. This volume continues the custom established for this series of publications by presenting chapters that take either a global or regional perspective.

GLOBAL THEMES

The chapters in this section are related to investigations that adopt as their starting point a global or international perspective. In Chapter I, Lehmann employs four case vignettes to expand upon the architecture and design principles of international information systems. Pauleen, in Chapter II, reports on lessons learned about the operation of distributed teams functioning across multiple cultural boundaries. In Chapter III, Mason investigates whether knowledge management systems are able to respond to culturally diverse users. Evaristo, in Chapter IV, continues the theme of distributed projects in proposing a cross-cultural distributed project management model. In Chapter V, Lim employs the cultural dimension of power distance to explore how users work with software agents. Huang, Yen, Lin, and Huang, in Chapter VI, propose a conceptual framework for the design of future Internet-based education systems for virtual universities. In Chapter VII, Ganesh, Madanmohan, Jose, and Seshadri discuss strategies for firms involved in B2B marketplaces. Chow, in Chapter VIII, investigates success factors for the use of extranets in the supply chain. Finally, in Chapter IX, Koh, Lee, Yen, and Havelka document how the skill requirements of information systems professionals change throughout their careers.

REGIONAL THEMES

The chapters in this section include investigations that initially involve one or a few countries, which present conclusions that may be applied globally. In Chapter X, Peterson and Kim investigate the perceptions of risk, related to project failure, held by information systems designers in the United States, Japan, and Korea. Karlsbjerg, Damsgaard, and Scheepers, in Chapter XI, present a taxonomy of intranet implementation strategies based upon the experience of firms in Denmark and South Africa. In Chapter XII, Singh, Zhao, and Hu investigate Web site localization by comparing the domestic and Chinese Web sites of companies based in the United States. Kaarst-Brown and Wang, in Chapter XIII, report how small investment firms competed globally while based in the Turks and Caicos Islands. In Chapter XIV, Brown, Stander, Hoppe, Mugera, and Newman compare how the national environment of Singapore and South Africa affects the adoption of Internet banking. Li and Chang, in Chapter XV, employ a multinational China-based company to propose a framework for e-business strategy. In Chapter XVI, Poók and Pence report on the information infrastructures of the Czech Republic, Hungary, Poland, and Slovakia in relation to their joining the European Union. Chen and Ching, in Chapter XVII, survey Taiwan financial service companies' use of customer relationship management systems to gain competitive advantage. Finally, in Chapter XVIII, Liu, Marchewka, and Ku propose a privacy-trust-behavioral model of intention regarding electronic commerce Web sites of companies based in the United States and Taiwan.

Acknowledgments

We thank the authors of the chapters in this volume. Your dedication to the topic of global information management and research expertise has made a significant contribution. We would also recognize the members of Idea Group Inc. for their assistance in this project. We would especially like to thank Dr. Mehdi Kosrow-Pour and Jan Travers for their valuable guidance.

We dedicate this book to our children Robb and Jeff (and Brandi) Hunter; and Shaune, Daniel, and Samantha Tan. You continue to challenge and inspire us as you become the next generation of global citizens.

Thank you all very much.

M. Gordon Hunter
Felix B. Tan
September 2004

Section I:

Global Themes

Chapter I

An Object-Oriented Architecture Model for International Information Systems?

Hans Lehmann, Victoria University of Wellington, New Zealand

ABSTRACT

Analysing four case vignettes in a grounded theory approach, this exploratory paper investigates the architecture and design principles of international information systems. A two-dimensional topology for international information systems — suggested in previous research — was confirmed as a useful architecture paradigm. In its terms, international information systems are configured from two elements: 'Core' systems (common for the whole enterprise) on the one hand and 'Local' systems (different for each site) on the other. The interface between the two is a third component. One case vignette in particular highlights the logical and organisational difficulties in defining these systems elements. Object orientation as the fundamental design principle is investigated as an approach to provide a solution for this problem. Because it enables implementation differentiation and flexibility for future functional changes, it is conjectured that object technology is an optimal — technical — development strategy for international information systems. Directions for further research are outlined.

INTRODUCTION

The notion that globalisation is the only key to survival in a rapidly shrinking world has been a hackneyed cliché for many businesses since the early 1980s. Equally, the pivotal importance of information technology as a key business driver has not been seriously questioned in 30 years. Yet the obvious fusion of these two truisms, the application of information technology throughout global operations, is still widely ignored by academics (Gallupe & Tan, 1999) and largely misunderstood by practitioners. As a result, international information systems projects over the last 20 years have often been downright disastrous. Research into why these applications are difficult and how they could be mastered should be of high priority, but is not: the ABI/INFORM database lists 32,919 papers with "information systems" as a keyword between 1985 and 2000. For the same time period, keywords to do with international information systems occur in 234 papers, i.e., in two-thirds of one percent.

Furthermore, the sparse research efforts by the academic community have been sporadic and dispersed over many, disjointed and often irrelevant topics (Palvia, 1998; Gallupe & Tan, 1999). This has left the field devoid of a firm theoretical base and framework from which to advise practitioners and to direct further applied research.

This exploratory paper validates a generic architecture common to international systems. Selecting and building an appropriate IT architecture is considered an important building block for the successful development of any complex system (Earl, 1989). Because such an *a priori* architecture simplifies the design process, it has the potential to make the development of international information systems faster and less risky.

A two-dimensional topology of systems' elements has been postulated some time ago. Using three case vignettes from the author's own experience in conjunction with a more recent, large case validates this two-dimensional architecture model. It also points to the difficulties with the definition of each of the topology dimensions as a major factor in the failure of IIS projects. The difficulties stem mainly from the fact that it is organisationally difficult — and politically sensitive — to achieve agreement for what systems and processes should be under central control and what may remain within local autonomy. Object orientation, with its emphasis on accommodating disparities in common entities, seems a suitable vehicle to deal with the 'local' variations to 'central' systems elements required by the architecture model.

Validation of the architecture model establishes a useful framework for further research into the nature of IIS. Combining the two-dimensional topology with the notion of object-oriented analysis, design and development of IIS provides a method for building IIS which is clearer, would avoid failure through destructive politics and thereby removes a large portion of the risk associated with these systems. It furthermore establishes the flexibility required for the ever-changing information systems and technology environment within multinational enterprises.

The paper is organised as follows: After a review of the (sparse) literature, previous research into the architecture of IIS is summarised. Next, the methodological backdrop to the approach used in the study — qualitative research, using Grounded Theory principles, especially the use of ideational concepts for sketching out a theory and its validation by re-casting these concepts — is introduced. The case vignettes are then described and their IIS structure expressed in terms of the postulated architecture model. Finally, the use if the architecture model as a framework for analysing, designing and building IIS is brought out and the benefits of an object-oriented approach are set out and demonstrated on examples drawn from the vignettes.

'INTERNATIONAL INFORMATION SYSTEMS' IN THE LITERATURE

The literature does not clearly identify a generally accepted term for information systems technology applied across borders. Often "global" is used (e.g., by Ives et al., 1991), but "transnational" is also in general use (e.g., King et al., 1993, 1999) for such systems. The first inevitably invites associations of vast enterprises covering the planet, whereas "transnational" is open to possible confusion with the precise use of the term coined by Bartlett and Ghoshal (1989) for describing one specific style of a firm's operation in more than one country.

In this paper, therefore, the term "international" is used. Furthermore, to distinguish international information systems (IIS) from other distributed systems, in this paper they are defined as:

The distributed information systems (within one firm) that support similar business activities in highly diverse environments, commonly found across country boundaries.

Most reviews of international information systems in the literature tend to agree that past research into IIS is sporadic and spread over a wide array of topics. Only in the last few years have researchers begun to direct their attention to the design and development of IIS. Some of this recent research focuses on the structure and architecture of IIS (e.g., Gibson, 1994; Burn et al., 1996; Targowski, 1996; Grover et al., 1996; Peppard, 1999) and the role of an information technology infrastructure (Weill, 1992, 1993; Weill et al., 1994a & b, 1995, 1998; Broadbent et al., 1997). This increasing interest in IIS, however, seems to be confined to academics in the information system discipline. IISs are still a very peripheral concern for international business researchers. Similarly, there is very little evidence in the practitioner literature.

FRAMEWORKS AND ARCHITECTURE OF INTERNATIONAL INFORMATION SYSTEMS

A number of researchers of IIS architectures use a framework for the classification of enterprises operating in more than one country that was developed by Bartlett and Ghoshal in 1989. Their model focuses on the level of global control versus local autonomy:

- 'Global' means high global control while 'Multinationals' have high local control.
- 'Internationals' are an interim state, transiting towards a balance of local and global.
- 'Transnational' organisations balance tight global control while vigorously fostering local autonomy. This strategy of "think global and act local" is considered optimal for many international operations.

Figure 1. Classification and migration through global business strategies

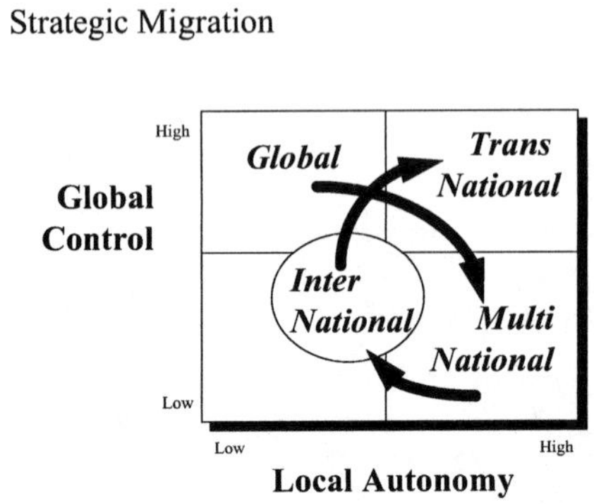

Butler Cox (1991) also put a developmental perspective on the Bartlett and Ghoshal framework. In Butler Cox's terminology, companies seem to become active internationally first as 'Exporter' of their goods or services — usually applying a *'Global'* business strategy. Increased activity in any one location encourages autonomy for local operations, taking on the role of 'National Adapter,' similar to the *'Multinational'* classification. In the next phase this degree of autonomy is counterbalanced by some global control as 'Central Co-ordinator,' i.e., an *'International'* firm. Finally, as global operations mature, firms move toward status of 'Global Co-ordinator' (equivalent to *'Transnational'*). Figure 1 demonstrates the Bartlett and Ghoshal (1989) classification and shows the 'migration paths' suggested by Butler Cox (1991).

A number of researchers have developed models of IIS with a direct, one-to-one relationship between Bartlett and Ghoshal's global business strategies and IT architectures as described in Table 1.

It seems that just as the '*international*' business strategy is an intermediary stage, so are the corresponding global information technology configurations. If these replicated/inter-organisational/intellectually synergy structures are regarded as embryonic 'integrated' architectures, then three mutually exclusive generic architectures could be defined, namely:

- centralised;
- decentralised (including autonomous and independent); and
- integrated.

While the centralised and decentralised structures have been researched over a number of years and are by now well understood, the nature of the 'integrated' architecture has rarely been an object of empirical study. However, the very nature of the systems under investigation makes it possible to establish some common sense, *a-priori* postulates as to the shape of their architecture.

International information systems support common functions across a number of local sites. The common sense deduction from this is the obvious requirement that such

Table 1. Linkage between global business strategy and IT architecture

	Architecture			
Bartlett & Ghoshal (1989)	Butler Cox (1991)	Karimi & Kosynski (1996)	Sankar et al. (1993)	Jarvenpaa et al. (1994[1])
Global	Centralised	Centralisation	Centralised	Headquarters-driven
Multinational	Autonomous	Decentralisation	Decentralised	Independent
International	Replicated	Inter-organisational	*(Undefined)*	Intellectual Synergy
Transnational	Integrated	Integrated	Integrated	Integrated

[1] Their classification is also supported by a series of case studies by Ives et al. (1991, 1992, 1994)

systems would have parts that are common to all sites and other parts which are specific to individual localities. The basis of this concept, i.e., the need for variation in international systems to accommodate differing local circumstances, has been established by Keen et al. as early as 1982, when a paradigm of a 'common core' of information systems applications with 'local' alterations was first articulated. There has been little further development of this model as far as the functionality of application systems is concerned, and what researchers concluded 10 years ago, namely that "the literature offers little guidance for...local versus common applications" (Ives et al., 1991), still holds today.

However, one exception is a set of conjectural papers published in the mid-1990s. Building on 'lived' experience in the development and implementation of IIS, Lehmann (1996, 1997) proposed a two-dimensional topology as an architecture model for international information systems. The topology consists of a 'common core' and 'local variations' of the system, linked together by a 'core/local interface,' as shown in Figure 2. Each local implementation of the IIS consists of the core and a varying configuration of local systems (as shown by the shaded 'Local Variations').

RESEARCH METHODOLOGY

The dearth of IIS research makes qualitative, theory-building methods an appropriate choice. Such methods are well established in organisational research and are becoming accepted in information systems research too (Galliers et al., 1987; Yin, 1989; Lee, 1989; Orlikowski et al., 1991). In particular, Eisenhardt (1989) describes the process of building theory, focusing especially on its inductive nature. In Sociology, Glaser and Strauss (1967) had already developed a specific inductive method, which they termed the 'Grounded Theory' approach, where theory is left to 'emerge' from the data — in which it is 'grounded.' Turner (1983) was one of the first to apply the Grounded Theory approach to management studies. Since 1984, Grounded Theory had been used in a number of business studies (Glaser, 1995). Orlikowski (1993, 1995) has pioneered Grounded Theory in Information System (IS) Research. Yoong (1996) and Atkinson (1997) are recent studies that use the GT approach.

Glaser and Strauss (1967) emphasise that not just large research projects, but also small informal cases and 'anecdotal comparisons,' as long as they are the researcher's 'lived experience,' are a legitimate source for the conceptualisation of theoretical constructs (*ibid.* p. 67). In a later monograph Glaser and Strauss (1994) set out in more explicit terms how to use the Grounded Theory method with case histories. The approach

Figure 2. Conceptual architecture model for international information systems

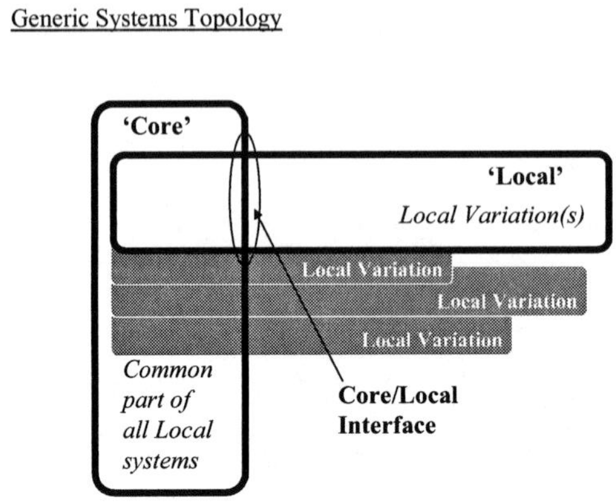

selected for the study of case vignettes in this paper is based on these methodological foundations.

The following sections thus first contain brief descriptions of the case vignettes, followed by an interpretation of the case facts using the postulated generic architecture (shown in Figure 2). Glaser and Strauss (1967) recommend this to see if a theoretical concept 'works,' i.e., if it can—just as well, or better—explain the case from the different perspective introduced by the postulated theoretical construct.

CASE VIGNETTE A: THE LEASING SUBSIDIARY OF AN INTERNATIONAL BANK

This UK firm with branches in the USA, the UK and three European countries was acquired by an Asian international bank to give them a base in the leasing business. The central computer into which all offices were linked was in New York. It soon transpired that systems were inadequate for expansion and the hardware was obsolete. A complete re-design of their information systems technology was therefore needed.

The business in the various countries was firmly limited to installment credit transactions and differed mainly in magnitude. This premise of business being the same did, however, not translate into systems terms. While the lease set-up process was similar, the lease administration part, which makes up more than two-thirds of the system, differed significantly from country to country.

To cope with this diversity, the new international system had two main components:

- A common leasing module dealt with credit and exposure management; it would set up the lease deals and the receivables stream.
- Local country modules would receive the basic data for input into the country-specific receivables and marketing modules, to cope with local languages as well as differing business practices.

Figure 3. Balanced topology of the leasing company

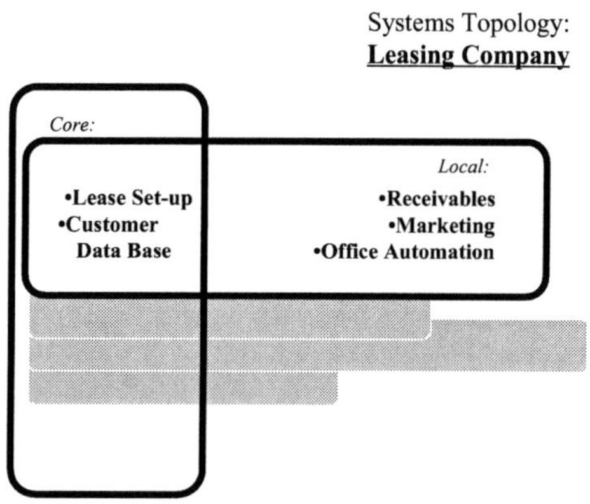

The topology of the new systems was thus characterised by:

- a 'core' of central application software for those key functions which need to be globally controlled,
- freedom for 'local' information technology applications, apart from stringent data interface standards; and
- technical people at the centre and business-oriented applications managers at the 'local' sites.

In terms of global business strategy, the leasing company balances strong global control with wide local autonomy to use the local environment to best advantage. This classifies them as Global Coordinator/*Transnational.*

CASE VIGNETTE B: AN AUSTRALASIAN MERCHANT BANK

This New Zealand-based merchant bank with substantial branches in Australia and London had expanded rapidly. The business of the bank consists mainly of money market dealing, investment banking and stock brokering. Information systems technology at the head office was fragmented and the branches ran odd assortments of software and equipment, loosely linked by public networks. A newly appointed central Treasurer set out to install global systems and controls.

There were two levels of information and systems needs:

- a comprehensive management control system to monitor exposure and risk internationally as well as for each local firm; and
- operations, specifically in the money market and stock broking activities required more systems support.

The resultant international system consisted thus of a narrow 'core' which only contained detailed and stringent data interface and communication standards for input into the Treasurer's own monitoring system. As an incentive to comply with the standards, an electronic mail and bulletin board service was offered on the 'core.' The 'local' systems were seen as entirely composed of the best suited local software, adapting to the diverse environments for money market and stock broking operations, provided they could comply with the data and information interface requirements. The branches would rely entirely on local, external technical and applications support.

The merchant bank gives thus high autonomy to its local offices and imposes only a reporting regime from the centre. It is low on global control and classified as a National Adapter/*Multinational.*

CASE VIGNETTE C: A NEW ZEALAND COMMODITY EXPORTING BOARD

The Board has a virtual monopoly in the purchase of fruit from producers. It owns packing houses and cool stores in New Zealand, operates its own charter fleet and runs a number of European sales offices, while North American sales are controlled through an agency. Asia is an important target for development in the near future.

The Board's systems strategy focuses strong production systems with some marketing modules integrated into them. The international part is, however, the smallest part of the system and consists mainly of common messaging formats. The shipping system is designed to send shipment details and forecasts to the branches who return sales statistics back to the centre. Some of the overseas sales offices use sales order entry systems specific to their environment.

Figure 4. Dominance of local Ssystems in the Merchant Bank's topology

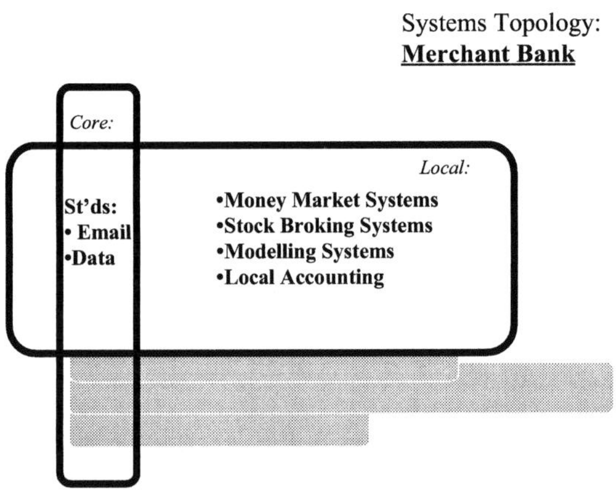

The architecture consists of:

- a large 'core' of extensive and sophisticated systems in the production, logistics and marketing (forecasting and decision-support/modelling) areas; and
- the messages and the independent sales and administration support systems at the branch offices are thus the only 'local' parts in the topology of this international system.

The commodity firm is strong on global control, leaving relatively little autonomy to the local sales offices. This classifies them as Exporter/*Global.*

CASE VIGNETTE D: AN AUSTRALASIAN FOOD PRODUCER'S CO-OP

Similar to the previous vignette, the Australasian Food Producer's Co-operative (the 'Co-op') is a statutory monopoly for the export of its produce from the home country. The UK was its main market until that country joined the EC and severely restricted the Co-op's access. In response, however, the Co-op rapidly built a global network of sales offices. This helped it first to become a global exporter and later, as it included local manufacturing resources, it turned into a truly international operator of considerable size: it ranks 25th among the world's food industry giants. Organised into regional enterprises with great autonomy, it found it difficult, however, to implement the global branding, which its competitors began to use to good effect during the last decade. Some five years ago, a new CEO began a strategic migration towards a 'transnational' global business strategy, balancing the regions' power with a corresponding measure of central control over global product, marketing and branding policy.

Figure 5. Systems topology of the "co-op"

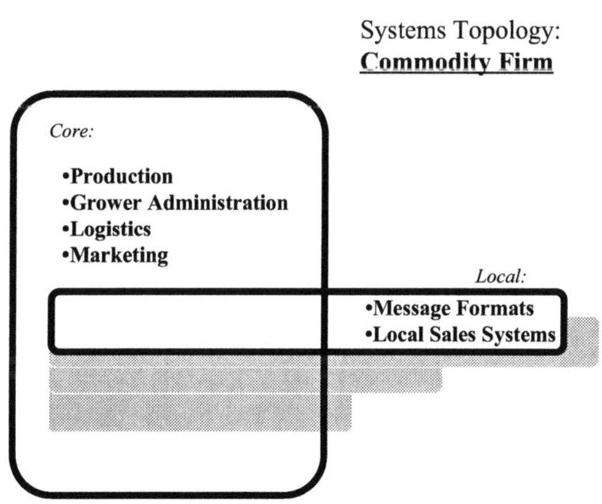

Figure 6. Globally standard IIS proposed by the IT people

The Co-op's IT people misunderstood this move as a return to a 'global' strategy, i.e., one aimed at minimising regional autonomy. For this reason, they set out to implement a globally standard IIS, with little local input and with hardly any information systems under local management — discarding the substantial regional information systems in existence. Figure 6 depicts this architecture.

Regional/local management had a different view of an appropriate IIS for the Co-op. Their vision was based on their existing systems, loosely linked by a common framework for production and marketing planning (Figure 7).

They furthermore interpreted the IT proposal as a covert way to increase central control and resisted this attack on the independence of their local 'fiefdoms' vigorously.

Figure 7. Dominance of local systems in the Merchant Bank's topology

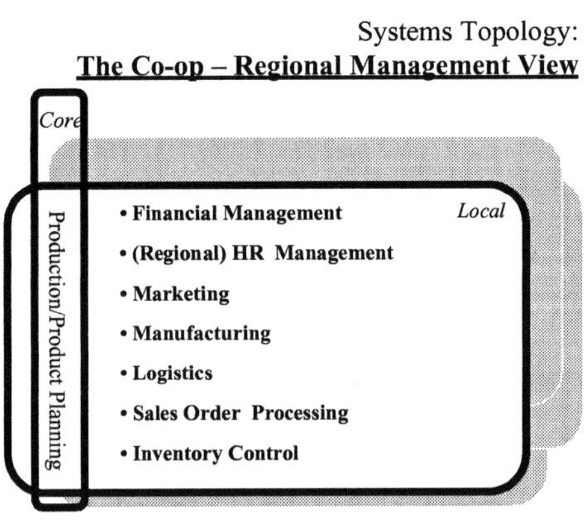

The IT people responded to this strong resistance from regional business management with political maneuvers. This made the communication between the business and IT more and more antagonistic. Because of the diagonally juxtaposed visions of the IIS architecture most appropriate for the Co-op, they went through several iterations of a cycle of business rejection and political reaction from the IT people.

In the end, the CEO terminated the project and re-allocated the team's people and technology resources.

SUMMARY OF THE CASE VIGNETTES

Table 2 summarises the topology and systems element characteristics of the four cases. The usefulness of the architecture model can now be assessed from two viewpoints:

a) as a *framework*; firstly for *describing* the structure of a wide variety of international information systems supporting an equally wide variety of multinational businesses; and, secondly, whether it leads to an enhanced *understanding* of the nature of international information systems; and
b) as a *blueprint* and/or *template* for the *design* of international information systems, especially with regard to its potential for *economies* in the development and/or implementation of international information systems.

The following paragraphs assess and discuss both framework and blueprint perspectives in more detail.

A FRAMEWORK FOR DESCRIPTION AND UNDERSTANDING

The two-dimensional topology is flexible, as a variety of systems architectures can be accommodated. Variety can occur not only in terms of 'core' versus 'local' across the

Table 2. Summary of the case vignettes

Architecture Elements	*LEASING FIRM*	*MERCHANT BANK*	*COMMODITY BOARD*	*CO-OP IT Vision*	*CO-OP Business Vision*
'Core'	'Medium core'; Selective Customer Data Base; Central Leasing Application for 'Deal-Set-up'	'Thin core'; Data & Information standards and communication standards (stringently enforced)	'Large core'; Large central system at the production sites and head office	All key business and control functions would reside and be managed centrally from Co-op HQ	Only central co-ordination functions would be managed by Co-op HQ
'Locals'	Local variations of the Leasing Receivables and Marketing systems	Locally selected technology (mostly out-sourced) for all key business functions	Message formats and protocols for interfacing to smaller, independent local systems	Only local sales support functions would be managed at regional/local level	All key business and control functions would be managed at regional/local level

board, but each 'core'/'local' mix and interface can be defined, maintained and changed in precise response to individual local site requirements

The two-dimensional topology can therefore implement all the architectures cited above:

- *centralised* architectures have a small 'local' part;
- *decentralised* architectures have a small 'core' part;
- *integrated* architectures have a varying 'core' to 'local' ratio for the IIS used in each location.

There is also no problem for describing the migration (proposed by Butler Cox, 1991) between global business strategies in the terminology of the 'core/local' topology:

- The large 'core' of an *Global* firm would
  - shrink as the autonomy of the parts of an *Multinational* increases;
  - re-establishing global control for the *International* would increase the size of the 'core' again and decrease 'local' parts until they reach equilibrium in the *Transnational* with specific 'core/local' ratios for each business unit.

The global business strategy, i.e., the degree of balance between local autonomy and global control, of the case vignettes seems to be correlated to the topology of their international information systems. This confirms the influence of strategy on information systems structure conjectured in the literature (especially Simon & Grover, 1993) and summarised in Table 1. A topology of large 'local' technology, compared to thinner 'core' components, seems to correlate to the higher degree of local autonomy and low level of global control as reflected in the Merchant Bank's organisational strategy. The balance in global vs. local control in the structure of the leasing company is reflected in a medium-to-large 'core' and equally sized 'local' systems. High global control and little autonomy for the branches in the commodity firm describe an architecture characterised by a large 'core' of systems at the centre and only a thin smattering of 'local' systems. The contradictory visions of what would be a suitable IIS structure held by the Co-op's IT people and local/regional management shows up as diametrically opposed. Figure 8 shows the position of the four case vignettes with respect to their topology and global business strategy.

DESIGN ECONOMIES FROM THE MODEL

Designing and developing an international system with the two-dimensional architecture would involve the three generic design domains:

- the 'Core' of common systems (infrastructure and applications);
- the 'Local' systems;
- the 'Core'/'Local Interfaces.'

These systems parts could be developed in three distinct steps and by independent teams within overlapping, or, in the case of 'local' systems acquisition projects, parallel

Figure 8. Link between global business strategy and the architecture (topology) of the international information systems described in the four case vignettes

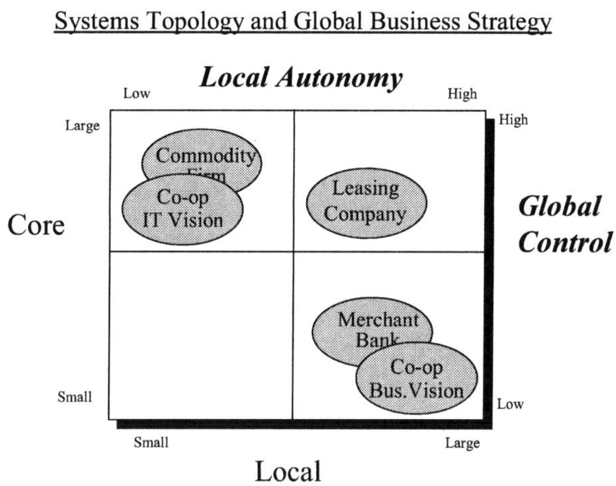

time frames. Firstly, the global business strategy would guide the definition of the 'Core' parts. Determination of the infrastructure elements and delineating which functional applications are to be rigid across local sites would be the major design parameters. This would be the foundation for the definition of 'Local' systems that have to complement 'core' applications. Conversely, this would also define which 'Local' applications should follow entirely their own, individual, specifications — albeit within the confines of the technology infrastructure. The definitions of the 'Core' and 'Local' functional content would then allow a detailed specification of the 'Core/Local Interfaces,' in terms of technical standards for the infrastructure elements and in terms of data/information standards for the functional application systems parts. This process, with its high degree of parallelism, would:

- break down and simplify the design task, reducing complexity;
- spread, and thereby reduce, the development risk; and consequently
- increase the predictability of project outcomes.

Two steps in this design process are crucial: first the definition of the 'Core' (infrastructure as well as functionality) and then the subsequent determination of the 'complementing Local' applications. Despite guidance from the global business strategy, designing and specifying the common parts of international information systems in the operational detail required for building and installing them turns out to be very difficult to do in practice. Equally tricky can be the determination of where, and how, the complementing Local applications differ from the 'Core' functionality. The following paragraphs discuss this process and suggest a direction for simplifying and error-proofing this process.

TRADITIONAL APPROACHES TO THE DEFINITION OF GLOBAL AND LOCAL SYSTEM ELEMENTS

CSC (1995) — in a case study of the way in which a number of multinational firms deal with the issues of global information systems versus local business requirements — summarises that building systems to satisfy a multiplicity of diverse business needs can take two different forms:

- The 'core' system is formed around the '*lowest common denominator*' of all the requirements (i.e., the sum of all local business system needs) in system building terms; however, this can be a disappointingly small proportion of the overall information system.
- The opposite stratagem, in CSC terms '*the grand design,*' attempts to specify a system which contains all the requirements of all local and global business units and agglomerates them into one information system; in mathematical terms this may be called the 'lowest common multiple' — and just as such a number can be alarmingly large, so can information systems built along this principle; some of the more spectacular information systems failures fall into this category: during the systems development time, the business changed so much that there could never be a 'final version' of the software.

In mathematics, however, there is a third possible stratagem for finding common elements among divergent number sets — multiples of common prime factors. In systems terms, these would be 'components' in the form of building blocks that would be used to assemble systems. The 'components' would carry the *global* standards, but their assembly could then follow individual *local* requirements. Information systems built in this way would satisfy both 'common' and 'local' needs and would avoid the conflicting trade-off stance altogether.

Such 'prime factors' for the establishment of global commonality can be implemented in three ways:

1. as infrastructure to enable common basic applications (such as email in the case vignette of the *Merchant Bank*); in this way, global standards are implemented in a form which would be immediately useful for the local business unit;
2. as a 'design template,' i.e., a set of design outlines and specifications for the global standard part of an application, from which the individual local systems can be built; the case vignette of the *Leasing Subsidiary* is a variation on this theme, with the template fairly firmly embedded in actual software templates;
3. as software components.

Both design templates and actual software components will consist of data and processes—which clearly and unambiguously define the resulting conglomeration as an *Object*.

Figure 9. Standard LEASE-SETUP 'message' acts on all objects, taking country variations into account

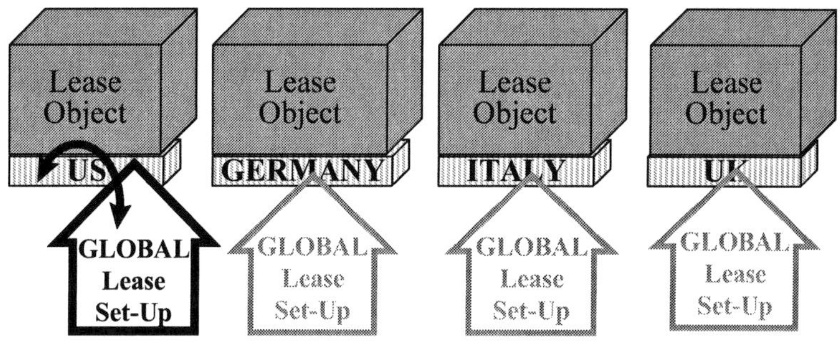

Figure 10. PAYMENT transaction is applied differently in each country although the accounting module/object is a global standard

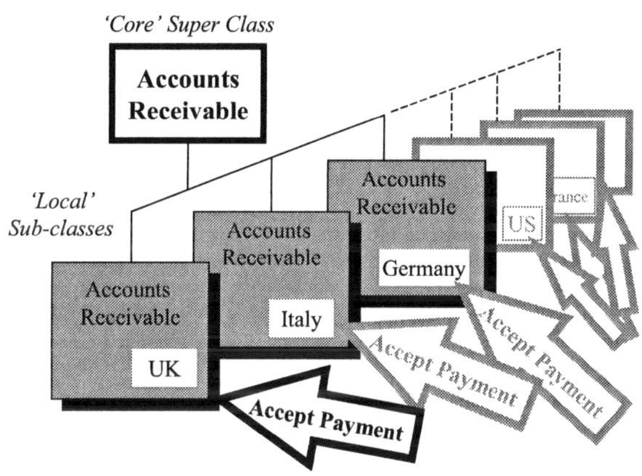

THE BENEFIT OF 'OBJECT' QUALITIES FOR 'CORE' SYSTEMS ELEMENTS

Three key qualities of object orientation with respect to the common/local issue in international information systems are discussed:

1. Objects are defined as *encapsulating* both data and processes/functions in one unit. This combination makes them very useful for vehicles of 'global'; standards, incorporating both data/information standards as well as 'prescribing' standard ways of operating.
2. Objects communicate with other objects using 'messages.' *Polymorphism*, defined as the capability of objects to deal differently with identical messages, is an essential quality for implementing 'local' requirements onto standard processes.

Examples are:

(a) The application of direct/sales taxes where the local applicability rules, rates, etc. are kept with the local accounting object, reflecting the different local statutory and tax regimes.

(b) If the 'Lease-Set-up' part of global standard information system in the case vignette *Leasing Subsidiary* were an object-oriented system, such a globally enforced message would have been the compulsory calculation and reporting of the profitability of each lease deal. Again, the processing modules in the 'message' would act on 'local' and individually different data (e.g., local cost-of-money rates, tax-rebate rules, etc.). Figure 8 depicts this.

3. *Inheritance* is the quality of objects to structure themselves hierarchically into 'super-classes and sub-classes pass 'down' characteristics (data and/or processes). This has two main uses in the global/local dichotomy:

(a) Consider 'Payments' transactions in the *Leasing Subsidiary* case vignette: while the gist of payment processing (application into a ledger, cash-book/bank reconciliations, etc.) is common, the operational detail of the payment process is not; each 'local' object would inherit the common 'Core' processes from a standard Accounts Receivable module (the 'super-class'), but implement typical local payment types (e.g., Direct Debits in the UK, Bank-Account-Transfers in Germany, negotiable promissory notes in Italy, etc.) in the 'Local' sub-class (denoted as the white squares); Figure 10 illustrates this.

(b) The second use would be the introduction of new functionality across the organisation such as new or updated global standards or new operational software developed in one site but potentially useful elsewhere. In the *Co-op* case vignette, the UK subsidiary developed a system of vendor-managed-inventory with a large supermarket chain who would pay for goods sold on the basis of their own point-of-sale records, without any orders or invoices involved (the white 'VMI' box in the 'UK' sub-class. Implemented in object-oriented form, this functionality could have been incorporated as an attribute in the super class of the Sales & Inventory object. In this way, this functional module would have been instantly available to all other 'Local' sub-classes through *inheritance* from the super-class of the Sales & Inventory object. In the actual case, it was the Malaysia local office that used it to strong competitive advantage. This is illustrated in Figure 11.

The advantages of using an object-oriented approach to the design/definition of the common and local parts of an international information system are, however, not restricted to the building of the system. As Butler Cox (1991) postulates, the business style of multinational enterprises is fluid and changes with their development. Moreover, King and Sethi (1993) demonstrated that multinational enterprises are hardly ever homogenous — they work at the same time in different modes and at differing degrees of 'penetration' into the 'Local' systems of different countries (e.g., applying a 'global' style in small subsidiaries and a 'transnational' style in larger, more sophisticated local environments). The ease and flexibility with which an object-oriented information systems architecture can be maintained and changed would certainly seem to make such an object-oriented approach an essential design consideration.

Figure 11. Once the vendor-managed-inventory (VMI) system functionality (developed locally in the UK) becomes part of the Sales&Inventory *super-class, it is instantly available to any 'local' sub-class — here demonstrated on Malaysia*

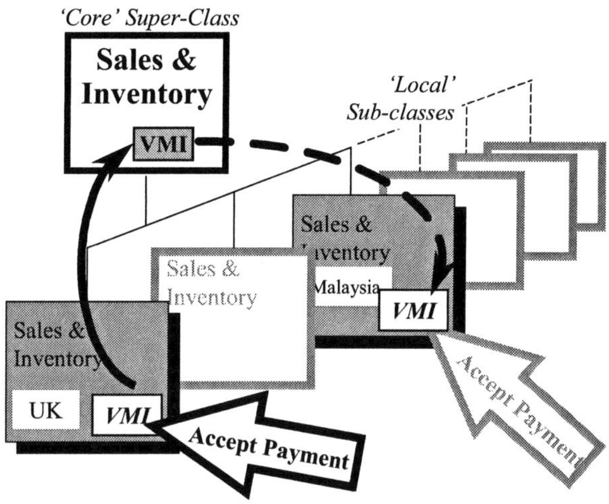

CONCLUSION

Previous research into the nature of international information systems has led to the proposition of a generic architecture paradigm, consisting of a two-dimensional topology. The model configures any international information system as consisting of a 'Core' of information systems common for all and 'Local' systems used in subsidiary sites. This generic structure model has been confirmed by applying it to the information systems of a set of four different multinational enterprises. It could be shown to be a practical and flexible tool to describe and understand the nature of international firms' information systems. It may therefore be concluded that the postulated architecture model may well be of use as a conceptual framework for future research into the structure of international information systems.

Because the architecture model prescribes a way of structuring international information systems, it could also have significant implications for the modus of developing them. Using the notion of the two-dimensional topology, linked by a standard interface, as a framework for the building and implementation of international information systems would allow in the first instance a systematic accumulation of a body of knowledge about this process. Secondly, it enables a modular and parallel systems building approach, as 'Core' and 'Local' systems and their interfaces could all be developed with significant independence. This could make the development of international information systems more predictable, shorter and less risky. The two-dimensional structure *eo ipso* could also provide for in-built flexibility for gradual future enhancement.

The 'goodness' of the design for an international information system seems to hinge on how well the 'Core' systems (technology and applications) are designed, as this determines to a large extent how easy it will be to apply, maintain and change the global standards of the enterprise. 'Local' systems — and their interfaces — are contingent on, and complement the 'Core's' technology. Object Orientation, as a base paradigm for the design of 'Core' elements, was found useful. In particular, the principles of *encapsulation, polymorphism* and *inheritance* ensure that the 'Core' systems can be implemented in differing degrees of 'penetration.' Furthermore, an object-oriented design is supposed to foster ease of maintenance, enhancements and other changes as future business needs and the evolution of the international firm itself dictate. It is conjectured that Object Orientation should be the preferred modus of analysis, design and development for international information systems.

To be of practical use, however, this architecture model now needs to be validated on a much larger, diversified and altogether more representative scale. More empirical research aimed at analysing the structure and architecture of international information systems, for which a Grounded Theory approach seems appropriate, is needed.

REFERENCES

Applegate, L.M., McFarlan, F.W., & McKenney, J.L. (1996). *Corporate information systems management: Text and cases* (Chapter 12, 4th ed.). Chicago, IL: Irwin.

Atkinson, G.J. (1996). *A study of perception of individual participants of a client group undertaking a series of meetings supported by a group support system.* Unpublished Doctoral Thesis, Curtin University of Technology, Perth, W.A., Australia.

Bartlett, C.A., & Ghoshal, S. (1989). *Managing across borders: The transnational solution.* Boston: Harvard Business School Press.

Broadbent, M., & Weill, P. (1997). Management by maxim: How business and IT managers can create IT infrastructures. *Sloan Management Review, 38*(3), 77-92.

Burn, J.M., & Cheung, H.K. (1996). Information systems resource structure and management in multinational organisations. In Palvia, P.C., Palvia, S.C., & Roche, E.M. (Eds.), *Global information technology and systems management - Key issues and trends* (pp. 293-324). Ivy League Publishing.

Butler Cox (1991). *Globalisation: The information technology challenge.* Amdahl Executive Institute Research Report. London.

Collins, R.W., & Kirsch, L. (1999). Crossing boundaries: The deployment of global IT solutions. *Practice-Driven Research in IT Management Series*™, Cincinnati, OH.

CSC Research and Advisory Services. (1995). *Globalisation and localisation: Implications for IS.* CSC Foundation Final Report. London.

Deans, C.P., & Karwan, K.R. (Eds.) (1994). *Global information systems and technology: Focus on the organisation and its functional areas.* Hershey, PA: Idea Group Publishing.

Earl, M.J. (1989). *Management strategies for information technology.* London: Prentice-Hall.

Eisenhardt, K.M. (1989). Building theories from case study research. *Academy of Management Review, 14*(4), 532-550.

Flowers, S. (1996). *Software failure: Management failure.* New York: John Wiley & Sons.

Galliers, R.D., & Land, F.F. (1987). Choosing appropriate information systems research methodologies. *Communications of the ACM, 30*(11), 900-902 .

Gallupe, R.B., & Tan, F.B. (1999). A research manifesto for global information management. *Journal of Global Information Management, 7*(3), 5-18.

Gibson, R. (1994). Global information technology architectures. *Journal of Global Information Management, 2*(1), 28-38.

Glaser, B.G. (Ed.) (1995). *Grounded theory: A reader* (Vols. I and II). Mill Valley, CA: Sociology Press.

Glaser, B.G., & Strauss, A.L. (1967). *The discovery of grounded theory.* Hawthorne, NY: Aldine Publishing.

Glaser, B.G., & Strauss, A.L. (1994). Case histories and case studies. (Reprinted from Glaser, B.G., & Strauss, A.L. (1970). *Anguish.* Mill Valley, CA: Sociology Press). In Glaser, B.G. (Ed.), *More grounded theory methodology: A reader* (pp. 233-245). Mill Valley, CA: Sociology Press.

Glass, R.L. (1992). *The universal elixir and other computing projects which failed.* Bloomington, IN: Computing Trends.

Glass, R.L. (1998). *Software runaways.* Upper Saddle River, NJ: Prentice Hall.

Grover, V., & Segars, A.H. (1996). The relationship between organizational characteristics and information system structure: An international survey. *International Journal of Information Management, 16*(1), 9-25.

Hamelink, C.J. (1984). *Transnational data flows in the information age.* Lund, Sweden: Student-litterature AB.

Ives, B., & Jarvenpaa, S.L. (1991). Applications of global information technology: Key issues for management. *MIS Quarterly, 15*(1), 33-50.

Ives, B., & Jarvenpaa, S.L. (1992). Air Products and Chemicals, Inc. *International Information Systems,* (April), 77-99.

Ives, B., & Jarvenpaa, S.L. (1994). MSAS Cargo International: Global freight management. In Jelass, T., & Ciborra, C. (Eds.), *Strategic information systems: A European perspective.* New York: John Wiley & Sons.

Jarvenpaa, S.L., & Ives, B. (1994). Organisational fit and flexibility: IT design principles for a globally competing firm. *Research in Strategic Management and Information Technology, 1*(39).

Karimi, J., & Konsynski, B.R. (1996). Globalisation and information management strategies. In Deans, C. & Jurison, J. (Eds.), *Information technology in a global business environment* (pp. 169-189). London: Boyd & Fraser.

Keen, P.G.W., Bronsema, G.S., & Auboff, S. (1982). Implementing common systems: One organisation's experience. *Systems, Objectives and Solutions,* 2.

King, W. R., & Sethi, V. (1993). Developing transnational information systems: A case study. *OMEGA International Journal of Management Science, 21*(1), 53-59.

King, W. R., & Sethi, V. (1999). An empirical assessment of the organization of transnational information systems. *Journal of Management Information Systems, 15*(4), 7-28.

Lee, A.S. (1989). A scientific methodology for MIS case studies. *MIS Quarterly, 13*(1), 32-50.

Lehmann, H.P. (1996). Towards a common architecture paradigm for the global application of information systems. *Proceedings of the IFIP WG 8.4 Working Conference on the International Office of the Future,* (April) Tucson, Arizona.

Lehmann, H.P. (1997). A definition of research focus for international information systems. *Proceedings of the 30th Annual Hawaii International Conference on Systems Sciences,* (January) Maui, Hawaii.

Lehmann, H.P. (2000). The fatal politics of multinational information systems: A case study. *Journal of Information Technology Cases & Applications, 2*(3).
Orlikowski, W.J. (1993). CASE tools as organisational change: Investigating incremental and radical changes in systems development. *MIS Quarterly,* (September), 309-337.
Orlikowski, W.J. (1995). *Organisational change around groupware.* Working Paper CCS 186, Massachusetts Institute of Technology.
Orlikowski, W.J., & Baroudi, J.J. (1991). Studying information technology in organisations. *Information Systems Research, 2*(1), 1-28.
Palvia, P. C. (1998). Global information technology research: Past, present and future. *Journal of Global Information Technology Management, 1*(2).
Peppard, J. (1999). Information management in the global enterprise: An organising framework. *European Journal of Information Systems, 8*(2), 77-94.
Sankar, C., Apte, U. & Palvia, P. (1993). Global information architectures: Alternatives and trade-offs. *International Journal of Information Management,* 13, 84-93.
Sethi, V., & Olson, J.E. (1993). An integrating framework for information technology issues in a transnational environment. In *Global Issues in Information Technology.* Hershey, PA: Idea Group Publishers.
Simon, S. J., & Grover, V. (1993). Strategic use of information technology in international business: A framework for information technology application. *Journal of Global Information Management, 1*(2), 29-42.
Targowski, A.S. (1996). *Global information infrastructure - The birth, vision and architecture.* Hershey, PA: Idea Group Publishing.
Tractinsky, N., & Jarvenpaa, S.L. (1995). Information systems design decisions in a global versus domestic context. *MIS Quarterly, 19*(4), 507-534.
Turner, B.A. (1983). The use of grounded theory for the qualitative analysis of organisational behaviour. *Journal of Management Studies, 20*(3), 333-348.
Van den Berg, W., & Mantelaers, P. (1999). Information systems across organisational and national boundaries: An analysis of development problems. *Journal of Global Information Technology Management, 2*(2).
Weill, P. (1992). *The role and value of information technology infrastructure: Some empirical observations.* Working Paper No. 8, University of Melbourne. Melbourne, July.
Weill, P. (1993). The role and value of information technology infrastructure: Some empirical observations. In Banker, R., Kauffman, R., & Mahmood, M.A. (Eds.), *Strategic information technology management: Perspectives on organizational growth and competitive advantage.* Hershey, PA: Idea Group Publishing.
Weill, P., & Broadbent, M. (1994a). Infrastructure goes industry specific. *MIS,* (July), 35-39.
Weill, P., & Broadbent, M. (1998). *Leveraging the new infrastructure: How market leaders capitalize on information technology.* Boston: Harvard Business School Publishing.
Weill, P., Broadbent, M., Butler, C., & Soh, C. (1995). Firm-wide information technology infrastructure investment and services. *Proceedings of the 16th International Conference on Information Systems,* Amsterdam.
Weill, P., Broadbent, M., & St. Clair, D. (1994b). I/T value and the role of I/T infrastructure investments. In Luftman, J. (Ed.), *Strategic alignment.* Oxford University Press.

Yin, R.K. (1989). *Case study research: Design and methods.* Newbury Park, CA: Sage Publications.

Yoong, S. P. (1996). *A grounded theory of reflective facilitation: Making the transition from traditional to GSS facilitation.* Unpublished Doctoral Thesis at the Victoria University of Wellington, New Zealand.

Yourdon, E. (1997). *Death march: The complete software developer's guide to surviving "mission impossible" projects.* Upper Saddle River, NJ: Prentice Hall.

Previously published in the Journal of Global Information Management, 11(3), 1-18, July - September 2003.

Chapter II

Lessons Learned Crossing Boundaries in an ICT-Supported Distributed Team

David J. Pauleen, Victoria University of Wellington, New Zealand

ABSTRACT

Most so-called virtual teams are virtual by degree only. The dichotomy of face-to-face versus virtual teams is a research-led phenomenon. Many teams are using combinations of face-to-face communication and ICT, and can be considered transitional or semi-virtual teams. Many of these teams cross multiple boundaries, including those of time, distance, organization and culture. This paper takes an in-depth look at one such ICT-supported distributed team and investigates some of the key issues that arose as it operated across multiple boundaries, in particular ethnic cultural boundaries. This paper answers recent calls for more in-depth and creative cross-culturally framed information systems research, and the findings suggest that while traditional notions of culture are still valuable for lending insight into team member behavior, team culture may also be seen as emergent and negotiated based on a wide range of contextual elements. The paper concludes with an analysis of the important practical lessons learned in this case. The implications for cross-cultural IS research are briefly discussed.

INTRODUCTION

This paper seeks to contribute to the general field of cross-cultural research in information systems by looking at the issues that arose in a multiple boundary crossing ICT- (information and communication technology) supported distributed team. It responds to the need for more research on how team leaders and members effectively respond to increasingly 'virtual' interactions. The paper takes an in-depth look at how one team leader, who led a transitional ICT-supported team, perceived the wide variety of challenging issues associated with the completion of a critical multi-organizational task in a distributed team environment. This case focuses particular attention on the selection and use of communication channels in building relationships and completing tasks across multiple boundaries, particularly ethnic cultural boundaries. This paper responds to recent calls for more in-depth and creative cross-culturally framed information systems research (Hunter, 2001; Meyers & Tan, 2002).

The case presented in this paper is unique not only for the insights gained into transitional teams and boundary crossing, but also in the nature of the research methodology, grounded action learning. This methodology allowed the team leader in this case the opportunity to reflect on and evaluate his experiences (along with other team leaders). His intention was to assist his organization in gaining a greater understanding of transitional ICT-supported teams and developing strategies to increase their use and effectiveness in the organization while at the same time gauging the consequences of their use across organizational and ethnic cultures.

This section will look at the use of ICT in organizational teams, the notion of cross-cultural relationship building, and leadership in ICT-supported teams. This is followed by a description of the methodology and its appropriateness in cross-culturally framed information systems research. The issues encountered by the team and the team leader's reflections and observations are then presented in a narrative case form, illustrated with the team leader's words. The paper concludes with an analysis highlighting the important practical lessons learned in this case, as well as the implications for the organization featured in this case and other organizations at large. The implications for cross-cultural IS research are briefly discussed.

Transitional ICT-Supported Teams

Fulk and DeSanctis (1995) discussed five key developments that have led to the possibilities of richer, more complex communication in organizations: increased speed of communication, dramatic reduction in communication costs, increased communication bandwidth, vastly expanded connectivity, and integration of communication with computing technologies. These developments have made the use of transitional ICT-supported teams not only possible, but also prevalent in today's organizations.

The term transitional ICT-supported team refers to organizational teams that have traditionally operated in a face-to-face context and are now increasingly engaged in virtual communications supported by a wide range of ICT including e-mail, teleconferencing, mobile technology, the Internet, videoconferencing and the like. These technologies enable team members to communicate more effectively and efficiently. For example, e-mail can allow team members to improve response times; e-mail attachments and Internet/intranet-based Web pages can disseminate information more quickly, and Web-based asynchronous discussion forums can extend face-to-face meetings, host problem-solving discussions and serve as information/knowledge repositories. ICT also allows traditional co-located teams to reach out beyond traditional organizational boundaries to access web-based information and network more freely with a wider variety of people. These virtual or semi-virtual teams often also function across time and distance, which can present significant challenges to team leaders and members, team processes and ultimately team outcomes.

The ability to keep up with technological developments has been identified as a critical factor in organizational success (Anison & Miller, 2002). These transitional teams face a range of challenging issues as they increasingly go virtual, particularly when the team is composed of members from different organizations and cultures. These issues include technology alignment, organizational policies and support and cultural preferences for particular communication channels and team protocols.

One model that might be useful in characterizing transitional team projects considers the degree to which virtual team members are geographically dispersed and the degree to which the organizational affiliation of team members is dispersed (Katzy, Evaristo & Zigurs, 2000). The first dimension is the familiar one of 'place'. The second describes the extent to which team members have the same or different affiliations to organizations, cultures or nations. The authors claim that examining a team project through these dimensions provides interesting insights into differences of the effects of such factors such as organizational culture, communication patterns and skill sets, among others.

Another important issue in ICT-supported teams is how information and communication technologies are chosen in support of team functions. Media choice theories attempt to explain why individuals use certain channels in certain situations and what the outcomes are of such use (Ngwenyama & Lee, 1997; Karahanna, 1995). Information richness theory (Daft & Lengel, 1986) states that different communication channels vary in their capacity to transmit rich information. "Rich" channels, such as face-to-face and video conferencing, enable rapid feedback, the use of varied social cues and personalization of use, helping to resolve equivocality (Ngwenyama & Lee, 1997), while 'lean" channels such as e-mail are more useful for reducing uncertainty (Daft & Lengel, 1986). However, this point of view has come under increasing scrutiny as some argue that media richness is a subjective rather than objective characteristic (Fulk, Schmitz & Steinfield, 1990) or that other media qualities such as synchronicity exert influence on media choice (Dennis, Valacich, Speier & Morris, 1990). These theories and many others (e.g., social presence theory, symbolic interactionism theory, critical mass theory, channel expansion theory, etc.) attempt to predict user choice of communications channels. Some researchers have simply concluded that appropriate media choice is more a function of preference, convenience and cost than of task media fit (Johansen, Vallee & Spangler, 1979). However, none of these theories explicitly or comprehensively deal with the influence of boundary crossing, particularly cultural boundaries, on media choice.

Cross-Cultural Relationship Building

The study of cross-cultural issues in information technology has been described as embryonic (Cukier & Middleton, 1996). Several recent articles have been written summarizing the literature and challenges of cross-cultural information systems research (Hunter, 2001; Karahanna, Evaristo, & Srite, 2002; Meyers & Tan, 2002) and this paper will only provide a brief overview of the subject. Essentially, there are a number of theories and models that have informed cross-cultural research, both methodologically and philosophically. Many of these are centered on the concept of national culture and are based on dichotomies or continuums of values, such as individualism-collectivism (Hofstede, 1980), high and low context (Hall, 1976) and monochronic-polychronic (Lewis, 1996). These value-based models predict individual and group attitudes and behaviors based on national culture. For example, when it comes to relationship building between individuals or groups in a business context, Hofstede's dimensions of uncertainty avoidance would predict that certain cultures would require extensive face-to-face contact between the parties over a long period of time to establish sufficient trust before any serious business can take place, while other cultures would only require a signed contract.

Indeed, the notion of culturally determined relationship building and trust have figured prominently in these cultural models and include models based on high-low trust (Fukuyama, 1995), social action-maintaining relationship (Newman, Summer & Warren, 1977) and merit based-relationship based (Newman et al., 1977) among others. In addition, the concept of relationship building is a chief characteristic of many other cultural models including Hall's high-low context model and Hofstede's uncertainty avoidance model as seen.

Communication, relationship building and the nature and development of trust in virtual teams are intimately connected and have been the subject of an increasing number of studies (e.g., Maznevski & Chudoba, 2000; Jarvenpaa, Knoll, & Leidner, 1998; Pauleen & Yoong, 2001a, 2001b). Stronger relational links have been associated with higher task performance (Warkentin & Beranek, 1999) and the effectiveness of information exchange (Warkentin, Sayeed, & Hightower, 1997). Maznevski and Chudoba (2000) found that deliberately addressing relationship building to develop shared views and trust across all types of boundaries characterize more effective virtual teams. According to Lau, Sarker, and Sahay (2000), effective communication is the key to successful virtual teams. Communication is an essential human behavior and one that is necessarily influenced by culture (Chidambaram, 1992). Differences may include initiation and coordination mechanisms, temporal orientation and modes of reaching decisions (Webb-Collins & Kirsch, 1999). These differences are sure to affect communication among groups from different cultural backgrounds. Cultural disposition toward communication is likely to affect both relationship building and the choice of communication channels.

Leadership in ICT-Supported Transitional Teams

Recent research into virtual teams and team leadership has split along the lines of a face-to-face versus virtual dichotomy, and leadership in transitional or semi-virtual teams remains an area of potentially fruitful research (Griffith & Meader, forthcoming). In the virtual team literature, team leaders are often seen as the nexus of the team, facilitating communications, establishing team processes, and taking responsibility for task completion (Duarte & Tennant-Snyder, 1999), and doing so across multiple boundaries. In virtual teams, technology becomes the crucial and ever-present link between team members (Lipnack & Stamps, 1997), one that team leaders must manage skillfully. Recent research has begun to look at leadership issues in virtual teams. Kayworth and Leidner (2001-2002) point out that leadership roles of social facilitation and communications processing will be even more important in virtual teams than in traditional co-located teams. They also indicate that cross-cultural issues and the need for proficiency with a wide range of ICT present particular challenges in virtual teams. This recent research suggests that the trend toward physically dispersed work groups necessitates further systematic inquiry into the changing role and nature of team leadership. This article looks at how these issues manifest in a transitional or semi-virtual team.

In a recent editorial in the *Journal of Global Information Management*, Hunter (2001) calls for the use of innovative research methods when looking at issues relating to cross-cultural research and the information systems field. The following section briefly explains the grounded action learning methodology used to gather the data for the case and its appropriateness in cross-culturally framed information systems research.

THE CASE

This case features the negotiation of historical claims between the New Zealand government and an indigenous Maori group. It involved many team members, organizations, and cultures, with large amounts of money and land at stake, and all taking place within a historical and political context. Table 1 lists the core Crown team members, the Claimant team members, extended team members and various stakeholders in this case. All together, there were approximately eight members on the core team, ten extended team members and several key stakeholder groups.

The Office of Treaty Settlements (OTS) is a part of the Ministry of Justice of the New Zealand government. The OTS's key responsibility is the settling of historical Treaty of Waitangi[1] claims[2] made by indigenous (Maori) groups. This project involved the negotiation and settlement of Treaty claim by Ngati Mea[3]. The project was essentially a negotiation between two parties, the Crown (represented by the New Zealand Government) and the Claimants (Ngati Mea). This project was unique because although two sides were involved in the negotiation, the Crown and the Claimant side, it was basically a collaborative rather than a confrontational negotiation. So in a real sense the project consisted of two teams working toward a common goal, a settlement. The wider Crown team, which consisted of a number of government departments including OTS, the Department of Conservation, the Ministry of Fisheries, Treasury as well as consultants

Table 1. Extended team membership

Core Team Members (Crown)	**Claimants**	**Extended Team Members**	**Key Stakeholders**
Representatives from Office of Treaty Settlements, Department of Conservation, Treasury	Principal Negotiator, other negotiators appointed by the mandated body	Other departments (Fisheries, Education), Lawyers, Specialists, Consultants	Ministers, Claimant community, Local authority, the New Zealand public

had access to a greater range of virtual technologies. Representatives of the Claimant side generally had the use of phones, fax and e-mail.

Within OTS itself, all permanent staff is located in Wellington. The Office also works with numerous consultants, such as negotiators and environmental specialists, from around the country. There is no policy regarding working virtually, but in practice what happens is that these consultants will come down to Wellington for an initial meeting and then work via phone and e-mail. Once-a-week meetings may occur when negotiations are moving very fast. According to BC (the team leader and research participant featured in this case), his department appears to be generally unaware of the influence that ICT may be currently having in its organization.

METHODOLOGY

The case presented in this paper is one of several cases gathered in a larger research project that focused on the experiences that a group of professional business people in their capacity as virtual team leaders found important. That research was guided by the research question, "What are the issues facing virtual team leaders and how do they manage them?" and involved seven virtual team leaders, each leading their own team within the larger context of their organizations and the rapidly evolving ICT environment. BC's case, featured in this article, is one of seven developed from the larger study. Although presented here as an individual case, it is important to understand that BC's case was an integral part of the larger study, just as BC's and the other team leaders' participation in the study was critical to the action learning program discussed below and the nature of the data collected.

Grounded Action Learning

The research methodology used to gather the data presented in this article derives from a combination of two research methodologies, action learning and grounded theory. An action learning program was used to gather data, while grounded theory techniques were used to analyze the detail. This methodology has been discussed in some detail in two previous papers (Pauleen & Yoong, 2001a, 2001b). Here, the main points of the methodology are briefly summarized.

Action Learning

Action learning is closely linked to action research and is cited as one of the several streams of action research (Lau, 1999). Action learning is described as the process by which groups of people work on real organizational issues and come up with practical solutions that may require changes to be made in the organization (Revans, 1982). Action learning is a practical group learning and problem-solving process where the emphasis is on self-development and learning by doing. The group, known as the action learning set, meets regularly and provides the supportive and challenging environment in which members are encouraged to learn from experience, sharing that experience with others, having other members critique and advise, taking that advice and implementing it, and reviewing with those members the action taken and the lessons that are learned (Margerison, 1988).

To ensure participants had experiences to talk about, a specially designed action learning-based virtual team training program was developed that provided participants with the knowledge and skills to both implement and lead a virtual team as well as the opportunity to be able to reflect on and talk about them. This training program functioned as a learning space for the participants and the researcher, allowing for structured, yet flexible training, semi-structured interviewing and freewheeling discussions. No particular hypothesis was being tested in this research design, but the research question was directed at how virtual team leaders implement and manage virtual teams. The grounded theory approach was expected to produce a set of constructs and a description of their relationships relevant to the experiences of the participants.

Grounded Theory

Traditional grounded theory is a methodology for developing theory that is grounded in data systematically gathered and analyzed in which theory emerges during actual research, doing so through the continuous interplay between analysis and data collection (Strauss & Corbin, 1990). Central features of this analytic approach include the general method of (constant) comparative analysis, theoretical sampling, theoretical sensitivity and theoretical saturation (Glaser & Strauss, 1967).

There have been recent studies in IS that have made use of traditional grounded theory methods and those which have used grounded theory. A number of IS studies, particularly in interpretive inductive studies requiring the development of meaningful categories, have combined various elements of grounded theory with other research methods. The most commonly borrowed elements from traditional grounded theory are the grounded theory coding techniques (open, axial and selective) used to analyze data. The grounded action learning approach used in this study was inspired by the grounded action research method introduced by Baskerville and Pries-Heje (1999) in which data gathered in an action research project was analyzed using grounded theory methods.

Data were collected during the action learning program and included semi-structured interviews and discussions between the researcher and the participants and informal participant reports, as well as the researcher's journal, and to various degrees: participant notes, organizational documentation and copies of electronic conversations, that is, e-mail.

BC and Participation in the Action Learning Program

BC, the participant featured in the case highlighted in this paper, took part in the action learning program. He is a senior policy analyst at the Office of Treaty Settlements (OTS). He can be considered, informally, a leader of his team, a large collaborative-type team that included many government and non-government members. The OTS, in which BC works, is a part of the Ministry of Justice of the New Zealand government and a key stakeholder in the team featured here. BC used the action learning program he participated in to look into the structural, procedural and technical issues of ICT-supported distributed teams, particularly with regards to cross-cultural issues. He was able to discuss his experiences and refine his ideas in discussions with the other action learning participants.

Table 2. Conditions present at start up of BC's virtual team

Key Issues	Related Issues	Specific Conditions
Team Issues	Project Goal Time Frame Team Membership	• Complex - Negotiate a historical Treaty settlement • Long-term with a deadline • Members from different organizations, • Virtual team experience - some with experience of electronic communication channels, others with none
Boundary Crossing	Organizational Culture/language	• Claimants and government; different government departments • Different cultures, different languages on occasions
Organizational Policies and Resources (including technology)	Policies Resources	• Policies on e-mail, Quality Assurance policies on letters sent outside the office • Unaware of what electronic resources were available to Claimants

BC made careful observations and reflections of the Crown officials (representing the New Zealand government) and Claimants' (Ngati Mea) sides and their ongoing interaction. These were focused by the issues raised in the action learning program, and they led BC to a number of insights on the issues that face a leader in a complex, cross-cultural negotiation in which the two sides are essentially striving to achieve a joint objective (BC's observation).

When the training program began in October 1999, Crown negotiations with Ngati Mea were already in progress and working against deadlines. Although BC was a participant in a virtual team study, it was never his intention to attempt a full-scale virtualization of this particular negotiation with Ngati Mea, as he did not believe it was the time to sit down and play with new technology that not everyone was familiar with or, in some cases, equipped with.

However, he had hoped to be able to implement "virtualness" around the edges of the negotiation process, by encouraging the use of virtual communication channels where possible. In the end, he was well placed to observe the part ICT played in this team in relationship to all the various team members and their organizational and ethnic affiliations. For example, in the first two weeks of the action learning program, BC related how he started to become aware of the way communications were being conducted within the negotiation process.

Finally, the grounded action learning methodology allowed both the research subject and the researcher to revisit their findings two years later to investigate whether they were still relevant to the functioning of teams in OTS, particularly in its planning for the selection and use of ICT in furthering the use of transitional or fully virtual teams within the organization and in future negotiations between the Crown and Claimants. This mixed methodology allowed for an in-depth field investigation of the subject and the integration of both the researcher's and the research subject's perceptions and has firmly grounded the data within the research context.

FINDINGS

This case highlights significant issues often found in diverse, boundary-crossing teams: ICT, and organizational and ethnic culture all figure prominently in this project (Table 2). These are discussed in the following section.

Boundary Crossing, Relationship Building and Use of Face-to-Face Communication

The defining condition in this project was the amount of boundary crossing that needed to be addressed–not only the cultural differences between the government and claimants, but also the organizational differences on the government team, which consisted of a number of government departments including the Office of Treaty Settlements (BC's office), the Department of Conservation, the Ministry of Fisheries and the Treasury, as well as numerous consultants and specialists. It was the effect of this boundary crossing in all its aspects that BC made special efforts to evaluate. Although face-to-face communication was critical in these negotiations, several other channels were also used and toward the end of the project, assumed a critical role. This section explains the driving factors in the selection and use of communication channels. These are summarized in Table 3.

Table 3. Driving factors in the selection and the use of communication channels

Communication Channel	Driving Factors for Selection and Use by Crown Side	Driving Factors for Selection and Use by Claimant Side
Face-to-Face	-Get core business done -Discuss and resolve key issues -Get people to do things -Convey negative responses to Claimants -Allow for credible, but inconclusive dialogue	-Build relationships -Get core business done -To raise key issues -Get Crown to focus on Claimant's issues
Letters	-Provide formal official response (must be Quality Assured)	-Provide formal, written records of every step of the process -Move things along and generate responses
Telephone	-Deal with matters under some urgency -To discuss issues informally (before or after sending a formal letter)	-To check on progress and keep up momentum -To discuss issues informally (before or after sending a formal letter)
E-mail	-Used primarily within the Crown side -Inter-departmental email discussion on issues relating to the negotiations. -Confirm and clarify details. -Possesses speed of a phone call with the paper trail of a written response. -Used to quickly transfer working documents back and forth in template form.	-E-mail was not used between OTS and the Claimants until the end of the negotiations when time constraints forced its use.

The reasons for using face-to-face meetings were varied and appeared to be strategically used by both the Crown and the Claimants. At the start of the negotiation, both parties were keen to build a relationship. BC recalled that before the negotiation process officially began, the claimants came down to Wellington and had a face-to-face meeting with the Office of Treaty of Settlement side, although no meeting was required. They reported on what they had been up to and asked a few questions. In retrospect, BC concluded:

I guess they just wanted to meet with us. It was more than was required and more than what any other claimant groups had done. But it built up some sort of rapport. It was just a little bit unusual at the very outset. Face-to-face was much more relevant than might otherwise have been the case. It's the way that Maori operate, by talking and looking at you.

In the initial meetings the Claimants wanted to learn about what the Crown's requirements were for negotiations, and the Crown wanted to consider whether it was feasible to begin negotiations. The two sides discussed the Crown's requirements for mandating negotiators. During these meetings, the claimants would often report on steps that they were taking in working towards formal negotiations. Often, it would not really have been necessary to meet to discuss these matters, but it was useful for the purpose of keeping in touch and building relationships. These face-to-face meetings were a regular aspect of this negotiation and provided an opportunity to bridge cultural differences. By the end of the action learning program, BC was convinced that these face-to-face meetings had been instrumental in building up the high levels of trust that were ultimately needed to successfully conclude the negotiations. He related:

There were two cultures that we needed to bring together, the claimant negotiators and the key government negotiators. Through a whole lot of face-to-face meetings over two years of interaction we built up a huge level of trust. This resulted in a close cultural melding.

Cultural melding, in the sense BC refers to, requires a high level of relationship building: a level of trust, between the Claimants and the government side, in which the claimants believe their interests would be looked after by the government.

Later, the negotiations entered into a more formal stage, with for example, face-to-face meetings with the Minister in Charge of Treaty of Waitangi Negotiations, to discuss the total financial quantum of the settlement. In this stage there were also many face-to-face meetings with the Crown negotiating team to work through other key issues to be negotiated (for example, what redress the Crown could provide in relation to sites of cultural and historical importance to the claimants).

At these meetings, key issues needed to be discussed face to face and people needed to be prodded to do things, something BC thought would be difficult in a completely virtual environment. At the same time, many minor issues and the exchange of technical information were being handled by post or fax. However, often the parties took advantage of the face-to-face meetings to also discuss or exchange supplemental information. According to BC, Ngati Mea called face-to-face meetings because they did not believe the OTS moved quickly enough or were giving the claimants the priority they thought they should receive.

OTS also used meetings strategically to convey negative responses to claimants. According to BC, this allowed the Crown the opportunity to explain *why* it could not do things. BC said it was easier to talk things through and gain immediate acceptance. Saying 'no' in a letter could lead to entrenched positions. Verbal, face-to-face communication allowed for indistinct, but still credible discussions to be held. BC elaborates:

When we are talking, we can explain our understanding of what the Crown policy is, for example, on a certain point. ...when you're talking although you might be focusing on the key point, with your phrasing and the exact nature of what you are saying you can get away with a lot more, maybe in terms of the exactness of what you are saying (compared to when we are writing when we are careful to cover all the conditions and provisos of what we are writing about).

The Crown and the Claimants both used letters to convey formality and official positions. The claimants tended to send a lot of letters. BC believed them to be very "written" oriented, because in this way they could compile a formal record of what had been done. Formal letters were also used to get things moving and to generate responses. BC reflects on why the claimants may have preferred letters:

I don't know exactly why but part of the reason is they are under a lot of pressure to get us performing. It also means that at the end of the day, if things boil down into disputes, they have a good record to show that they have done everything they can.

The OTS tended to respond verbally because they were not so concerned with recording every step. Also it gave them an opportunity to explain things and then process their responses in a controlled QA (quality assurance) protocol. In the end they would respond in a formal letter. According to BC:

Letters are quite interesting. A lot of times we use the mail to give something a bit more significance, a bit more finality. In our office, the official record is a hardcopy.

Using ICT

While the core business of the negotiation was done in face-to-face meetings, telephones and e-mail were often used to set up and confirm meeting times. The telephone was also used to deal with issues under negotiation when something had to be dealt with quickly. A lot of important work was done on the telephone. The telephone allowed for spirited talk and even argument.

E-mail was often used to confirm details, particularly within the Crown side. According to BC, e-mail possesses the speed of a phone call with the paper trail of a written response. E-mail was also used to quickly transfer working documents back and forth in a template format, which is used to build up and agree on components that involve a settlement. Currently the template is only e-mailed internally on the OTS side.

At the outset of negotiations, the Crown was unaware that the Claimants had access to e-mail as they did not use it until the final weeks of negotiations. It was in the final stages of the negotiations, when the wording of the key document needed to be agreed upon, that telephones and e-mail began to play a crucial role. Although much of the discussion around the document's wording was around detail, significant issues emerged from these discussions. However, as a date had already been set for the signing of the document, there was no time to set up face-to-face meetings to discuss these issues. These significant issues needed to be discussed and settled virtually, and it became evident to BC that when the final deadline was upon the two parties, they were able to use electronic channels to settle some final, difficult issues. Again, BC believed that this was possible because they had already developed trust and strong personal relationships. He elaborated:

The last four weeks before the signing of the Heads of Agreement was frantic and involved a different way of working together virtually. After a series of critical face-to-face meetings to work out some difficult points and with just a couple of days to go, communication took place primarily by phone and email. Important issues, normally dealt with face to face, were resolved virtually and they performed admirably on their side under that regime. This kind of tells me that although the preference might be there for face-to-face, when the costs are too high they work fine with other scenarios.

It seems that in this high stakes, complex project, there really was a need to develop a high level of relationship between the team members. In this case, the claimants took the lead in relationship building by insisting on regular face-to-face meetings, whether or not tasks actually had to be done. The benefits of this concerted effort at face-to-face relationship building eventually led to greater use of virtual communication channels as the negotiations progressed and eventually concluded.

By the middle of the action learning program, BC was able to inventory the ICT at the team's disposal and assess how the communication channels were being used and to what effect. Other than e-mail, no Internet-based technologies were used. According to BC, the Crown was unaware of whether the claimants were familiar with any Internet-based tools and in any case most of the OTS staff were unfamiliar with those tools. He believed that using real-time electronic meetings (such as video conferencing) was not appropriate because the two sides were geographically close enough to make the effort to meet face to face and would get more done that way. He also mentioned that telephones work much better in a one on one situation anyway:

I thought about the online conference. In this case, a videoconference would be less efficient at getting through stuff I think then a face-to-face meeting or even the telephone. It wouldn't be appropriate in this circumstance.

The other disadvantage of on-line meetings for OTS is that they would potentially leave an electronic recording of exactly what was discussed, which they do not want. Not only would there be a loss of frankness in the discussions, according to BC, but everything that is put down on paper in the OTS must be QA'd, so they are assured that what they are saying is correct. BC explains that the written confirmation of spoken positions is a key process in OTS:

In fact when we do make commitments at meetings, we actually follow that up on paper. It's the reviewed position that they have to go by, not what we say.

Other barriers to more fully virtualizing communications between the two sides had to do with issues of accessibility. One of the most significant according to BC is that Ngati Mea do not appear "to be up on the Internet". They lack the equipment as well as the training, and perhaps even the inclination at this stage to work more fully through virtual channels. BC pointed out that the Claimants are not set up commercially or in a general business sense and that "they have to come up to speed on lot of stuff, not just the technical and communications stuff."

The situation is further complicated by the fact that it would be completely inappropriate, for historical reasons, for OTS to implement any sort of policy to try and encourage the claimants to virtualize their communication processes. A lack of electronic tools does not disadvantage claimants in their negotiations with the Crown, and the Crown cannot make any assumptions about what tools are available to claimants. The Crown must be sensitive to their preference of tools. As BC states:

We are careful being the Crown. We do not want to force certain things on the claimant communities. Obviously, if they have it, it's an advantage, but we don't really push them on that sort of thing.

The next section summarizes the selection and use of communication channels across boundaries in relationship building and task completion in light of media choice theories. The impact of culture in this process is also discussed.

DISCUSSION

This case involved a highly complex task involving much cultural and organizational boundary crossing. The usual array of communication channels was available to the Crown side and to a lesser degree to the Claimant side. The key factors restricting the full use of communication channels appeared to have much to do with cultural (including organizational) preferences for particular channels, and to a lesser degree with familiarity and skill in using certain channels. Based on the actions of the Claimant group and the observations of BC, it is apparent that a high level of trust-based relationship was needed to successfully complete this negotiation.

Most of the conditions at the start of this project supported the selection of face-to-face meetings to build relationships: specifically, a long timeframe, close physical proximity of the parties, the lack of high technology on the Claimant's side and their lack of experience with it, as well as their strong preference for traditional channels such as face-to-face and formal letters.

BC was able to conclude that in this negotiation process, there was a hierarchy of communication practices determined in part by Office of Treaty Settlement protocols and Ngati Mea preferences, and in part by whatever seems to work best in a given situation. Letters represented the most formal and permanent communication channel on both sides. Face-to-face meetings were used to build relationships, to work out important issues and to get tasks done, often through the use of subjective, multi-layered dialogues, which were influenced by relationships, history and status. Telephones were also be used to discuss important issues and get agreement. E-mail was officially used to confirm details and also to transmit documents quickly and efficiently. Unofficially, e-mail was used to discuss important matters, although there remain unresolved organizational issues regarding its use in this manner. The use of videoconferencing is problematic as the technology is not available to all parties and is not particularly reliable. There are also security concerns.

After the last few hectic weeks before the negotiations were successfully concluded, BC had additional observations and insights on the way communication channels were used, and how organizational and cultural factors influenced this use. He came to realize that relationship building was an important facet of the negotiation process, particularly for Ngati Mea.

But even within the Crown side, it became clear to BC that a sound relationship was the foundation of successful virtual communications, and that face-to-face meetings were the key to building and maintaining good relationships. Building rapport through face-to-face meetings was demonstrated by OTS's excellent relationship with the Department of Conservation. The two offices held regular relationship meetings together that would include discussions on protocols for working with each other. This contrasts with OTS's poorer communication with some other departments where there was less effort on behalf of either party to build relationships.

The claimants continued to demonstrate a clear preference for face-to-face meetings throughout the negotiation process and particularly at critical times. Toward the end of the negotiations when things became critical, the claimants wanted face-to-face meetings, to as BC put it:

...nut everything out, which indicated to me that they have a huge preference for face-to-face meetings.

But it also became evident to BC that when the final deadline was upon the two parties, that they were able to use virtual channels to settle some final, difficult issues. BC believed that this was possible because they had already developed trust and strong personal relationships. In a very real sense then, the team featured in this case was a transitional ICT-supported team, geographically close enough to meet face to face to accomplish the all-important relationship building and adopting technology on an as-needed basis to assist task processes.

Limitations

There are several obvious limitations to this study. Foremost is the fact that only one point of view is put forth, that of BC. In a study such as this with its cross-cultural focus, it would have been very valuable to have the insight of other critical players in this case, particularly from the Claimant side. However, there is still significant value in these findings, as BC was a critical player in this case and his thoughtful reflection on the negotiation process and the other partied involved has provided valuable insight into the workings of an multiple boundary-crossing ICT-supported team involved in a complex task, as well as the influence of both organizational and ethnic culture on the team process and outcome.

IMPLICATIONS FOR RESEARCH AND PRACTICE

This case highlights many of the issues involved when working across multiple boundaries in ICT-supported teams. BC's team consisted of the Claimants and a number of Crown departments and ministries, as well as consultants and specialists. The Claimants were very different from the Crown members of the team. They were an indigenous cultural group with their own beliefs, values and protocols (Hall, 1976; Hofstede, 1980). These differences affected team interaction (Chidambaram, 1992),

Apparent cultural preferences on the part of the Claimants for particular communication channels might be explained by media choice theories. The Crown's use of communication channels may be likewise understood. The Claimant's preference for face-to-face communication might be explained by social presence theory (Short & Christie, 1976), while a desire for formal letters could be understood using symbolic interactionism theory (Trevino, Lengel & Daft, 1987). The Crown's use of e-mail and telephone (within its own ministries and departments and consultants) to set and confirm appointments, send documents, etc., might be understood in terms of information richness theory (Daft & Lengel, 1986) and/or media synchronicity theory (Dennis et al., 1998). Clearly as Ngwenyama and Lee (1997) state, media choice theories can be used to explain why individuals use certain channels in certain situations.

Researchers looking at cross-culturally framed information systems research should take note. While traditional value-based cross-cultural theories such as Hall's high/low context theory and Hofstede's uncertainty avoidance dimension would seem to have validity in explaining the Claimant's preferences for face-to-face communication and the high need for relationship building, the findings from this case demonstrate that other factors are at work. Deadlines, the need to work more quickly and the availability of alternative communication channels and team members' comfort with such channels did, especially toward the end of the negotiation process, play a large part in getting the parties to use these technologies in ways that traditional cross-cultural theories would not necessarily predict, but which media choice theories may.

The findings in this study seem to raise the challenge — how do we understand the selection of communication channels when communicating across cultures and other boundaries. It may be that these findings tend to support the relatively new notion that culture may be seen as emergent and negotiated based on a range of contextual elements (Holden, 2002), some of which may be explained by cross-cultural and media choice theories. However, these theories don't appear to tell the whole story. There is still much here for researchers to investigate and for which to develop more comprehensive theory.

In the next section, the implications for practice and organizational policy are discussed. BC then offers his further reflections two years after participating in the AL program.

ICT-Supported Teams in Practice

The OTS and the other New Zealand departments and ministries involved in this negotiation are making ever-increasing use of ICTs within their organizations. This case brings to light both the opportunities and difficulties when using these technologies in ICT-supported transitional teams both within the government and in negotiations with

Table 4. Key lessons learned in this action learning study

Lesson Learned	Organizational Response	Relevant Theories/Models
The importance of relationship building in multiple boundary crossing teams	Allow sufficient relationship building time and activities in project timelines	Pauleen & Yoong, 2001b
The need to consciously select and use appropriate ICT channels	Provide access and training to a wide range of ICT channels	Chidambaram, 1992
	Negotiate appropriate channel selection and use with team members	Various Media Choice theories (e.g., Daft & Lengel, 1986; Dennis et al., 1990; Trevino et al., 1987)
The usefulness of value-based cultural models in guiding cross-cultural interaction *with* the caveat that ultimately individual and group behavior may be flexible and negotiable based on a range of contextual elements	Provide member training in multiple boundary crossing interaction	Hall, 1976; Hofstede, 2000; Webb-Collins & Kirsch, 1999
	Foster cross-cultural learning and participation within the organization	Meyers, & Tan, 2002
	Use trained cultural facilitators in complex multiple boundary crossing project work	Holden, 2002

those outside the government. Table 4 highlights the lessons learned in this action learning study as they relate to multiple boundary team-based activity in OTS and how organizations, including OTS, might respond. Relevant theories and models are included.

Although virtual teams are becoming more prevalent, for most organizations like the OTS, the reality is that traditional co-located teams will remain the norm for the foreseeable future. Barriers exist that hinder a more complete virtual negotiation process, including those of access and skills to technology and the political, cultural and historical context, which make it difficult for the Crown side to push too strongly.

For the OTS, the 'virtualizing' of both the office and the negotiation process is not a matter of policy, rather it simply reflects trends in the wider business community. By adopting ICT to expand their communication and information sharing capabilities, traditional co-located teams can serve as exemplars in the organization and serve as a critical link on the way to fully virtual teams. BC has been pushing the virtual envelope in the OTS. His team is making as much use as possible of virtual technology. The key, according to BC, is the development of trust between team members, a conclusion that tends to support previous research in this area (e.g., Jarvenpaa et al., 1998; Lau et al., 2000; Nandhakumar, 1999; Traumuller, 1995).

By the end of the action learning program, after reflecting on his experiences, BC was able to see the need for OTS to have an ICT-supported transitional team strategy. He believed such a strategy would consist of at least a couple of components: one, a face-to-face meeting at the start of the team so members would be clear and comfortable about what each of the members were doing; the second component would be the use of short time frames for finishing blocks of work so team members would get positive feedback that would give them the idea that they were on track.

Further Reflections from BC

Since the action learning program concluded two years ago, BC has informally discussed his reflections within the OTS, and has also been involved in another set of negotiations with another Crown-Claimant team. Through this process, BC has validated his earlier reflections and has also clarified some aspects. He continues to believe that the complex nature of negotiations requires a great deal of face-to-face meetings, especially for sensitive issues (e.g., issues at the core of the historical grievances, as opposed to straightforward commercial issues) and at critical turning points in negotiations — or example, overcoming a difficult impasse. BC has confirmed that as negotiations become more intensive, more work can be done virtually if relationship building has already occurred and the fundamental underlying issues have been resolved face-to-face.

As far as the use of communications channels, BC has noticed that the use of audio-conferencing has now been established for some negotiating meetings, especially where the issues are second-order. Both the Claimants and the Crown are now using e-mail as an efficient way for dealing with transactional issues such as arranging meeting times and project management communications (e.g., reminders about task timeframes) and for widely distributing versions of formal quality-assured letters.

It is important to note that many of the Claimant negotiators (although Maori) may be used to operating in a pakeha (European New Zealand) environment, and could potentially be more accustomed to virtual communications than OTS staff. In such cases, the need for face-to-face meetings would still arise, primarily in relation to issues that are complex, intractable, or that relate to cultural matters or are closely linked with the historical grievances under negotiations.

A key lesson is that not all aspects of negotiations are of such a complex nature as to require face-to-face meetings. Furthermore, given that any negotiations will involve many face-to-face meetings, those meetings could lead to a sufficient development of trust and relationship building to enable a significant amount of second-order issues to be addressed through less-rich communications channels. It is clear that effective teamwork will involve a mixture of channels. A key challenge for OTS is to balance the various communication channels, to try and address second-order issues using virtual channels (which are cheaper) so that face-to-face meetings can focus on resolving key outstanding matters.

ACKNOWLEDGMENTS

The author would like to thank Bede Clarke, Senior Policy Analyst for the Office of Treaty Settlements, New Zealand for his participation in this study and his help with this paper.

REFERENCES

Anison, L. & Miller, P. (2002). Virtual teams: A virtue for the conventional team. *Journal of Workplace Learning 14*(4), 166-173.

Baskerville, R. & Pries-Heje, J. (1999). Grounded action research: A method for understanding IT in practice. *Accounting Management and Information Technologies*, 9, 1-23.

Chidambaram, L. (1992). The electronic meeting room with an international view. In R.P. Bostrum, R.T. Watson, & S.T. Kinney (Eds.), *Computer augmented teamwork: A guided tour*. New York: Van Nostrand Reinhold.

Cukier, W., & Middleton, C. (1996). Cultural diversity and group decision support systems. In E.J. Szewczak & M. Khosrowpour (eds.), *The human side of information technology management* (pp. 294-316). Hershey, PA: Idea Group Publishing.

Daft, R.L. & Lengel, R.H. (1986). Organizational information requirements, media richness and structural design. *Management Science*, 32(5), 554-571.

Dennis, A.R., Valacich, J.S., Speier, C., & Morris, M.G. (1990). Beyond media richness: An empirical test of media synchronicity theory. *Proceedings of the 31st Hawaii International Conference on Systems Sciences*, IEEE.

Duarte, N. & Tennant Snyder, N. (1999). *Mastering virtual teams: Strategies, tools, and techniques that succeed*. San Francisco: Jossey-Bass.

Fukuyama, F. (1995). *Trust: The social virtues and the creation of prosperity*. London: Hamish Hamilton.

Fulk, J. & DeSanctis, G. (1995). Electronic communication and changing organizational forms. *Organization Science*, 6, 338-349.

Fulk, J., Schmitz, J., & Steinfield, C. (1990). A social influence model of technology use. In J. Fulk & C. Steinfield (Eds.), *Organizations and communication technology* (pp. 117-142). Newbury Park: Sage.

Glaser, B. & Strauss, A.L. (1967). *The discovery of grounded theory: Strategies for qualitative research*. New York: Aldine De Gruyter.

Griffith, T. & Mead, D. (forthcoming). Prelude to virtual groups: Leadership and technology in semi-virtual groups. In D. Pauleen (Ed.), *Virtual teams: Projects, protocols, and processes*, Hershey, PA: Idea Group Publishing.

Hall, E.T. (1976). *Beyond culture*. New York: Anchor Press.

Hofstede, G. (1980). *Cultural consequences: International differences in work related values*. Beverly Hills, CA: Sage.

Holden, N. (2002). *Cross-cultural management: A knowledge management perspective*. Harlow, UK: Financial Times-Prentice Hall.

Hunter, M.G. (2001). Cross cultural research issues (Editorial). *Journal of Global Information Management, 9*(4), 3-5.

Jarvenpaa, S.L., Knoll, K., & Leidner, D.E. (1998). Is anybody out there? Antecedents of trust in global virtual teams. *Journal of Management Information Systems, 14*(4), 29-64.

Johansen, R., Vallee, J., & Spangler, K. (1979). *Electronic meetings: Technological alternatives and social choices*. Reading, MA: Addison-Wesley.

Karahanna, E. (1995). Electronic media in the workplace: An integrated model of media choice. *The Journal of Management Systems, 7*(2), 1-28.

Karahanna, E., Evaristo, R., & Srite, M. (2002). Methodological issues in MIS cross-cultural research. *Journal of Global Information Management, 10*(1), 48-55.

Katzy, B., Evaristo, R., & Zigurs, I. (2000). Knowledge management in virtual projects: A research agenda. *Proceedings of the 33rd Annual Hawaii International Conference in Systems Science*, Hawaii. IEEE.

Kayworth, T. & Leidner, D. (2001-2002). Leadership effectiveness in global virtual teams. *Journal of Management Information Systems, 18*(3), 7-40.

Lau, F. (1999). Toward a framework for action research in information systems studies. *Information Technology & People, 12*(2), 148-175.

Lau, F., Sarker, S., & Sahay, S. (2000). On managing virtual teams. *Healthcare Information Management Communications Canada,* 14, 46-53.

Lewis, R.D. (1996). *When cultures collide: Managing successfully across cultures.* London: Nicholas Brealey Publishing.

Lipnack, J. & Stamps, J. (1997). *Virtual teams: Reaching across space, time, and organizational boundaries*. New York: John Wiley & Sons.

Margerison, C. (1988). Action learning and excellence in management development. *Journal of Management Development, 7*(5), 43-55.

Maznevski, M. L., & Chudoba, K.M. (2000). Bridging space over time: Global virtual team dynamics and effectiveness. *Organization Science, 11*(5), 473-492.

Meyers, M.D., & Tan, F. (2002). Beyond models of national culture in information systems research. *Journal of Global Information Management, 10*(1), 24-33.

Nandhakumar, J. (1999). Virtual teams and lost proximity: Consequences on trust relationships. In P. Jackson (Ed.), *Virtual working: Social and organizational dynamics* (pp. 46-56). London: Routledge.

Newman, W.H., Summer, C.E., & Warren, E.K. (1977). *The process of management.* Englewood Cliffs, NJ: Prentice Hall.

Ngwenyama, O.K. & Lee, A.S. (1997). Communication richness in electronic mail: Critical social theory and the contextuality of meaning. *Management Information Systems Quarterly, 21*(2), 145-167.

Pauleen, D. & Yoong. P. (2001a). Facilitating virtual team relationships via Internet and conventional communication channels. *Internet Research: Electronic Networking Applications and Policies, 11*(3), 190-202.

Pauleen, D.J., & Yoong, P. (2001b). Relationship building and the use of ICT in boundary-crossing virtual teams: A facilitator's perspective. *Journal of Information Technology, 16*(4), 205-220.

Revans, R. (1982). *The origins and growth of action learning*. Chartwell-Bratt, Bromley.

Short, J., Williams, E., & Christie, B. (1976). *The social psychology of telecommunications*. New York: John Wiley.

Strauss, A.L. & Corbin, J. (1990). *Basics of qualitative research: Grounded theory procedures and techniques*. Newbury Park: Sage.

Traunmuller, R. (1995). Enhancing office systems with the CSCW-function. In Bots, P., Glasson, B., & Vogel, D. (Eds.), *The international office of the future: A problem analysis. Faculty of Systems Engineering, Policy Analysis and Management* (pp. 27-44). Delft University of Technology, Delft, The Netherlands.

Trevino, L.K., Lengel, R., & Daft, R.L. (1987). Media symbolism, media richness, and media choice in organizations: A symbolic interactionist model. *Communication Research, 15*(5), 553-574.

Warkentin, M. & Beranek, P.M. (1999). Training to improve virtual team communication. *Information Systems Journal,* 9, 271-289.

Warkentin, M. E., Sayeed, L., & Hightower, R. (1997). Virtual teams versus face-to-face teams: An exploratory study of a web-based conference system. *Decision Sciences,* 28, 975-996.

Webb-Collins, R. & Kirsch, L. (1999). *Crossing boundaries: The deployment of global IT solutions.* Cincinnati, OH: Pinnaflex Educational Resources.

ENDNOTES

1. The Treaty of Waitangi is New Zealand' founding document, which was signed in the 19th century between representatives of the British Crown and the Maori, the indigenous people of New Zealand. The Treaty sets out the rights of the two groups.
2. Currently, there are over 800 claims being made by Maori groups against the Crown. These usually involve land and resources believed to be unjustly taken from the Maori by the Crown over the last 150 or so years.
3. The Maori group has been called "Ngati Mea" to protect its anonymity. The group resides within driving distance of Wellington, where most of the Crown team was located.

Previously published in the Journal of Global Information Management, 11(4), 1-19, October - November 2003.

Chapter III

Culture-Free or Culture-Bound? A Boundary Spanning Perspective on Learning in Knowlege Management Systems

Robert M. Mason, Florida State University, USA

ABSTRACT

Knowledge management systems (KMSs) have been criticized as having a North American bias. The cultural dimension of KMSs, particularly the relationship of learning and culture in KM projects, are rarely discussed. This paper addresses these concerns in a review of the conceptual foundations for KM and by examining implementations of KM projects. Despite the evolutionary changes in how KM is viewed, KMSs, as they have been designed, implemented, and reported, do not appear to provide for cultural diversity among users. Instead, the reports of KMSs indicate that such systems seek to create and maintain a homogeneous organizational culture, and the adoption of such a shared culture appears to be a prerequisite for success. The paper discusses KMSs as systems that exhibit boundary spanning objects and processes in three different categories, and an analysis of reported projects reveals that boundary spanning across national and ethnic boundaries is rare.

INTRODUCTION: INFORMATION TECHNOLOGY AND KNOWLEDGE MANAGEMENT

Global enterprises increasingly turn to knowledge management systems (KMSs) to raise productivity and remain competitive. Although there is considerable evidence that applications of information technology (IT) for storage and improved access of information help create value, some observers believe that KMSs are limited in their utility because they have been designed with a North American bias (Nonaka, 1995). To understand why this bias may be of concern, it is useful to consider KM programs in the context of the resource-based view (RBV) of the firm (Penrose, 1959).

The dynamic economic environment and the ever-increasing innovative capabilities of global organizations have renewed interest in the RBV. In the opinion of many writers, the RBV has had a significant impact on how information systems and strategy are viewed. The RBV is closely linked with strategy and sustainable competitive advantage (Barney, 1994, 1996), plays a major role in how the modern economic theory of the firm has developed (Madhok, 2002), and perhaps has become the most influential framework for the theory of strategic management and sustained competitive advantage (Barney et al., 2001). The knowledge-based view (KBV) of the firm, foreseen by Drucker (1988), is a special case of the RBV with a focus on knowledge as an organizational resource (Grant, 1996a,1996b; Spender, 1996).

As in the more general case of the RBV's influence on strategic management and competitive advantage, the KBV provides the conceptual foundation for much of the research and design efforts that link information technology (IT) and systems, strategic IT, organizational learning, and knowledge management systems. The KBV has shaped the discussion of KMSs and the role of information technology in strategy and competitive advantage (Huber, 1991; Mata, 1995; Kogut, 1996; Alavi, 2001). While the KBV is the foundation for IT support of knowledge management, most observers agree that knowledge by itself is not the source of a competitive advantage. Instead, organizations use knowledge to gain a competitive advantage through learning (Stata, 1989), by the development of competencies (Rebentisch & Ferretti, 1995; Alavi, 2001), and through knowledge integration (Grant, 1996b; Kogut, 1996). In other words, the value of KM programs depends not only on the application of IT but on the individual and organizational learning and knowledge integration that comes from revealing and using all the intellectual assets of the organization. This requires a mix of technology and organizational processes.

From the perspective of the KBV, a global enterprise that has members from distinct ethnic and cultural backgrounds would appear to have a potentially large asset in the rich source of tacit and experiential knowledge of its members. KM efforts would be one way to access this potential asset. This paper addresses the question of how and to what extent firms may be taking advantage of the knowledge asset represented by having members with diverse backgrounds.

In the remainder of the paper, we will use the terms "KM" and "KMS" interchangeably; both will refer to the set of activities directed toward knowledge asset management. Our approach is in the spirit of socio-technical systems, thus KM and KMS include physical resources (e.g., computers and communication infrastructures), conceptual resources (e.g., repositories of expressed knowledge, tacit knowledge), and the social and organizational processes associated with the use of these resources.

The paper contributes to the discussion of KM by drawing attention to the importance of culture in KM efforts and proposing that KMSs may be viewed as boundary spanning systems. The cultural boundary is important for two reasons. First, ethnic backgrounds and national cultures represent a potential knowledge asset of the enterprise. Second, culture is a significant factor in how people learn, and learning is required for the organization to take advantage of the potential intellectual assets in the organization. The paper uses Carlile's boundary spanning framework (Carlile, 2002) to review the functions of KMSs and to examine how these systems handle culture.

In the following three sections, this paper provides a synopsis of knowledge management approaches and learning from three perspectives: (a) a summary of the goals and expected benefits of KMSs; (b) the role and importance of culture and ethnic background in learning; and (c) reports on the implementation and use of KMSs and the role of culture in these implementations. The fourth section reviews these KMS implementations using Carlile's framework of practical boundary spanning systems to gain a better understanding of the relative importance of culture in the design and use of a KMS in a particular organization. Based on this review, most KMSs appear to fulfill the boundary spanning role primarily at two of Carlile's three levels. The KM activities at the third level are not directed at spanning boundaries between or among national or ethnic cultures but rather at spanning hierarchical boundaries. As a consequence, benefits that might accrue from the surfacing of differences in tacit cultural knowledge are unlikely to be realized. The final two sections discuss the contribution of this paper, its limitations, and its potential significance for research and practice.

KMS DIRECT AND INDIRECT BENEFITS: THE VALUE OF CULTURE AND LEARNING

Early studies of KMSs viewed such systems as the application of IT to improve the information value chain (Rayport, 1995) by providing an organizational memory (Walsh & Ungson, 1991) and supporting organizational learning (Huber, 1991). A recent review of KM elaborates on this basic value chain model and summarizes prior studies as being focused on four basic processes in knowledge work: creating, storing/retrieving, transferring, and applying knowledge, with each of these processes being subjected to more detailed analyses (Alavi, 2001, p. 114). This conceptual model views KMSs as systematic attempts to make visible the collective and individual knowledge in an organization, develop a knowledge-intensive culture, and support this culture through an IT-enabled infrastructure (Davenport & Prusak, 1998; Alavi, 2001). Specifically, a KMS is viewed as a class of information systems applied to managing knowledge (Alavi, 2001), and IT is seen as an enabler of the basic processes of creating-storing/retrieving-transferring-applying knowledge.

While much effort is devoted to creating repositories of explicit knowledge, KM efforts also seek to identify sources of tacit knowledge, to surface and improve access to this knowledge, and to enable collaborative learning and integration across functional areas (Alavi, 2001; Leonard, 2002; MacCormack, 2002; Fulmer, 2003). In their study of

31 KM projects, Davenport et al. (1998), identified four broad types of objectives: (1) create knowledge repositories, (2) improve knowledge access, (3) enhance knowledge environment, and (4) manage knowledge as an asset.

Reports of tangible benefits have included cost savings (arising from, among other things, reduced time required for knowledge access by having ready access to needed information) and increased revenue and profit. These benefits are recognized as both direct and indirect. Direct benefits are those associated with having the collective knowledge readily accessible to all organizational members: a corporate memory (e.g., best practices, repositories of projects, compilations of data about clients and suppliers, etc.) and the facilitation of communication among the organization's members (a shared structure and communication infrastructure, meta-information that enables members to locate others that may have the information they require), and knowledge transfer (Fulmer, 2003). These direct benefits are antecedents to indirect benefits: knowledge creation and innovation in products and processes, improved collaboration and the synthesis of tacit knowledge from multiple entities in the organization, the creation and maintenance of core competencies, and improved competitiveness. The indirect benefits require learning and knowledge integration across functional boundaries and may be even more valuable than the direct benefits. However, as noted by many of the cases, it is easier to see the impact of KM efforts on the direct benefits, and firms are able to assign values to these direct benefits more easily than to indirect benefits (Alavi, 1997; Davenport et al., 1998; Hansen, 2002; MacCormack, 2002; Ng, 2002; Soo, Devinney, et al., 2002). The difficulty of measuring indirect benefits may be overcome by measuring their impact. For example, Soo et al. (2002) suggest measuring changes in the innovative outputs that arise from the KM effort.

A global firm comprises organizations that have individuals from many different ethnic and cultural backgrounds. From the KBV, knowledge from this diverse set of employees represents a potential intellectual asset. Appropriating the benefits from this asset, however, requires learning. The next section reviews the relationship between culture and learning.

LEARNING: THE ROLE OF CULTURE

Learning is defined as "a relatively permanent change in the ability to exhibit a behavior; this change occurs as the result of a successful or unsuccessful experience" (Klein & Mowrer, 1989, p. 2). This definition is consistent with the experiential learning model (Kolb, 1984) and with a prevailing view in the IT field that "an entity learns, if, through its processing of information, the range of its potential behaviors is changed" (Huber, 1991, p. 89). Although these definitions are consistent — each relates to ability or capability to demonstrate behavior, for example — they have differences: Klein and Kolb emphasize experience as the basis of learning; Huber, as an IS researcher, views the learner as an information (or knowledge) processing entity. Huber also posits that an organization learns if "...any of its units acquires knowledge that it recognizes as potentially useful to the organization" (Huber, 1991, p. 89).

Although a discussion of the similarities and distinctions between individual and organizational learning could be interesting and enlightening (e.g., Dixon, 1994), this paper accepts that individual learning is critically *important to* (if not, as Huber posits, *equivalent to*) organizational learning. Consequently, the following discussion focuses on individual learning and looks particularly at contemporary theories about culture-based learning.

The study of learning processes has provided a rich field for psychologists for well over 100 years (Mowrer, 1989). Earlier views of learning focused on the stimulus-response relationship and later (particularly Skinner's work) associated reinforcement (rewards) with behavior. Contemporary learning theory, in contrast, has moved away from global theories and looked more closely at concept formation, problem solving, language, and other cognitive aspects of learning. Even the biological aspects of learning are getting new attention (LoLordo, 1989; Mowrer, 1989; Zull, 2002).

Table 1. Characteristics and critical success factors (CSFs) of sample KMSs

Source	Context/Users	Technology and Methods	CSFs (author's summary)	Support for multiple cultures?
(Massey, 2002)	Nortel New Product Development (NPD) teams	Electronic Performance Support System	Financing Standardized process Codified format	No
(MacCormack, 2002)	Siemens ShareNet—multiple (sales, developers, managers)	Knowledge Repository User Network for query-response	Structured inputs Query-Response component Marketing of system use; incentives	No
(Fulmer, 1999; Fulmer, 2000; Fulmer, 2003)	Buckman Labs K'Netix	Monitored but unmoderated Forums Email	Principles Code of Ethics Adaptability Dedicated staff	Sensitive to culture Tried multiple languages; settled on English only
(Ng, 2002)	PwC KnowledgeCurve; 99% of 150,000 world-wide consultants	Intranet; Lotus Notes databases	KM content process (editing/vetting) Users' view as integrated system Access from any location	No
(Leonard, 2002)	NASA/JPL: LLIS, APPL and KSI (face-to-face program); Project Libraries; *Know Who* directory; *Technical Questions* DB; legacy reviews; personal knowledge organizers (oral histories)	Customizable portal; DocuShare, leadership development; Internet; DBs	Resources Culture and commitment (varied from low to high)	No (Focus is on US organization)
(Rukstad, 2001)	DaimlerChrysler Knowledge Management Strategy	CAD/CAM Product DB EBOKs TechClubs	Support Link to performance Top management consensus on required capabilities	No
(Chard, 1997; March, 1997)	E&Y: EYKnowledgeWeb	LotusNotes; multiple platforms	Culture and commitment; built into performance reviews that encouraged use	No
(March, 1997; Hansen, 2002)	Andersen Consulting: Knowledge Xchange	LotusNotes; divergent and specialized forums (GBP); search capabilities	Annual review and incentives to encourage sharing KMs with specialized roles Pre-filtering, context	No
(Alavi, 1997)	KPMG: Kweb	Web-based intranet; links to external	Integration into business processes; incentives to embrace culture	No
(Davenport, Long, et al., 1998)	Multiple (see Table 2)	Multiple	K-friendly culture, change in motivational practice, clear language, 5 others	None mentioned

Because learning is related to language and language use, and language rules may be "hardwired" as Chomsky proposed in the 1980s (see Baker, 2001, for recent renewal of these arguments), one might conclude that the basic biological foundation for learning is genetic and therefore would not vary from culture to culture. A common biological base does not prevent the development and evolution of learning skills and abilities along different paths in distinct ethnic groups and cultures.

Indeed, for over 70 years researchers have examined the link between culture (and ethnicity) and learning. Hall (1959) summarizes the relationship between learning and culture by saying, "once people have learned to learn in a given way, it is extremely difficult to learn in any other way [...C]ulture reflects the way one learns." In 1930, Vygotsky and Luria proposed a culture-centered approach to learning, and others developed this approach further (Vygotsky et al., 1987; Forman et al., 1993; Kozulin, 1998). In this perspective, culture is a source of differences in cognition as cognitive processes are formed through sociocultural activities. Cole and others developed this into a contextual theory of cognitive functions (Cole, 1971), which has as a foundation the idea that different cultures have different systems of mediated learning experience

Table 2. Examples from "successful knowledge management projects" (adapted from Davenport et al., 1998)

Context/Users	Technology and Methods	Notes
HP (9 projects)	Electronic Sales Partner Lotus Notes DB	Highly enthusiastic managers ("most successful implementation of software I have seen in 20 years")
Sequent Computer	Sales oriented document respository	Highly enthusiastic managers
BPX	Videoconferencing Document scanning/sharing Education, support	Anecdotal success stories
Microsoft	Breakout of K competencies Link K competencies to staffing and HR	
National Semiconductor	Intranet Lotus Notes	Engineers preferred Intranet; sales and marketing preferred Notes
TelTech Resource Network	Referral of experts (technology not specified) Customer feedback Records of K resources used in proposals and projects compared with "wins"	Basis for business model
Sematech	Formal K transfer practices • K transfer organization • K transfer sessions • Client "assignees"	Face-to-face most effective for K transfer
Consulting firm	Structured K base	Contributions to K base a significant factor in compensation (not entirely successful)
Automobile co.	Specific K application guidelines Decision audits to assess use of Knowledge	Success not established
Automobile co.	Grapevine: interprets, stores, routes competitive intelligence to managers	
Skandia	Intellectual capital audit	
Dow	Review and manage patents	Expect $100 million more in licensing revenue

(MLE). Such systems are important to cognitive development, and differences in development (which can arise from different MLEs) become evident when a learner makes a transition from one system to another.

Some have argued "cross-cultural differences tend to disappear under the influence of systematic exposure to formal schooling" (Kozulin, 1998, p. 110, citing Cole, 1990). However, recent research confirms Hall's general statement and provides empirical evidence for Cole's theory that individual cognitive abilities can develop differently in different cultures. Even within the U.S., studies have indicated that some groups exhibit distinct behavioral styles through which they express their abilities to learn. Individuals who have experienced culturally distinct environments while growing up tend to use the skills gained in these environments even after they are adults (Hilliard, 1992). In studies of young adults who have completed schooling in one culture and move to another culture, the results indicate that the nature of the initial formal schooling makes a difference. This difference is not simply a difference in knowledge base, but rather is associated with the basic skills by which one learns new concepts. The young adults exhibited specific difficulties associated with coding schema, concepts, and graphic and symbolic devices used in communication of ideas (e.g., tables, ordering, plans and maps). The difficulties extend to cognitive activities such as the ability to identify or define problems (that is, the ability to apply their knowledge to a set of data and infer the implicit question or issue), the ability to work with multiple sources of information, etc. In short, the young adults are missing cognitive antecedents that would enable them to excel in their new environment (Kozulin, 1998). Kozulin concludes, "...cross-cultural differences in cognition are most probably related to learning practices characteristic of different cultures and subcultures..." and "Two major determinants of cognitive prerequisites are conceptual literacy and facility with other symbolic psychological tools, and a mediated learning experience responsible for the integration of these tools into the cognitive system of the student" (Kozulin, 1998, p. 129). His work showed that intervention could help learners develop the basic skills that would enable them to learn effectively in the new environment.

Research on western-style education has included considerable recent interest in problem-based learning (PBL) as a distinct pedagogy. The potential benefit of PBL is that it provides a better match of formal learning experience with environmental demands after the completion of formal schooling. While some research reports mixed results, there is considerable enthusiasm for this approach in the sciences and medicine (Culatta, 1994; Boud & Feletti, 1998; Jones et al., 2001). The use of PBL to develop competencies — abilities that enable persons to continue to learn — is suggested by Tien et al. (2003), who compare the competencies identified as goals by the educational systems in Australia, the UK, the U.S., Canada, and Taiwan. Significantly for this review, the competencies in different countries are similar but not identical in wording. For example, Australia and the UK emphasize instrumental competencies more so than the U.S., Canada, and Taiwan. Taiwan lists "self-understanding and potential development" as one of its ten competencies, a skill not articulated by any of the other countries' educational systems (Tien, 2003). Tien's study illustrates that even with many shared goals, the educational objectives of even industrialized countries have differences; the differences across a wider range of countries might be even greater.

Kozulin's studies suggest that cultural backgrounds can perform a critical role in the development of problem definition and problem solving abilities. Moreover, intervention to develop missing skills may be required for a learner to make a successful transition from one learning system to another. Since many KMSs are intended to support group problem solving and learning across functional boundaries, a PBL approach may be a useful way to bridge cultural differences, as suggested by Tien (2003). However, there are practical difficulties in designing PBL interventions for non-western and non-English speaking students (Walker, 1996; Allen & Rooney, 1998).

KMS IMPLEMENTATIONS AND CULTURE

This study examined published reports of KMSs in the literature and focused particularly on the role of culture in KMS and KM project success. For individual cases, this section draws heavily on teaching cases (business school cases of individual implementations) and other summaries of actual implementations (e.g., Davenport et al., 1998). Table 1 summarizes the findings, including this author's summary of what appear to be critical success factors in each case. Table 2 provides additional examples of the implementations summarized by Davenport et al. (1998).

The results of the studies summarized in Tables 1 and 2 are striking in three ways. First is the emphasis by so many of the implementations on standardization, both technical standards and the format of the content. Second is the frequent mention that an organizational culture of knowledge sharing is a correlate of success. Third is the prevalent, though not universal, use of incentives to change behavior and encourage system use. National culture and ethnic background of the users are rarely mentioned.

Only one case directly discusses the importance of national culture. In the Buckman Laboratories case, national culture and non-English speakers were handled explicitly (Fulmer, 2003). K'Netix, as it evolved, encouraged contributions to the knowledge base in whatever language the contributor felt most comfortable. A key component of the system was a group of translators hired by the firm. The translators translated into English the contributions selected by the forum monitors, making these contributions more accessible to employees whose native language was not English. As the system evolved, separate regional forums began, including a Spanish language case database on best practices. In 1997 there were 1,787 cases in English and 685 in Spanish (Fulmer, 2003). At the last report, however, the firm had standardized on English as the common language for contributions, rather than having separate language forums (Fulmer, 2000). Consequently, the only case that gave attention to national culture appears to have "regressed to the mean" and decided on a standardized approach to knowledge sharing.

This review of KM contrasts with the attention paid to culture in studies of other information systems (e.g., Walsham, 2002), and in other studies of international management, (e.g., Erez & Earley, 1993). The tendency to ignore cultural background in KM efforts suggests that KMS designers may be implicitly adopting the "culture-free" hypothesis as a basis for design. The culture-free hypothesis expresses the thought that there is universality to organizational design and structures — that organizations are micro social entities that can exist without reference to their immediate societal environment. This is in contrast to the culture-bound view that organizations match their structures to fit their societal environment (Maurice, 1976). For a wide range of production firms, the culture-free hypothesis is supported by a meta-analysis (Miller, 1987). There is some evidence supporting the culture-free hypothesis in banks (Birnbaum,

1985), but research that examines this hypothesis in other professional service work is missing. Perhaps the lack of attention to (national) culture in KM efforts is understandable, particularly given the experience and decision of Buckman Laboratories, but there is no research that supports "culture-free" as a normative approach to knowledge work (or other professional service work).

KMSS AS BOUNDARY SPANNING SYSTEMS

Boundary spanning has been recognized as a necessary component in processes that require coordination and translations among diverse groups (Star & Greisemer, 1989) and different functional groups or "thought worlds" (Dougherty, 1992). Product development is an example of a process requiring such coordination. KM and learning are often discussed in the context of new product development (NPD) processes; (Chapman & Hyland, in press; Verganti, 1998; Ramesh and Tiwana, 1999; Alavi, 2001).

Carlile's (2002) study of boundary spanning objects in an NPD process suggests a useful framework for examining the functions of KMSs. Carlile looked at the four primary functions involved in the creation of a new product (sales/marketing, design engineering, manufacturing engineering, and production) and examined how the new product development team dealt with the specialized knowledge of each area. Each of the four functional areas had different and specialized (in Carlile's terms, "localized and embedded") knowledge, structured in a way that made sense to the group. This knowledge specialization presented a barrier to the effective operation of the NPD team — the team found it difficult to exchange and synthesize knowledge as necessary for the successful development of a new product. Carlile observed that the team overcame this barrier by using boundary spanning objects that operated at three different levels: syntactic, semantic, and pragmatic.

At the syntactic level, *repositories* enabled communication of facts and agreed-upon tasks and actions. At the semantic level, *standardized forms* and *methods* enabled not only communication of facts, but also provided a way for the different groups to clarify differences in meaning. The objects at this semantic level (standard forms and methods) enabled the team to translate the knowledge embedded in one group so that other groups contributing to the product development could understand it. At the pragmatic level, *objectives, maps,* and *models* enabled each group to transform embedded knowledge into knowledge that the team (and others not in the group) could understand.

Earlier, Brown and Duguid (1998) pointed out different roles of boundary spanning activity, noting particularly the need for translators between communities. In commenting on Carlile's model, Brown (2002) suggests that Carlile's three levels correspond to three different levels of knowledge ambiguity among communities of practice. At each level, different types of boundary objects are necessary for communication, knowledge transfer, and learning.

At the syntactic level, the differences across the boundaries are explicit, clear, and stable. A shared syntax is a necessary (but not necessarily sufficient) condition for knowledge sharing under these conditions. Taxonomies and classification (e.g., shared databases) provide this syntax and enable the sharing and transfer of knowledge among groups that have a clear understanding of their differences and understand that these differences are relatively stable.

At the semantic level, the differences are neither clear nor necessarily stable (Brown, 2002), and the solution is to provide a means of translating meanings across boundaries. At this level, Carlile (2002) observed the use of standardized forms and methods as boundary objects.

At the pragmatic level, Brown (2002) notes that the knowledge of one group is not neutral to another's. The communities may have different values and/or power relationships, and this level of difference requires boundary objects that provide additional capability beyond the first two levels. At this level, the groups must transform their knowledge and create new (shared) knowledge rather than simply exchanging or transferring knowledge. Resolution of the group differences requires objects such as models and maps, objects that enable the sharing of methods of thinking and the surfacing of assumptions and values. Only when both groups understand these differences can the two communities be co-creators of knowledge that did not exist before the differences were discussed. It is at this level that one might observe the kind of knowledge integration and synergy anticipated by the KBV of the firm.

Considering KM efforts and KMSs as processes to span the boundaries of groups that have different knowledge and experience, one can map the examples from Tables 1 and 2 onto Carlile's framework. At the syntactic level, shared databases (e.g., repositories of best practices, evident in all the cases) provide a way for groups to transfer knowledge. At the semantic level, standard forms and practices (highlighted in Nortel, Siemens, PwC, EYKnowledgeWeb, KPMG's KWeb and Andersen's Knowledge Xchange) provide the means of representation of different knowledge in familiar formats and a means of learning from groups that have a different knowledge base.

At the pragmatic level, where differences in value would be apparent, Tables 1 and 2 provide fewer examples. The use of principles (Buckman Laboratories) as a boundary-spanning object between upper management (the owner) and employees could be considered a pragmatic level object, helping transform the owner's values into practice. An interesting distinction is that the principles are not developed in collaboration with the employees but rather presented to them as a standard — a codified set of behavioral principles that enable them to understand the goals and objectives of the firm. Similarly, one may consider the incentives and efforts to change corporate culture as boundary spanning activities at the pragmatic level.[1] These activities may also be viewed as procedures and routines that serve to translate values from the executive level to managers and other employees — procedures that might also be viewed as semantic. In other words, these processes span a hierarchical boundary, not an ethnic or national cultural boundary.

Table 3 summarizes Carlile's levels, Brown's (2002) view of the community differences for which spanning is required, and the corresponding boundary spanning techniques used by KMSs. Note that all of the KM efforts report use of syntactic boundary objects (repositories).

Table 4 presents the examples in Table 1 in terms of the different boundary objects in Carlile's model. Most objects and processes span communities of practice (e.g., different functional groups) at the syntactic or semantic level, and the majority are operating at the syntactic level — that is, most of the KMSs are set up to facilitate the exchange of knowledge at the level where the knowledge differences are, in Brown's (2002) words, "clear and stable." (The leadership development process at NASA/JPL, assuming it includes dialogue among individuals with different values, is an example of

Table 3. Levels and functions of boundary spanning processes with KMS examples

Level (Carlile, 2002)	Community Differences (Brown, 2002)	Boundary Object Function or Solution (adapted from (Brown, 2002)	KMS Boundary Objects	Example KMS
Pragmatic	Knowledge is not neutral; values and power may differ across boundaries	Transformation Exercise of power	Objectives Maps Models Incentives Forums	Buckman Labs Siemens Andersen E&Y KPMG Others (Davenport, Long et al., 1998) Buckman Labs NASA/JPL
Semantic	Unclear	Translation	Principles Standard methods and procedures Standard, codified forms	BuckmanLabs K'Netix Nortel E&Y Andersen PwC
Syntactic	Clear, stable	Communication Shared knowledge and data base	Repositories	All

a process operating at the pragmatic level, but the literature provides little detail.) All behavioral change processes and exercise of power (incentives, top management support, etc.) are classified as "pragmatic" processes that span the boundary between executives and other employees — a hierarchical boundary spanning process. Only the forums of Buckman Labs explicitly recognize the role of national culture and provide the opportunity for knowledge to be shared across ethnic boundaries and national backgrounds. The latest report suggested that these forums now are using a single language (English) for sharing knowledge.

CONCLUSIONS AND DISCUSSION

The intellectual foundation for KM is the KBV of the firm. For global enterprises, the tacit and experiential knowledge of staff from different cultures is a potential knowledge asset, one that a KM effort might seek to exploit. However, reviews of KMSs and KM efforts revealed little attention directed toward the cultural or ethnic backgrounds of staff. National and ethnic culture are not important considerations in most KM efforts. There is, however, a universal emphasis on creating and maintaining an organizational culture that supports knowledge exchange and the use of the KMS. Indeed, the success of these KM efforts seems linked to establishing and rewarding a shared organizational culture. The research identified only one system — Buckman Labs — that explicitly recognized the potential value in different national cultures. The latest report on this system indicates that the decision has been made to emphasize English and thus provide a common language base for all employees, so the sole example of a KM that

Table 4. KMSs as boundary spanning systems

KMS, KM effort	Boundary Objects or Processes	Level in Carlile's Framework		
		Syntactic	Semantic	Pragmatic
Nortel New Product Development (NPD) teams	Financing	X		
	Standardized process	X	X	
	Codified format	X	X	
Siemens ShareNet—multiple (sales, developers, managers)	Structured inputs	X	X	
	Query-Response		X	
	Incentives			X
Buckman Labs K'Netix	Forums	X	X	X*
	Code of Ethics		X	
	Principles		X	X
PwC KnowledgeCurve; 99% of 150,000 world-wide consultants	Users' view as integrated system	X		
	KM content process (editing/vetting)	X	X	
NASA/JPL: LLIS, APPL and KSI (face-to-face program); Project Libraries; *Know Who* directory; *Technical Questions* DB; legacy reviews; personal knowledge organizers (oral histories)	Resources	X		
	Portal	X		
	DocuShare	X		
	DBs	X		
	Leadership Development		X	X
DaimlerChrysler Knowledge Management Strategy	CAD/CAM	X		
	Product DB	X		
	EBOKs	X		
	TechClubs	X	X	
E&Y: EYKnowledgeWeb	Lotus Notes	X	X	
	Performance reviews			X
Andersen Consulting: Knowledge Xchange	Pre-filtered respositories	X	X	
	K Managers with specialized roles		X	
	Annual reviews		X	X
KPMG: Kweb	Intranet	X		
	Integration into business processes	X	X	X
	Incentives		X	X
Synthesis of Multiple Successful KM Projects (Davenport, Long et al. 1998)	Link to performance, value		X	
	Standard, flexible K structure	X	X	
	"K-friendly" Culture			X
	Clear language	X	X	
	Change in motivational practice			X
	Multiple channels for K Transfer	X	X	
	Senior Management Support			X

had national culture as a significant dimension has discarded this factor in its latest incarnation.

When examined using the lens of Carlile's pragmatic view of boundary objects, current KMSs and knowledge management efforts have emphasized the syntactic (knowledge transfer) and semantic (knowledge translation) levels of boundary spanning. Boundary spanning at the pragmatic (knowledge transformation and learning) level has focused on internal cultural and behavioral change (related to administrative power) rather than capturing and exploiting knowledge from diverse national or ethnic cultures.

Present-day KM efforts are richer in vision and technique than earlier KM implementations that had a primary emphasis on computer-based repositories of readily available data and knowledge — they include processes and forums in which people exchange knowledge face to face, for example. However, even current KM efforts have not explicitly recognized the potential value of objects and processes that operate at the pragmatic level to span cultural boundaries. It may be that boundary spanning at this level is required to realize the full potential of knowledge integration and co-creation of new knowledge from culturally diverse groups.

Based on this review, it appears that current KM efforts have developed consistent with the "culture-free" hypothesis. That is, by not taking into account multiple national cultures or ethnic backgrounds, these KM efforts tacitly adopt the position that a single organizational culture is appropriate without regard for the societal environment.

This finding raises several issues. The "culture-free" hypothesis is not consistent with other studies that examine culture in IT applications such as group support systems (Watson, 1994) and software production (Walsham, 2002). This raises the question of whether KM is going through a temporary phase and will evolve to a more culturally sensitive form of management, or if KM, as organizational processes and structures seem to be doing for production organizations, will appear the same regardless of cultural environment or the cultural composition of the organization. The stated goal of KM efforts is to achieve knowledge integration and benefit from the collective knowledge of the organization through learning. Because learning is so dependent on cultural experience, it remains an open question if the current culture-free approach is better. Can a culture-free approach enable a KM effort to effectively surface, or otherwise benefit from, the tacit knowledge of its staff from different cultures, the component of the "collective knowledge assets" that comes with the staff? More fundamentally, if KM is intended to exploit the entirety of the organization's intellectual assets, why do current systems appear to have ignored this ethnic and cultural component of the organization's collective knowledge? If organizations truly value all components of the collective knowledge, then KM efforts should pay more explicit attention to the use of objects and processes at the pragmatic level to span cultural boundaries.

By reviewing contemporary learning theory and showing the relationship of culture to learning in the context of knowledge management and knowledge management systems, this paper contributes to the KM discussion by directing attention to the cultural aspects of KM. The paper also highlights what may be a limitation of current KM efforts by viewing them as boundary-spanning systems that often operate at two of the levels in Carlile's model — the syntactic and the semantic — but not always at the pragmatic level.

The research in this paper has been limited to secondary reports of KM efforts taken from the management and information systems literature. Consequently, it has all the limitations that come with depending on the reports of others. Constraints of time and space, and the author's judgment on dealing with these constraints mean that the research may have missed significant reports from other fields. The paper also is limited by the scope of the reviewed literature. It represents a review of the recent KM literature and, while it examined the strategic links of KM to the KBV of the firm, the paper did not critically examine the fundamental assumptions in knowledge ontology and epistemology. The paper did not review, for example, ERP systems and other strategic systems that impose structure on entire organizations. A comparison of such systems with KM efforts might lead to different conclusions and different insights. Moreover, the paper did not examine the philosophical, linguistic, and semiotic foundations for the "syntactic, semantic, and pragmatic" levels in Carlile's boundary spanning model. Such an investigation might provide a richer set of explanations and models with which to improve our understanding of the complex practice of sharing and developing knowledge in organizations.

IMPLICATIONS FOR KM RESEARCH AND PRACTICE

The findings raise several questions about the current designs of KMSs and the evolution of KM. Missing in current discussions of KMSs is a discussion of the potential knowledge assets represented by culturally diverse staff in global organizations. Despite the KM goals of tapping into all parts of the collective knowledge in the firm, this cultural aspect of the organization's knowledge appears, from published reports, to be neglected. Without research on its value, one cannot say that this culturally based set of assets has sufficient value to deserve more formal recognition in KM efforts.

KM efforts appear to be adopting, without debate or research support for its efficacy, a culture-free design. Research is needed to determine if this culture-free approach is most appropriate for KM, or if other, more culturally sensitive approaches to KM would enable an organization to realize even greater benefits than it can realize using current practices. As researchers, we may want to look toward the critical review of our models of KM (e.g., Schultze, 2002), and to consider the suggestion that we re-frame our systems as supporting emergent knowledge (Markus et al., 2002), in order to think "outside the box" of the prevailing view of KM as applications of IT to improve the information value chain. If a culture-free approach is judged to be appropriate, then research is needed to understand if an intermediate step is necessary to enable individuals from non-western (non-North American) cultures to adapt to these approaches.

From a practical viewpoint, the research suggests that current KM efforts may need to give added attention to the learning dimension of their portfolio of activities. Contemporary learning theory shows a strong relationship between learning and cultural experience — individuals learn based on how they have learned in the past, and early ethnic and cultural experiences provide a base of models and abilities that enhance and constrain an individual's capabilities in new situations. The implication of these theories for KM practice is two-fold. First, staff from different ethnic and cultural backgrounds will not necessarily share a basic set of models when they begin with their new firm, and these staff may benefit from efforts to bridge the gap in meta-learning skills and models. The frequent reports of KM efforts to reward use of a KMS (conform to a corporate norm) may reflect one approach to bridging this gap, but typically these have been reported simply as efforts for corporate cultural change. Second, if a firm is to benefit from the tacit knowledge these staff members bring with them, it may need to incorporate processes and techniques that are not evident in current examples of KM efforts.

ACKNOWLEDGMENTS

This paper is significantly different from an earlier version, and the author thanks Roberto Evaristo and three anonymous reviewers for thoughtful and insightful comments that led to this revision.

REFERENCES

Alavi, M. (1997). *KPMG Peat Marwick U.S.: One Giant Brain.* Cambridge, MA, HBS: Report No. 9-397-108, 21pp.

Alavi, M., & Leidner, D.E. (2001). Review: Knowledge management and knowledge management systems: Conceptual foundations and research issues. *MIS Quarterly, 25*(1), 107-136.

Allen, R., & Rooney, P. (1998). Designing a problem-based learning environment for ESL students in business communication. *Business Communication Quarterly, 61*(2), 48.

Barney, J., Wright, M., et al. (2001). The resource-based view of the firm: Ten years after 1991. *Journal of Management, 27*(6), 625-641.

Birnbaum, P. H., & Wong, G.Y.Y. (1985). Organizational structure of multinational banks in Hong Kong from a culture-free perspective. *Administrative Science Quarterly, 30*(2), 262-277.

Boud, D., & Feletti, G. (1998). *The challenge of problem-based learning.* London: Kogan Page.

Brown, J.S. (2002). An epistemological perspective on organizations and innovations. *Organizational Knowledge and Learning Conference 2002*, Athens.

Brown, J.S., & Duguid, P. (1998). Organizing knowledge. *California Management Review, 40*(1), 90-111.

Brown, J.S., & Duguid, P. (2001). Knowledge and organization: A social-practice perspective. *Organization Science, 12*(2), 198-213.

Carlile, P.R. (2002). A pragmatic view of knowledge and boundaries: Boundary objects in new product development. *Organization Science, 13*(4), 442-255.

Chapman, R., & Hyland, P. (in press). Complexity and learning behaviors in product innovation. *Technovation.*

Chard, A.M. (1997). *Knowledge Management at Ernst & Young.* Stanford, Graduate School of Business, Leland Stanford Junior University: 28.

Cole, M. (1971). *The cultural context of learning and thinking: An exploration in experimental anthropology.* New York: Basic Books.

Culatta, E. (1994). New modes for making scientists. *Science, 266*(5186), 875+877+883-884+888.

Davenport, T.H., Long, D.W.D., et al. (1998). Successful knowledge management projects. *Sloan Management Review, 39*(2), 43.

Davenport, T.H., & Prusak, L. (1998). *Working knowledge: How organizations manage what they know.* Boston: Harvard Business School Press.

Dixon, N.M. (1994). *The organizational learning cycle: How we can learn collectively.* New York: McGraw-Hill.

Dougherty, D. (1992). Interpretive barriers to successful product innovation in large firms. *Organization Science, 3*(2), 179-202.

Drucker, P.F. (1988). The coming of the new organization. *Harvard Business Review, 66*(1), 45.

Erez, M., & Earley, P.C. (1993). *Culture, self-identity, and work.* New York: Oxford University Press.

Forman, E.A., Minick, N., et al. (1993). *Contexts for learning: Sociocultural dynamics in children's development.* New York: Oxford University Press.

Fulmer, W.E. (1999). Buckman Laboratories (B). Cambridge, MA, Harvard Business School: Report No. 9-800-033, 2pp.
Fulmer, W.E. (2000). Buckman Labs (A) Teaching Note. Cambridge, Harvard Business School: Report No. 5-800-360.
Fulmer, W.E. (2003). Buckman Laboratories (A). Cambridge, MA, Harvard Business School: Report No. 9-800-160, 15pp.
Grant, R.M. (1996a). Prospering in dynamically-competitive environments: Organizational capability as knowledge integration. *Organization Science, 7*(4), 375.
Grant, R.M. (1996b). Toward a knowledge-based theory of the firm. *Strategic Management Journal, 17*(Winter), 109-122.
Hall, E.T. (1959). *The silent language.* New York: Doubleday & Co.
Hansen, M.T., & Davenport, T.H. (2002). *Knowledge management at Andersen Consulting.* Cambridge, MA: HBS.
Hilliard, A.G. III (1992). Behavioral style, culture, and teaching and learning. *Journal of Negro Education Africentrism and Multiculturalism: Conflict or Consonance, 61*(3), 370-377.
Huber, G. (1991). Organizational learning: The contributing processes and literatures. *Organization Science, 2*(1), 88-115.
Jones, R., Higgs, R., et al. (200*1).* Changing face of medical curricula. *The Lancet, 357*(9257), 699-703.
Klein, S.B., & Mowrer, R.R. (1989). *Contemporary learning theories: Instrumental conditioning theory and the impact of biological constraints on learning.* Hillsdale, NJ: Lawrence Erlbaum Associates.
Kogut, B., & Zander, U. (1996). What firms do? Coordination, identity, and learning. *Organization Science, 7*(5), 502-518.
Kolb, D.A. (1984). *Experiential learning: Experience as the source of learning and development.* Englewood Cliffs, NJ: Prentice-Hall.
Kozulin, A. (1998). *Psychological tools: A sociocultural approach to education.* Cambridge, MA; London: Harvard University Press.
Leonard, D., & Kiron, D. (2002). *Managing knowledge and learning at NASA and the Jet Propulsion Laboratory* (JPL). Cambridge, MA: Harvard Business School.
LoLordo, V. M., & Droungas, A. (1989). Selective associations and adaptive specializations: Taste aversions and phobias. In S. B. Klein, & Robert R. Mowrer (Eds.), *Contemporary learning theories: Instrumental conditioning theory and the impact of biological constraints on learning* (pp. 145-180). Hillsdale, NJ, Lawrence Erlbaum Associates.
MacCormack, A. (2002). *Siemens ShareNet: Building a knowledge network.* Cambridge, MA: Harvard. Report no. 9-800-160, 27pp.
Madhok, A. (2002). Reassessing the fundamentals and beyond: Ronald Coase, the transaction cost and resource-based theories of the firm and the institutional structure of production. *Strategic Management Journal, 23*(6) 535.
March, A. (1997). *A note on knowledge management.* Cambridge, MA, HBS: Report No. 9-398-031, 20pp.
Markus, M. L., Majchrzak, A., et al. (2002). A design theory for systems that support emergent knowledge processes. *MIS Quarterly, 26*(3), 179.

Massey, A.P., Montoya-Weiss, M.M., & O'Driscoll, T.M. (2002). Knowledge management in pursuit of performance: Insights from Nortel Networks. *Management Information Systems Quarterly, 26*(3), 269-289.

Mata, F.J., Fuerst, W.L., & Barney, J.B. (1995). Information technology and sustained competitive advantage: A resource-based analysis. *MIS Quarterly, 19*(4), 487.

Maurice, M. (1976). Introduction - Theoretical and ideological aspects of the universalistic approach to the study of organizations. *International Studies of Management & Organization, 6*(3), 3.

Miller, G. A. (1987). Meta-analysis and the culture-free hypothesis. *Organization Studies, 8*(4), 309-326.

Mowrer, R. R., & Klein, S.B. (1989). *A contrast between traditional and contemporary learning theory. Contemporary learning theories.* Hillsdale, NJ: Lawrence Erlbaum Associates.

Ng, P., & Farhoomand, A. (2002). *The PricewaterhouseCoopers Know-ledgeCurve and the spinoff of PwC Consulting.* Hong Kong, The Centre for Asian Business Cases, University of Hong Kong: HKU221; reprinted by HBS as 02/147C; 15pp.

Nonaka, I., & Takeuchi, H. (1995). *The knowledge-creating company.* New York and Oxford: Oxford University Press.

Penrose, E.T. (1959). *The theory of the growth of the firm.* Oxford: Blackwell.

Ramesh, B., & Tiwana, A. (1999). Supporting collaborative process knowledge management in new product development teams. *Decision Support Systems, 27*(1-2), 213-235.

Rayport, J.F., & Sviokla, J.J. (1995). Exploiting the virtual value chain. *Harvard Business Review, 73*(6), 75-85.

Rebentisch, E.S., & Ferretti, M. (1995). A knowledge asset-based view of technology transfer in international joint ventures. *Journal of Engineering and Technology Management, 12*(1-2), 1-25.

Rukstad, M.G., & Coughlan, P. (2001). *DaimlerChrysler knowledge management dtrategy.* Cambridge, MA, HBS: Report No. 9-702-412; 18.

Schultze, U., & Leidner, D.E. (2002). Studying knowledge management in information systems research: Discourses and theoretical assumptions. *Management Information Systems Quarterly, 26*(3).

Soo, C., Devinney, T., et al. (2002). Knowledge management: Philosophy, processes and pitfalls. *California Management Review, 44*(4), 129.

Spender, J.-C. (1996). Making knowledge the basis of a dynamic theory of the firm. *Strategic Management Journal,* 17, 45.

Star, S. L., & Greisemer, J.R. (1989). Insitutional ecology, 'translations' and boundary objects: Amateurs and professionals in Berkeley's Museum of Vertebrate Zoology, 1907-39. *Social Studies in Science,* 19, 387-420.

Stata, R. (1989). Organizational learning: The key to management innovation. *Sloan Management Review, 30*(3), 63-74.

Tien, C.-J., Ven, J-H., & Chou, S-L. (2003). Using the problem-based learning to enhance student's key competencies. *Journal of American Academy of Business, Cambridge, 2*(2), 454.

Verganti, R. (1998). Leveraging on systemic learning to manage the early phases of product innovation projects. *Journal of Product Innovation Management, 15*(5), 471-472.

Vygotsky, L.S., Rieber, R.W., et al. (1987). *The collected works of L.S. Vygotsky*. New York: Plenum Press.
Walker, A., Bridges, E., & Chan, B. (1996). Wisdom gained, wisdom given: Instituting PBL in a Chinese culture. *Journal of Educational Administration, 34*(5), 12.
Walsh, J. P., & Ungson, G.R. (1991). Organizational memory. *Academy of Management Review, 16*(1), 57-91.
Walsham, G. (2002). Cross-cultural software production and use: A structurational analysis. *MIS Quarterly, 26*(4), 359-380.
Watson, R. T., Ho, T.H., & Raman, K.S. (1994). Culture: A fourth dimension of group support systems. *Communications of the ACM, 37*(10), 44-55.
Zull, J.E. (2002). *The art of changing the brain: Enriching teaching by exploring the biology of learning*. Sterling, VA: Stylus Publishing, LLC.

ENDNOTE

[1] The author thanks an anonymous reviewer for pointing out that incentives and other behavioral change processes could be classified in this way.

Previously published in the Journal of Global Information Management, 11(4), 20-36, October - November 2003.

Chapter IV

Managing Distributed Projects Across Cultures

Roberto Evaristo, University of Illinois at Chicago, USA

ABSTRACT

A model of cross-cultural distributed project management is proposed. The model is based on recent theoretical developments regarding trust and team processes and suggests that trust relates to cultural differences between distributed members. Trust affects project performance in two key ways: through the traditional view of main effects on performance, and via moderating effects on other determinants of performance in distributed projects. We also use a task characteristics categorization, together with several mini-case studies, to create a set of guidelines for best practices for management of cross-cultural distributed projects.

INTRODUCTION

Virtual organizations are increasingly becoming the focus of attention in different literatures. In particular, much research has been carried out in the area of virtual teams and how groupware is able to support such teams. An essential component of the rise of the use of virtual teams is the geographical distribution afforded by the globalization of businesses as well as the availability of inexpensive, advanced information and communication technologies (ACT).

This geographical distribution increases the chance that members of such teams originate from different cultures. Differences in expectations and value-laden behaviors resulting from culture clash create performance and personal relationship issues.

Although some of these problems are addressed in the literature, one large frontier remains unexplored: the management of cross-cultural differences and their consequences that are inherent in distributed projects.

Project management knowledge and application has been devoted for a long time to single-site, single projects (Turner, 1993, 1995). Recently, several authors have started to discuss different issues occurring either when projects are composed of many subprojects — also called programs (Graham, 2000) — or are spread across different locations, in a "virtual" or "distributed" project. The latter situation creates several logistical and management problems that are exacerbated by cultural differences.

This manuscript discusses issues associated with managing distributed projects across different cultures. It is mostly conceptual, but it also includes descriptions taken from mini case studies conducted by the author in Japan (Evaristo & Scudder, 2000), USA (Suleiman, Evaristo, & Kelly, 2000), Germany (Evaristo, 2001), and Norway.

We will discuss in more detail what a distributed project is, its differences from a traditional project, and the resulting need for a customized way to manage it. We then present a proposed model to manage distributed projects, emphasizing the direct and moderating role of trust in the performance of cross-cultural distributed projects under cultural differences. Finally, based on the model, we offer recommendations on how to best manage distributed, cross-cultural projects.

DISTRIBUTED PROJECTS

The overwhelming number of projects described in the research-oriented literature as well as most of the practical and theoretical developments in this area is focused on single, stand-alone projects. This is also true for practitioner-oriented books on how to perform project management (Lock, 1996). A few exceptions describe multiple-site instead of single-site projects; for instance, sets of smaller projects that, although performed independently, need to be managed concurrently by the same management team. These sets are defined in the literature as "programs" (Van Der Merwe, 1997).

In fact, a careful analysis of other examples in the literature suggests that there are more types of projects than the single site type, as the multiple site dichotomy described above seems to suggest. For instance, Kumar (1996) describes a software development project where most of the developers were in India and the client was in the U.S. In this case, they were separated by thousands of miles, by 12 time zones, and by cultural and religious differences, but were still working on the same project, characterizing a single project involving multiple non-collocated sites. This type of project differs from the previous two examples because now the non-collocated stakeholders are working on the *same* project, and we will define it here as "distributed projects."

The fundamental differences in coordination and resource needs across sites in programs (little), or across sites on a single but distributed project, led Evaristo and Fenema (1999) to two conclusions: (1) qualitatively different management approaches would be critical for successful completion of such projects; and (2) before attempting to address such differences, it would make sense to inquire whether these three types of projects were, indeed, a complete set of project types. The resulting work developed a complete typology of projects that included exploration of alternatives (e.g., multiple distributed projects with shared locations). The crucial differences among them relate to both increasing difficulties in communication and coordination of interdependent shared resources. For instance, *collocated programs* have the lowest level of project interdependence resulting from related objectives, shared common deliverables, information, resources, and technology. On the other hand, in *multiple collocated programs*, the geographical dispersion not only creates communication problems, but also the sharing of resources requires more careful scheduling. We refer the interested reader to the original work for a full discussion.

In the next section, we present the proposed model with a discussion of the individual elements. The result is a set of guidelines on how to manage such projects. We conclude with recommendations for further research.

A MODEL TO UNDERSTAND THE MANAGEMENT OF CROSS-CULTURAL DISTRIBUTED PROJECTS

Managing distributed projects is extremely complex; this is particularly true when stakeholders originate from multiple cultures. The model presented here uses recent theoretical findings to fit observations and data collected by the author and by other sources to develop a new perspective on key issues these projects face. However, this model is not intended to be exhaustive or definitive, but an informative model (Figure 1) that will provide general guidelines to manage such projects.

The key variable in this model is trust as it is impacted by cultural differences or similarities. Separate authors have addressed different aspects of trust in the context of virtual teams [e.g., Jarvenpaa & Knoll, 1998; Solomon, 2001; Jarvenpaa & Leidner, 1998; Iacono & Weisband, 1997)]. The proposed model will expand the understanding of how trust affects the performance of distributed projects by adding a moderating effect to the more traditional direct or main effect. This distinction is based on the work of Dirks and Ferrin (2001), and it implies not only different theoretical rationales, but also different "methodological designs, different forms of statistical evidence, and different ways of using trust as a managerial intervention" (p. 451).

The main effects operate in a clear-cut manner: higher levels of trust are supposed to result in more positive attitudes, superior levels of cooperation and other forms of workplace behavior, as well as higher levels of performance. The moderating effect alternative explanation of trust, although not new, had received very little attention in the literature before Dirks and Ferrin (2001). They suggest that the presence of trust changes the effect of other determinants on work attitudes, perceptions, behaviors, and performance outcomes. Basically, trust enables an environment where more cooperation, higher performance, and other attitudes and perceptions are more likely. For simplicity, instead of providing a slew of performance-affecting determinants, we presented the team processes listed by Marks et al. (2001).

The other key difference on how our model treats trust in the context of distributed projects is the explicit recognition of cross-cultural differences as a factor in the formation of differences in trust. This model will allow us to generate guidelines on better ways to manage distributed projects whose stakeholders originate from different cultures. Additionally, we will expand on how culture is one of the antecedents of trust, as

Figure 1. Model of management of cross-cultural distributed projects

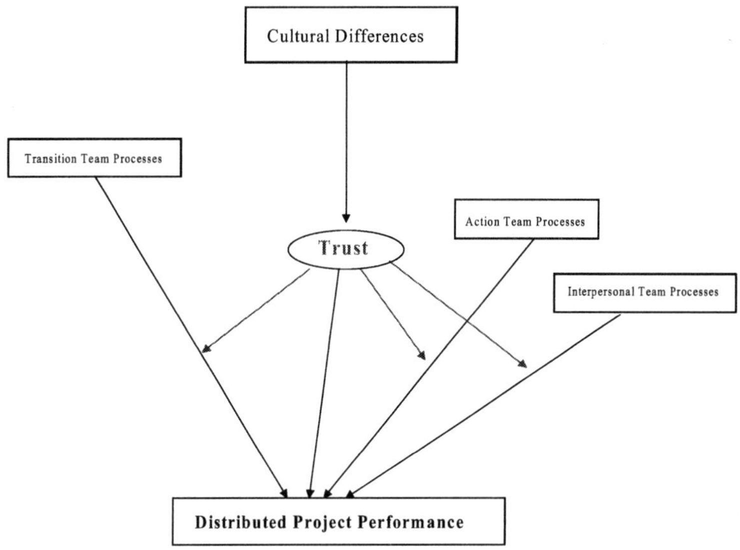

well as the concept of trust as both main and moderating effects on determinants of high performance in distributed projects. We will use examples from previous research to provide initial support for the model.

CULTURAL DIFFERENCES

One of the most commonly cited definitions of culture is that of Hofstede (1980). Hofstede defines culture as a collective programming of the mind. He further states that, "The collective level of mental programming is shared with some but not with all other people; it is common to people belonging to a certain group or category, but different among people belonging to other groups or categories" (p.15). Hofstede further defines culture by its dimensions: power distance, uncertainty avoidance, individualism, femininity, and time horizon. It should be noted that Hofstede's definition of culture is value-based, and that alternatives do exist (Straub, Loch, Evaristo, Karahanna, & Srite, 2002). Although there have been several challenges to Hofstede's operationalization of culture, and therefore his results (e.g., Erez & Earley, 1993), Hofstede's definition of culture and its dimensions is satisfactory for our purposes.

Power distance is a measure of the acceptance of the perception of inequality by members of a culture. For instance, Hoecklin (1994) mentions that "French managers thought that authority to make decisions came as a right of office," whereas Dutch ones had a "relaxed approach to authority," and Germans felt more comfortable in formal hierarchies.

Uncertainty avoidance, on the other hand, is the lack of tolerance for ambiguity and the need for formal rules. Lifetime employment, for instance, is more likely to exist in countries with high uncertainty avoidance like Japan.

Individualism/collectivism is how much a given individual puts his or her interests ahead of the interests of the group to which he or she belongs. Earley (1989), for instance, reports on a study where individualistic people performed best when operating individually and not anonymously, whereas collectivistic people performed best anonymously and in groups. The opposite was true for collectivistic individuals when they had to identify their work and present it individually. Individualistic people fared worst in their performance when working in groups and, therefore, anonymously.

Masculinity/femininity relates to the extent of emphasis on work goals and assertiveness as opposed to personal and family goals. In countries with a more feminine culture, quality of life is more valued than career advancement. One of the consequences is that people tend to spend more time with their family. Finally, time horizon relates to short-term vs. longer-term orientation. Western societies are typically shorter-term oriented than Eastern societies. In the next section, we will elaborate on how these cultural differences may affect trust.

Culture as One of the Trust Determinants

In this manuscript, we use Rousseau, Sitkin, Burt, and Camerer's (1998) cross-disciplinary definition of trust: a psychological state comprising the intention to accept vulnerability based upon positive expectations of the intentions or behavior of another. One of the reasons people may not initially have trust for one another is the lack of knowledge about rationale for past or present behaviors and intentions and, therefore, the lack of willingness to risk vulnerability to an unknown situation. Although this is likely in face-to-face situations, it is even more true in distributed environments, where people cannot observe rich cues (Daft & Lengel, 1986). When there are deep cultural differences coloring the perceptions of individuals across distributed groups, the most difficult situation develops. In this case, individuals may not even understand the cause and effect relationships that determine how certain behaviors originate from stimuli. As a result, behaviors intended to generate trust with the other distributed sites may be misinterpreted. Our model tries to emphasize this point by depicting "cultural differences" as a precursor of trust, and implies that the way in which these differences are managed can have a considerable impact on how trust is generated and, as a result, ultimately affect the performance of the project as a whole.

Team Processes and Associated Task Characteristics: A Gateway to Cultural Understanding in Distributed Projects

Another part of the model is the team processes borrowed from Marks, Mathieu, and Zaccaro (2001). In this section, we explore the connection between team processes and task characteristics and their combined fit with the cultural characteristics described above. Marks et al. (2001) proposed a taxonomy of team processes. Although intended for face-to-face teams, their taxonomy can be adapted and applied to distributed teams as well. These team processes involve tasks that can be characterized using one of the several task categorizations available (e.g., Evaristo, 1993). Crucial to the understanding of the model is the fit between the cultural characteristics of an individual and the task characteristics afforded by the team processes in distributed projects. Higher fit between cultural and task characteristics is also a likely cause of higher performance.

One task categorization schema presented by Evaristo (1993) includes the following two task characteristics: task formalization and task complexity. *Task formalization* is defined as the level of specification and structure afforded by the task. A highly structured task can be fragmented or subdivided more easily, with less need for interconnections and coordination across subparts. *Task complexity* actually includes three types of complexity (Wood, 1986): component complexity, coordinative complexity, and dynamic complexity. Component complexity depends on the number of distinct acts and distinct information cues necessary to perform a task. Coordinative complexity refers to the nature of relationships — such as sequencing and timing — between task inputs and task products. Finally, dynamic complexity is a function of changes in the states of the world that have an effect on the relationships between task inputs and products, and/or the task inputs and products themselves.

Distributed projects are likely to have more of each of these complexities than collated projects, sometimes overlapping and compounding each other, making the management of these projects unwieldy. For instance, task formalization affects the ease of divisibility of project parts that are developed in dispersed locations. Higher formalization is likely to imply higher cohesion within collated "modules" and lower coupling across dispersed "modules," one of the tenets of structured programming (Page-Jones, 1980). Modules overlapping locations are even more affected by coordinative and dynamic complexity because of need to share resources.

These task characteristics also have an intrinsic connection with the original team processes taxonomy proposed by Marks et al. (2001) and based on a meta-analysis of team process research in the last several decades (Table 1). They suggest that there are three types of team processes: transition, action, and interpersonal processes.

In the *transition processes,* the level of task structure or formalization is lowest. Task formalization is related to two cultural characteristics: uncertainty avoidance and, to some extent, individualism. A less formalized task, because of its lower structure, brings more uncertainty and, therefore, seems more appropriate in cultures with lower uncertainty avoidance. Moreover, when appropriate channels do exist in the mechanism of formalizing the task, strategies to deal with power distance may change as well. Both situations are magnified in distributed environments.

Action processes, on the other hand, afford a much higher level of task structure, to the point that one can prepare a checklist of action items and routinely go through them. Therefore, cultures higher in uncertainty avoidance can also excel at these tasks. An interesting caveat is that more formalized tasks can stifle creativity (Amabile, 1998), particularly for people in highly individualistic cultures.

Across both transition and action processes, task complexity is inherently task-specific and cannot be predicted based on process alone. Cultural values and task complexity are related in several ways. The timing of certain acts, as in coordinative complexity and the perception of waiting in between different tasks, is connected to the perception of short-term vs. long-term horizon. Something else is the "mañana" approach, or the fact that many cultures' tasks have very elastic impressions of urgency.

Table 1. Taxonomy of team processes (Marks et al., 2001)

Process Dimensions	Definition
Transition processes	
Mission analysis, formulation, and planning	Interpretation and evaluation of the team's mission, including identification of its main tasks as well as the operative environmental conditions and team resources available for mission execution
Goal specification	Identification and prioritization of goals and subgoals for mission accomplishment.
Strategy formulation	Development of alternative courses of action for mission accomplishment
Action processes	
Monitoring progress toward goals	Tracking task and progress toward mission accomplishment, interpreting system information in terms of what needs to be accomplished for goal attainment, and transmitting progress to team members
Systems monitoring	Tracking team resources and environmental conditions as they relate to mission accomplishment, which involves (1) internal system monitoring (tracking team resources such as personnel, equipment, and other information that is generated or contained within the team) and (2) environmental monitoring (tracking the environmental conditions relevant to the team)
Team monitoring and backup behavior	Assisting team members to perform their tasks. Assistance may occur by (1) providing a teammate verbal feedback or coaching, (2) helping a teammate behaviorally in carrying out actions, or (3) assuming and completing a task for a teammate.
Coordination	Orchestrating the sequence and timing of interdependent actions
Interpersonal Processes	
Conflict management	Preemptive conflict management involves establishing conditions to prevent, control, or guide team conflict before it occurs. Reactive conflict management involves working through task and interpersonal disagreements among team members.
Motivation and confidence building	Generating and preserving a sense of collective confidence, motivation, and task-based cohesion with regard to mission accomplishment.
Affect management	Regulating member emotions during mission accomplishment, including (but not limited to) social cohesion, frustration, and excitement.

Furthermore, team members originating from lower uncertainty avoidance cultures may be better at engaging in transition process typically with lower structure/higher uncertainty; on the other hand, members from higher uncertainty avoidance cultures will be better at action processes. Task complexity and culture are related on a case-by-case basis. In addition, interpersonal processes are better dealt with if there is the recognition of such cultural differences.

In summary, higher fit between the task characteristics in the team processes described above and cultural characteristics of the distributed project stakeholders will improve the likelihood of better performance in the distributed project. However, an important effect is that the levels of trust among the stakeholders moderate such a relationship. The next section will detail the moderation relationship. In the sequence, we will detail the main effect of trust on overall project performance.

Trust as Moderating Effect on Determinants of Performance in Distributed Projects

Dirks and Ferrin (2001) propose that trust generates two distinct processes that in turn either inhibit or foster positive outcomes in the relationship. One is that trust affects how people assess future behavior of partners; the other is that trust affects how people interpret past or present actions of the same partners. Taken as a set, trust in past, present, and future actions would tend to reduce some of the uncertainty or ambiguity inherent in relationships.

They propose that trust moderates the relationships between motivational constructs and workplace behavior and outcomes. In this view, instead of thinking of trust as a factor that causes individuals to engage in workplace behaviors, consider trust as an influencing factor on the type or degree of behavior. Therefore, whereas the main effect argument would suggest that trust creates cooperative behavior, the moderation argument suggests that trust helps facilitate the existence of cooperative behavior, because one believes that a partner is more willing to cooperate under high trust. In our model, we propose that this moderation effect changes the way in which the team processes proposed by Marks et al. (2001) affect the performance of distributed projects. The team processes themselves relate to culture as outlined earlier in the chapter, even though relationships are not shown on the model for parsimony of drawing.

The other side of the coin is that when trust is non-existent, cooperation is less likely to exist. For instance, in one experiment conducted in the fall of 1999, students in Norway played the role of analysts, interviewing "users" in Virginia and delivering system specifications for Colorado "developers," all in the same semester. Differences in culture made certain exchanges very difficult, with strong finger pointing and blaming behavior associated with perceived shirking behavior across locations. Moreover, teams that started with high trust tended to end with high trust as well, explainable by the self-reinforcing positive pattern generated. A similar result was also found by Kanawattanachai and Yoo (2002).

In fact, Dirks and Ferrin (2001) explicitly make an argument about what they call cooperative behavior in mixed-motive contexts, more likely to be the case in distributed projects with a high cross-cultural component. Ways in which this type of problem has been studied include negotiation, prisoner's dilemma, and social dilemmas (e.g., Komorita & Parks, 1995). Dirks and Ferrin, however, proposed to use the goal expectancy model to suggest that trust would moderate the relationship between an individual's goal and cooperative behavior and, therefore, performance. Cooperative behavior is included in the transition, action, and interpersonal processes.

Therefore, under low trust, cooperative goals are less likely to cause cooperative behavior. Moreover, in situations where there are competitive goals in addition to cooperative goals, the amount of trust may determine which one will be given extra weight. For instance, co-authors working on the same project may feel like slacking off if they perceive the other is likely to do the same.

The cooperative behavior is particularly critical for coordination of an action process. Coordination is the need to decide who does what when. If there is a single project manager or coordinator — or even a clear hierarchical chain — this is much simpler. Nevertheless, this is not always the case. Ambiguities or lack of structure or specification in the task may require a high amount of communication to clarify doubts.

This brings us to the connection of coordination with power distance, another cultural characteristic. Distributed projects have a natural tendency to flatten the perceived hierarchy, since direct managers become less visible. Moreover, it becomes more likely that people will connect with each other using different types of media that are inherently less formal and allow for direct communication across actual hierarchical lines, therefore bypassing formal reporting loops. That was the case where Evaristo

(2001) describes the development of a grant proposal in a distributed environment involving members from different cultures and hierarchical levels and some of the resulting strategies.

Trust as Main Effect on Performance in Distributed Projects

The basic tenet of trust as a main effect comes from Mayer et al. (1995), who suggested that one's beliefs about another individual's ability, benevolence, and integrity increase one's willingness to risk. Risk-taking behavior is associated with expectations for higher individual performance; in the case of groups, it is associated with higher unit performance (Larson & LaFasto, 1989). This was found in many of the seminal works in trust in virtual teams (e.g., the multiple Jarvenpaa et al. studies (1998) and Iacono and Weisband (1997)).

In fact, the creation of trust among members who may never meet face-to-face (Jarvenpaa & Knoll, 1998) was deemed critical for high distributed team performance. Ideally, advanced communication technologies play an important role in this trust creation, because, particularly in relatively low-budget or very short projects, there may never be enough resources to enable face-to-face meetings. This applies to all types of distributed projects described above and becomes more complex as the projects become more distributed in dimensions beyond a number of locations and projects (Evaristo & Scudder, 2000). For instance, if in addition to distance across locations, the cultural heterogeneity of groups is high, it is harder to develop trust among stakeholders.

An example is gleaned from Evaristo and Scudder (2000). They describe the Titus Project, a joint venture of U.S. West, Time Warner, Toshiba, and another Japanese partner with the objective of wiring several main Japanese cities with optical fiber for high bandwidth data and voice and image transmission. In the beginning, problems abounded. The job of the Denver office was to send plans on installation location of cables, repeaters, and other hardware. The local field personnel would survey the actual location and propose changes to send back to the U.S. A few mistakes on both sides created an uncomfortable level of mistrust, which was eventually solved via face-to-face meetings. One of the interviewees said:

There has been face to face contact with people in Denver — we made it a point that they know each other. Time Warner people here in Japan knew people in Denver originally, but they still also invited them to come to Japan to meet Japanese staff. We sent Japanese staff to Denver or Pennsylvania so they could meet each other and see each other's shop just to get a better feeling for what they do. They feel that's very important, because if there's a problem, it gets solved a lot faster if there's a relationship there, and to know what the other can physically handle. E-mails took a while because everyone had different types of communications software, and we had to get that consistent throughout.

However, after the distributed teams involved met face-to-face and got used to the repetitive procedure of exchanging corrected plans between the Tokyo and the Denver office, most of the exchanges boiled down to electronic file transfers. U.S. designers learned 50 or so different Japanese characters relevant to the maps and plans to help alleviate language barriers. A similar effect is described by Ridings, Gefen, and Arinze (2002) who mention that "trust had a downstream effect on members' intentions to both give information and get information through the virtual community" (p. 271).

Cultural heterogeneity affected the amount of immediate trust placed on the distributed partner, and interventions were designed to re-create trust that had been compromised. In this case, the main effect of the trust created between the groups directly increased the distributed team performance. Ultimately, although face-to-face interaction is ideal to initiate a project, this may not be possible. Careful ACT use should be planned as a close second best to facilitate initial trust formation, and should be the vehicle for continued interaction.

GUIDELINES TO MANAGE CROSS-CULTURAL DISTRIBUTED PROJECTS

The current manuscript does not attempt to present an exhaustive list of methods to manage distributed projects when they involve members from different cultures. The length of this topic more appropriately lends itself to a more detailed discussion in a full-length book. Therefore, the objective is to use our model to point out critical issues to which a practitioner may need to attend. Similarly, these issues may provide a fertile ground for further research.

The inspiration for the suggestions below is based on the following tenets: (1) fit between task characteristics embedded in team process with cultural characteristics of stakeholders improves performance, particularly in the presence of high trust (moderation effect); and (2) trust is impacted by cultural differences and, in turn, directly affects performance as well.

The first step is to analyze the level of heterogeneity of project participants related to the crucial cultural dimensions. Although in certain negotiation tasks cultural heterogeneity has not shown to be an issue (Anderson & Hiltz, 2001), in other situations cultural heterogeneity and corresponding background differences create situations where understanding of cause-effect relationships are fundamentally different (Evaristo & Scudder, 2000). As a result, trust tends to take longer to develop, consequently affecting project performance negatively. This effect happens not only directly (i.e., lower trust causes lower cooperation and, therefore, lower performance) but also indirectly (i.e., in absence of trust, there is less incentive to engage in cooperation, or, even worse, individuals are likely to engage in competitive behavior, therefore lowering performance).

The second step is to analyze the level of complexity and formalization in the key tasks in which the group will be involved in the context of both project type (which suggests resource interdependencies and coordination levels required) and the different team processes presented in the model.

The third step will be to cross such analysis with the cultural analysis from the first step. For instance, let us examine the most complex case of "multiple distributed projects with shared locations." The manager can start the process by analyzing how task formalization or complexity affects each of the crucial team processes — action, transition, and interpersonal. Next, the manager should cross this result with the cultural characteristics of the stakeholders — uncertainty avoidance (because of the ambiguity that the complexity will generate); power distance (How many bosses are there? Who has control over what?); individualism (How will other group members perform, and do I care?); and femininity-masculinity (Who will get the recognition?). The result should be a clear picture of the project and its susceptibility to trust and cultural variations.

Once this step is completed, it behooves the project manager to evaluate how cultural heterogeneity may affect the trust formation in that particular project and among team members, and be proactive about it. One of the potential solutions is trust development training (Beranek, 2000). Heterogeneous groups are not synonymous with "bad" groups; in fact, Mejias (1997) found that such groups generated more ideas, lending support to creation of synergies, as long as their energy is appropriately channeled.

Another key suggestion is to surface some of the cultural assumptions that may be influencing the appropriate behavior of several of the different team locations. Such strategy may help individuals begin to understand that differences do exist, and that these differences are often responsible for some of the reactions they may be observing.

CONCLUSION

In this chapter, we have used a theory-driven approach to provide a model that helps to understand the management of distributed projects with stakeholders originating from different national cultures. We used the model not only to reinforce the traditional view of trust's main effects on the performance of such projects, but also to explore the moderation effects of trust on the relationship between team processes and performance, something new in this context. Moreover, we explained how culture affects trust and, therefore, overall performance. Using the explanatory power of our model, we provided guidelines for the management of cross-cultural distributed projects.

The concepts discussed here provide a glimpse into research areas that are likely to generate fruitful results and, therefore, merit further exploration (e.g., the identification of formats and templates for structuring and splitting projects across locations, with the objective of fostering cultural synergies based on task analysis). In addition, further examination should be given to an understanding of critical aspects of the cultural makeup of teams that can decrease the incidents of cross-cultural relationship problems. And when cross-cultural relationship problems do occur, we need to better understand how to address them with the effective use of ACT. Another question of interest is trust development in a cross-cultural environment in general. Finally, a qualitative analysis of best practices may yield further improvements to the state of the art in the management of distributed projects in culturally diverse teams.

ACKNOWLEDGMENTS

The author would like to thank two anonymous reviewers and suggestions made by Michael Eierman and Fred Niederman.

REFERENCES

Amabile, T.M. (1998, September-October). How to kill creativity. *Harvard Business Review, 76*(5), 77-87.

Anderson, W.N., & Hiltz, S.R. (2001). Culturally heterogeneous vs. culturally homogeneous groups in distributed group support systems: Effects on group process and outcomes [Paper]. At *34th Hawaii International Conference on System Sciences*, Maui, Hawaii.

Beranek, P. (2000). The impacts of relational and trust development training on virtual teams: An exploratory investigation [Paper]. At *33rd Hawaii International Conference on System Sciences*, Maui, Hawaii.

Daft, R. L., & Lengel, R.H. (1986). Organizational information requirements, media richness and structural design. *Management Science, 32*(5), 554-571.

Dirks, K.T., & Ferrin, D.L. (2001). The role of trust in organizational settings. *Organization Science, 12*(4), 450-467.

Earley, C. (1989). Social loafing and collectivism: A comparison of the United States and the People's Republic of China. *Administrative Science Quarterly, 34*, 565-581.

Erez, M., & Earley, P.C. (1993). *Culture, self-identity and work.* New York: Oxford University Press.

Evaristo, J.R. (1993). *An empirical investigation of the effect of information technology and information characteristics on individual information load* [Unpublished thesis]. Minneapolis, MN.

Evaristo, J.R. (2001). Non-consensual negotiation in distributed collaboration. *Communications of the ACM, 44*(12), 89.

Evaristo, J.R., & Fenema, P.C.V. (1999). A typology of project management: Emergence and evolution of new forms. *International Journal of Project Management, 17*(5), 275-281.

Evaristo, J.R., & Scudder, R. (2000). Geographically distributed teams: A dimensional analysis [Paper]. At *33rd Hawaii International Conference on Systems Sciences*, Maui, Hawaii.

Graham, A.K. (2000). Beyond PM 101: Lessons for managing large development programs. *Project Management Journal, 31*(4), 7-18.

Hoecklin, L. (1994). *Managing cultural differences—Strategies for competitive advantage.* Workingham, UK: Addison-Wesley.

Hofstede, G. (1980). *Culture's consequences: International differences in work related values.* Newbury Park, CA: Sage.

Iacono, C.S., & Weisband, S. (1997). Developing trust in virtual teams [Paper]. At *HICSS-30*, Maui, Hawaii, USA.

Jarvenpaa, S.L., & Knoll, K. (1998). Is anybody out there? Antecedents of trust in global virtual teams. *Journal of Management Information Systems, 14*(4), 29-65.

Jarvenpaa, S.L., & Leidner, D.E. (1998, November-December). Communication and trust in global virtual terms. *Organization Science: A Journal of the Institute of Management Sciences, 10*(6), 791.

Kanawattanachai, P., & Yoo, Y. (2002). Dynamic nature of trust in virtual teams. *The Journal of Strategic Information Systems, 11*(3-4), 187-213.

Komorita, S., & Parks, C. (1995). Interpersonal relations: Mixed-motive interaction. *Annual Review of Psychology, 46*(1), 183-207.

Kumar, K., & Willcocks, L.P. (1996). Offshore outsourcing: A country too far? [Paper]. At *European Conference on Information Systems*, Lissabon, Portugal.
Larson, C., & LaFasto, F. (1989). *Teamwork*. Newbury Park, CA: Sage.
Lock, D. (1996). *Project Management* (Vol. 6). Hampshire, UK: Gower.
Marks, M., Mathieu, J.E., & Zacaro, S.J. (2001). A temporally based framework and taxonomy of team processes. *Academy of Management Review, 26*(3), 356-376.
Mayer, R.C., Davis, J.H., & Schoorman, F.D. (1995). An integrative model of organizational trust. *Academy of Management Review, 20*, 709-734.
Mejias, R.J., Shepard, M.M., Vogel, D.R., & Lazaneo, L. (1997). Consensus and perceived satisfaction levels: A cross-cultural comparison of GSS and non-GSS outcomes within and between the United States and Mexico. *Journal of Management Information Systems, 13*(3), 137-161.
Page-Jones, M. (1980). *The practical guide to structured systems design*. New York: Yourdon Press.
Ridings, C.M., Gefen, D., & Arinze, B. (2002). Some antecedents and effects of trust in virtual communities. *The Journal of Strategic Information Systems, 11*(3-4), 271-295.
Rousseau, D., Sitkin, S., Burt, R., & Camerer, C. (1998). Not so different after all: A cross-discipline view of trust. *Academy of Management Review, 23*, 387-392.
Solomon, C.M. (2001). Managing virtual teams. *Workforce, 80*(1), 60-64.
Straub, D., Loch, K., Evaristo, J.R., Karahanna, E., & Srite, M. (2002). Toward a theory based measurement of culture. *Journal of Global Information Management, 10*(1), 13-23.
Suleiman, J., Evaristo, R., & Kelly, G. (2000). Facilitating and coordinating distributed joint applications development. *Informatica, 24*(2), 159-166.
Turner, J.R. (1993). *The handbook of project-based management*. London, New York: McGraw Hill.
Turner, J.R. (1995). *The commercial project manager*. London: McGraw Hill.
Van Der Merwe, A. (1997). Multi-project management—organization structure and control. *International Journal of Project Management, 15*(4), 223-233.
Wood, R.E. (1986). Task complexity: Definition of the construct. *Organizational Behavior and Human Decision Processes, 37*, 60-82.

An earlier version of this chapter was presented at BITWORLD 2001, Cairo, then printed in the Journal of Global Information Management, 11(4), 58-70, October - December 2003.

Chapter V

The Role of Power Distance and Explanation Facility in Online Bargaining Utilizing Software Agents

John Lim, National University of Singapore, Singapore

ABSTRACT

Online transactions have become increasingly popular and deserve greater attention from a research perspective. Whereas there are various aspects of online transactions, this study specifically examined an online bargaining scenario utilizing software agents. User's performance and attitudes were studied in a 2x2 factorial-design experiment. The independent variables were power distance (a dimension of culture)-for reasons associated with increasing and irresistible globalization, and explanation facility-for its conjecturable benefits in helping users to better understand and work with their software agents. Results showed these factors to have an interaction effect on task performance; as well, explanation facility exhibited main effects on trust and satisfaction. The findings have implications for system designers and builders; they also help managers in tailoring their expectations on what technology can deliver-under which conditions.

INTRODUCTION

Bargaining activities occur in a wide variety of political, economic, and social settings. Recently, with the rapid advancement of e-commerce, people find themselves increasingly involved in cross-cultural dealings; the pertinence of cultural differences in negotiations is becoming noticeable (Tse et al., 1994; Brett and Okumura, 1998; Gelfand and Christakopoulou, 1999; Tinsley et al., 1999), and carries important implications for research. At the same time, the form of negotiation support is evolving, and is influenced by fields such as distributed artificial intelligence. The growing interest in autonomous interacting software agents and their potential application in e-commerce opens up exciting possibilities for "automated" bargaining. The potentials of agents in online bargaining have been recognized in single-issue as well as multiple-issue negotiations (Terry et al., 2000; Yan et al., 2000). The early work consists mainly of conceptual models or exploration of technical feasibilities. A "drawback" of automated agents lies in the fact that oftentimes recommendations or actions are unexplained to the user. This paper explores the incorporation of explanation facility, a feature of knowledge-based systems (KBSs), into automated agents. The lack of research efforts on *both* culture and explanation facility in the context of negotiation provides the primary impetus for the current study. The study contributes toward research and practice by identifying and examining a novel combination of two important aspects, power distance (an important cultural dimension) and explanation facility. It serves to highlight to practitioners the crucial conditions that would elicit effectiveness of the negotiation support technology.

Prior research has generally revealed culture's influence in IS research (e.g., Tan et al., 1993; Watson et al., 1994). At the same time, work in KBSs has shown the usefulness of explanations (e.g., Gregor, 1996). However, as pointed out earlier, efforts are wanting in jointly visiting and examining culture and explanation facility.

The current research focuses on impact evaluation involving software agents. One independent variable in this research is explanation facility, a unique characteristic of KBSs. We attempt to incorporate explanation facility into the software agent and examine its influence. Mindful of the role of culture in bargaining and other business activities, and how it may moderate the impact of explanation facility, power distance-an important dimension of culture-is examined as the other independent variable. Specifically, the current study addresses the following research questions: (1) What are the effects due to availability of explanation facility on negotiation outcomes? (2) How does power distance moderate the above relationships? These questions address new and important aspects of negotiation support, and have implications on both the design and use of negotiation technologies.

This paper is organized as follows. The next section provides an overview of the theoretical background. Subsequently, the research model and hypotheses are presented. The research method is deliberated, followed by the results section. A discussion of the findings is then provided.

THEORETICAL BACKGROUND

Negotiation Agents and E-Commerce

A main hindrance of effective bargaining can be attributed to the cognitive limitations of the human information processing capability as well as process losses associated with communication. At the individual level, a consequence of being capacity-limited is to "satisfice" (Simon, 1957) and employ cognitive heuristics that would produce decisions efficiently (Kahneman et al., 1982) though most often imperfectly (Kerr et al., 1996). At the group level (negotiation involves more than one person), process losses happen such as "concentration blocking," which refers to the phenomenon that fewer comments are made because group members concentrate on remembering rather than thinking of new ones, until they can contribute them, as a result of short-term memory limitations (Nunamaker et al., 1991– includes an elaborate discussion on gains and losses associated with group processes).

The limitations as mentioned lead to the concept of negotiation support systems (NSSs), which are interactive, computer-based tools intended to assist negotiating parties in reaching an agreement. In some sense, NSSs can be considered a subset of group support systems (GSSs), which can be used to support virtual teams (Huang, Wei, and Lim, 2003). NSSs offer the potential for enhancing the problem-solving process and for helping to alleviate the cognitive and socio-emotional stumbling blocks to successful negotiations (Anson and Jelassi, 1990; Foroughi and Jelassi, 1990; Jelassi and Foroughi, 1989). Research on NSSs has primarily focused on two key technological aspects: (1) group decision and/or conflict resolution models to help negotiators reduce discord and increase the chance of reaching consensus, and (2) providing rich communication media to enhance communication exchange between antagonists (Bui and Shakun, 1997).

While traditional Web-based NSSs are quite powerful tools, they require near-constant human input. The advance of e-commerce has highlighted the need for automated negotiation support, a process that involves one or more bargaining agents. These agents employ artificial intelligence techniques such as machine learning, case-based reasoning, rule-based formalism, neural networks, and predicate calculus. In addition, the implementations and the computational approaches employed by some NSSs are relevant to and suggestive of possibilities for electronic bargaining agents. One example of the use of computational techniques is a concession model of Matwin et al. (1991), which presents a general strategy of concession in multi-issue negotiation.

Identifying efficient solutions involves substantial information-processing activities-estimating of preferences, generating of alternatives, evaluating of alternatives, and so on. It cannot be realized without considerable information-processing capacity. However, negotiators often do not know each other's utility function. Even if a good guess is somehow made available, it is virtually implausible to process the myriad possibilities of solutions; a sub-optimal settlement is often the result. Negotiation agents, by negotiating with each other automatically with minimum user supervision, remove those human factors to a great extent. Notwithstanding the above, it is viable to treat negotiation as a multi-tiered phenomenon. At the top level are full-scaled negotiations such as union-management contract bargaining; the middle level consists of "defined type" bargaining involving a limited number of issues; at the bottom level are mini-negotiations regarding very specific issues. It is fair to suggest that negotiation support tools generally focus on the lower two levels; but more general communications tools could simply keep track of all issues, and calculate costs and benefits of potential arrangements and tradeoffs, as a support rather than substitute for the human negotiator.

Compared to the traditional NSSs, negotiation agents perhaps remove to a larger degree the cognitive heuristics of humans, as the agents are not quite bounded in their computational/information-processing ability. Hence, there are possibilities for them to make fully rational decisions. However, the complex nature of negotiation activity puts a high demand for extensive instructions to be fed to negotiation agents. The negotiation strategies specified, together with their modifications, form the centerpiece of an electronic negotiation agent. Their appropriateness has a great impact on the agent's subsequent performance vis-à-vis bargaining outcomes. Negotiation agents can offer a number of potential benefits. First, a system of autonomous agent-based negotiations could increase the efficiency of settlements even for semi-structured, multi-issue business bargaining problems, in which negotiators hold private preference functions and possess incentives to misrepresent their own preferences (Oliver, 1997). Second, agents decrease the opportunity costs of transactions. For automated agents, the modest computational costs associated with today's high-speed computers are only a fraction of traditional transaction costs. Third, people may utilize autonomous negotiation agents to accomplish the task in their stead, and avoid the hassle of bargaining with others face-to-face. It should be pointed out that substituting agents for humans does not make the decisions rational-it simply narrows the field of discussion. Given a variety of issues, it could in principle define a pareto threshold where all decision combinations are technically equivalent and let bargaining between humans occur along this threshold.

Some examples of agents are briefly discussed in the following. The MAGMA project, which provided market architecture for multi-agent contracting, was limited to simple offers that negotiated over price alone in a limited form of the Vickrey auction and did not support multi-attribute negotiation (Tsvetovatyy et al., 1997). Kasbah system was an object-oriented distributed artificial intelligent multi-agent system, using offers which competing buyer and seller agents created, rejected, and accepted; however, the attribute for negotiation was again price alone (Chavez and Maes, 1996). The Virtual Property Agency was developed to support users with information collection and channel to communicate with their counterparts (Yen et al., 2000).

Explanation Facility

Either the provider or the receiver of information may *initiate* explanations (Gregor and Benbasat, 1999). Provider of information may initiate (i.e., provide) explanations for purposes of clarifying, justifying, or convincing; an explanation used in this sense may be viewed in terms of rhetoric or argumentation (Toulmin et al., 1984). At the same time, receiver of information can initiate (i.e., request) explanations to resolve misunderstanding or disagreement (Gilbert, 1989; Ortony and Partridge, 1987; Schank, 1986). In this paper, we are more interested in the provision of explanations. Three main types of explanations have been identified (Chandrasekaran et al. 1988; Clancey, 1983, 1993; Southwick, 1991). *Reasoning trace* refers to a record of the inferential steps taken by the system to reach a conclusion. *Justification* is an explicit description of the causal argument or rationale behind each inferential step taken by the system. *Strategy* is a high-level goal structure that determines how the system uses its domain knowledge to accomplish a task.

Providing explanations for recommended actions is deemed one of the most important capabilities of expert systems or knowledge-based systems. Explanation facilities are used primarily because they can assist users with the understanding of unfamiliar terms and requests during data input, thus leading to greater accuracy of input.

Use of explanations improves the performance achieved with KBSs as an aid (Gregor and Benbasat, 1999). Use of feedback explanations increases the accuracy of decision making (Dhaliwal, 1993; Dhaliwal and Benbasat, 1996). Hayes-Roth and Jacobstein (1994) noted that explanations, by virtue of making the performance of a system transparent to its users, are influential for user acceptance of intelligent systems and for improving users' trust in the advice provided. A more detailed literature review can be found in Gregor and Benbasat (1999).

Nevertheless, much of the previous research on explanation facility has been done in the expert systems or KBS field. This paper argues for the extension or transfer of this idea to software agents in the e-commerce context. It is clear that agents are now being actively deployed in e-commerce, for example, to search on the Web for the trader selling a given product at the cheapest price (Jango: http://www.jango.com; PersonaLogic: http://www.personaLogic.com; Firefly: http://www.firefly.com; AuctionBot: http://auction.eecs.umich.edu). However, marriage of explanation facilities to software agents is not yet commonplace. Most agents just get specifications from users and accomplish the tasks on behalf of their owners-without ever providing owners with explanations of their actions.

Power Distance

Culture is the collective programming of the mind that distinguishes the members of one group or category of people from another (Hofstede, 1991). Hofstede identified four dimensions of culture: power distance, individualism, uncertainty avoidance, and masculinity, which have formed the foundation of much cultural research. Other dimensions proposed include communications context (Hall, 1976), monochrony (Hall and Hall, 1990), and time perspective (Hofstede and Bond, 1988). An example of recent empirical work involves a qualitative study to examine the linkage between information technology transfer and cultural factors that support or impede a successful transfer (Hill, Straub, and El-Sheshai, 1998). Other recent studies on culture include Rose and Straub (1998) and Straub et al. (2001, 2002).

This study focuses on *power distance*, which indicates the extent to which a society accepts the fact that power in institutions and organizations is distributed unequally (Hofstede, 1980).[1] The choice of this dimension is dominated by the relevancy of context: it is conceivable that people's attitude toward power distribution in society would be highly relevant to the bargaining context. Other dimensions are outside the scope of this work and remain for future investigations.

HYPOTHESES

Figure 1 depicts the research model. Three dependent variables are of interest: performance, trust in agent, and satisfaction with outcome. Performance is indicated by the user's total utility scores pertinent to the agreement reached. The model also looks at the degree of trust the user has placed in the software agent, as well as the satisfaction level of the user related to the outcome.

Power distance is positioned as a moderating variable: it moderates the relationship between availability of explanation facility and the outcome variables.

Figure 1. Research model

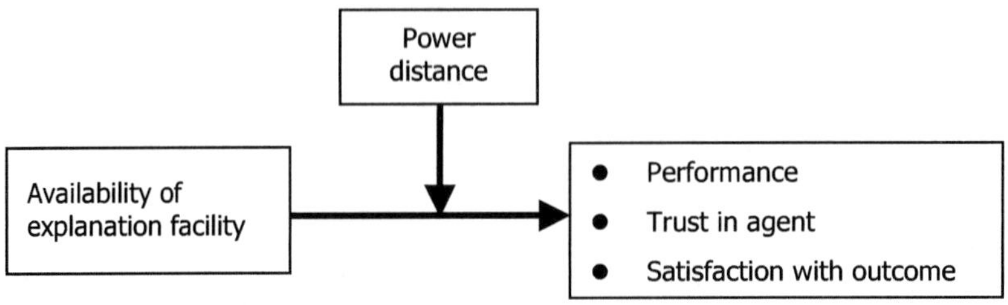

Effect on Performance

Effect of Availability of Explanation Facility

Use of explanations improves the performance achieved with a KBS as an aid (Gregor and Benbasat, 1999). Dhaliwal's (1993) study showed that the use of feedback (reasoning trace) explanations improved the accuracy of decision making. Gregor (1996) found that use of explanations of all types was related to improved problem-solving performance. De Greef and Neerincx (1995) reported that an "aiding" interface improved performance with a statistical program. Although these studies were done mainly on KBS, it is reasonable to anticipate benefits from explanations in the context of negotiation support. There are mainly two functions of explanation facility in negotiation agents: one is to describe what agents have done and why agents behave that way; the other is to provide users with the problem-solving knowledge. Explanations not only increase the comprehensibility of the analyses given by the negotiation agent, but also exert informational power over the negotiators (Kaplan and Miller, 1987), thus improving performance.

H1: Negotiation performance levels will be higher for those using explanation facilities in contrast to those not using them.

Moderating Effect of Power Distance

The positive effect of explanation facility is culturally bound; power distance should have a marked bearing on the use of explanation support. In a culture with large power distance, subordinates defer to superiors and do not question their authority. People with this kind of mentality are more likely to pay attention to explanations (offered by the negotiation system) which they view to be a source of authority. On the contrary, people coming from low power-distance background may care less about explanations, as they tend to make decisions based on their own knowledge and experience.[2] We thus predict an interactive effect, that the influence of explanation facility on negotiation outcomes is smaller with lower power distance than with higher power distance.

H2: The effect of explanation support on performance will be moderated by power distance, such that it is more positive for users from large power distance culture than those from small power distance culture.

Effect on Trust

Effect of Availability of Explanation Facility

Trust can be defined as the willingness to risk increasing "one's vulnerability to another whose behavior is not under one's control" (Zand, 1972, p. 230). The first rule of successful negotiation, according to Bazerman and Neale (1992), is to build trust. In a traditional buyer-seller negotiation, trust is referenced toward the opposite party. However, with the introduction of negotiation support, which acts as a form of mediation for two parties, trust is referenced toward the system. It is important to note this subtlety.

It is widely agreed that the trust is the key success factor of e-commerce (Keen, 1999). An increase in trust decreases transaction costs of relationships because individuals have to engage less in self-protective actions in preparation for the possibility of others' opportunistic behavior (Kramer and Tyler, 1996). Research has revealed a close relationship between trust and control mechanisms (Holland and Lockett, 1998; Luhmann, 1979). If people have full control over the transaction, they are more likely to have trust. It is evident that a lower level of sense of control is associated with the use of negotiation agents when explanations are not provided.

H3: Trust in agent will be greater for those using explanation facilities in contrast to those not using them.

Moderating Effect of Power Distance

To the extent that performance is moderated by culture (see earlier section), trust is similarly affected.

H4: The effect of explanation support on trust in agent will be moderated by power distance, such that it is more positive for users from large power distance culture than those from small power distance culture.

Effect on Satisfaction

Effect of Availability of Explanation Facility

We predict a generally elevated degree of satisfaction with outcome owing to the provision of explanations. Greater satisfaction is anticipated, mainly as a result of better understanding of the otherwise hidden process the software agents have undergone, the improved negotiation performance, and the increased trust in the software agents.

H5: Satisfaction with outcome will be greater for those using explanation facilities in contrast to those not using them.

Moderating Effect of Power Distance

To the extent that performance is moderated by culture (see earlier section), satisfaction is similarly affected.

Figure 2. Number of subjects per treatment (n=82)

		Yes	No	Total
Power Distance	High	25	22	47
	Low	17	18	35
	Total	42	40	82

H6: The effect of explanation support on satisfaction with outcome will be moderated by power distance, such that it is more positive for users from large-power distance culture than those from small power-distance culture.

RESEARCH METHOD

Research Design

A laboratory experiment with 2x2 factorial design was conducted. The independent variables were power distance (high vs. low) and explanation facility (present vs. absent). Eighty-two undergraduates from a large university served as subjects for this experiment. Two groups of students for investigating the power distance effects were used, consisting of Singaporeans and Mauritians. A marked difference in power distance was shown between the two peoples in Hofstede's (1980) analysis. Figure 2 shows the design and the breakdown of numbers in the four cells.

Task

Adapted from Jones (1988), the task involved negotiation over four issues of a purchase agreement for turbochargers, an engine subcomponent. The issues are unit price, purchase quantity, warranty period, and date of first delivery. The subject assumed the role of buyer, while computer simulated sellers. In this experiment, two software agents were used to aid negotiation. One is search agent, which gets requirements from its owner, then looks for potential sellers on the Internet. The other is negotiation agent, which uses genetic algorithm (GA) in carrying out negotiation process.

Genetic algorithms are search techniques inspired by the natural evolutionary processes of variation and selection. GAs can be viewed abstractly from many aspects: process description, searching, learning, and evolution. The most concrete view is to analyze the actual process of genetic operations. The operations perform an efficient search, which has an implicit parallelism, as different members of the population can operate independently. New offspring are continuously created and they search new areas. For details on GA, refer to Goldberg (1989).

Experimental Procedure

Upon arrival, the subject was given general instructions that outlined the procedures and the aim of the experiment. He/she then proceeded to read the case and gain background information about the company he/she was supposedly representing. The subject's preferences and utility function were elicited; the utility scores reflect the degree of importance one places on the ranges of values for each issue (Keeney and Raiffa, 1991). Since in this study power distance is one independent variable, a culture inclination questionnaire adapted from (Hofstede, 1991) was administered before the negotiation process. The subject also answered a demographics questionnaire. Next was training on how to use the system, after which the subject was given ten minutes to familiarize with the system, and given a chance to clarify doubts with the experimenter. The subject then proceeded to negotiate; with the use of the system, there were three specific activities he/she had to carry out: initialize negotiation agent, start search process, and decide on seller. For each condition, a log file was used to capture operations. After the negotiation, the subject completed a set of questionnaires on satisfaction and trust. See appendices for the various questionnaires.

Computation of Utility Scores for Measuring Performance

For each of the negotiation variables (unit price, purchase quantity, warranty period, and date of first delivery), several conditions (or values) exist. For the moment, suppose the negotiation only involves "unit price" and "warranty period," and that each has three values; thus, "unit price" assumes the values "high," "medium," and "low"; and "warranty period" assumes the values "short," "medium," and "long." If the buyer accords higher priority to "unit price" than "warranty period," there should be a way for this preference to be expressed. The approach is to give a higher weightage (e.g., 60) to "unit price" while a lower one (e.g., 40) to "warranty period"; the total weightage is 100. As well, one should be allowed to distinguish between the different values within a variable. For example, if the lowest unit price is hoped for, then this value of unit price will have a weightage of 60 (maximum); a higher unit price, say "medium," may have a weightage of 50, while "high" unit price may have a weightage of only 35. Similar specifications can be made of "warranty period," in which case "long" warranty period will be given a weightage of 40 (maximum).

Elicitation of one's utility functions involving all four issues (unit price, purchase quantity, warranty period, and date of first delivery) allows the carrying out of several types of analysis, such as deriving the total utility obtainable for a certain combination of variable values. The sum of utilities of the *agreed* values, then, represents the total utility score attained, or negotiation performance. See Lim (1999) for a detailed discussion on utility specification.

RESULTS

Control checks were performed on gender, working experience, experience in making business decisions, experience in using computer, and experience in using NSS. Results showed no difference across conditions.

A t-test was conducted on the power-distance scores between the "high" condition (Singaporeans) (mean=73.54, s.d.= 65.63) and the "low" condition (Mauritians) (mean=40.42, s.d.=55.35), and confirmed significant differences in the desired direction ($t=2.42$, $p<0.02$). This served as a manipulation check for the culture factor.

Table 1. ANOVA table for performance

Source	*Sum of Squares*	*df*	*Mean Square*	*F*	*p*
Power distance	94.27	1	94.27	6.51	0.01
Explanation facility	3.85	1	3.85	0.27	0.61
Power distance *Explanation	222.24	1	222.24	15.36	0.00

Performance

Performance was measured as the sum of utility scores obtained for a given settlement (i.e., points a negotiator earned), and ranged between 1 and 100; more points indicate better outcome.

The ANOVA table is shown in Table 1. A significant interaction effect is seen (F=15.36,p=0.00).

The interaction effect was further explored by studying the effect of explanation while holding the other independent variable constant (see Figure 3). In high power-distance situation, explanation served to boost performance significantly (t=5.48, p=0.00); the effect of explanation was not observed in low power-distance situation (t=0.72, p=0.48).

Hypothesis H1 was not supported; H2 was supported.

Trust in Agent

Table 2 shows the ANOVA results. A significant main effect due to explanation is observed (F=24.73, p=0.00). Subjects provided with explanation facility perceived greater trust (mean=3.96, s.d.=0.44) than those not provided (mean=3.44, s.d.=0.57). The questionnaire adopted a scale of 1 to 5.

Given the importance of the trust variable within the e-commerce context (McKnight et al., 2002), a content analysis was also performed to supplement the perceptual measure. Questions were placed on the system's screens at strategic points throughout the process; users' responses (in words) were coded into a scale of 1 to 5 by two independent coders. Cohen's Kappa (which corrects for chance agreement) was calculated to assess inter-rater reliability, and turned out as 0.68. Consistent with the perceptual measure, content analysis of trust showed a significant main effect owing to explanation (F=12.27, p=0.00). This supplementary analysis serves to strengthen the finding on the role of explanation vis-à-vis trust.

Hypothesis H3 was supported; H4 was not supported.

Satisfaction

The ANOVA results are shown in Table 3. Explanation exhibited a significant main effect on satisfaction (F=47.42, p=0.00). Subjects provided with explanation facility reported higher satisfaction with outcome (mean=3.64, s.d.=0.56) than those not provided (mean=3.15, s.d.=0.67). The questionnaire adopted a scale of 1 to 5.

Hypothesis H5 was supported; H6 was not supported.

Figure 3. Interaction effect on performance

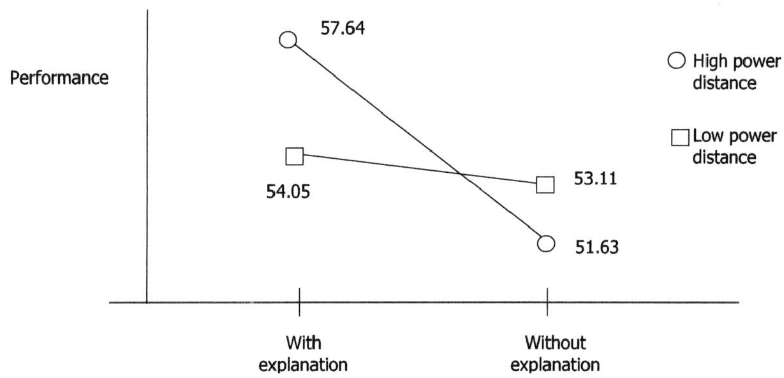

DISCUSSION

It is to be recognized at the outset that *all* users were supported with software agents, which conducted search and negotiation. In other words, this study is not about the effect of software agents per se, since the situation of not having agents has been earlier argued as inferior. Rather, this paper posited and found that users derived benefits from software agents to *various* extents. Particularly, people from high power-distance culture were more influenced by the explanation incorporated into the system and obtained higher performance. In contrast, low power-distance people, who plausibly deemed explanation less important, consequently reported lower performance. This manifestation of interaction between power distance and technological feature has practical implications. A direct implication would be the need to tailor the expectations that practitioners should have of potential benefits of technology, owing to the role of power distance. Its ramification in a globalized workplace and mixed-cultural workgroups must be appreciated.

Notwithstanding the interaction led on by power distance, we would still recommend the incorporation of explanation facility into software agents. The bettered performance in terms of negotiation settlement could be attributed to the user's (or negotiator's) acceptance of explanations given by the system. Heeding the explanations, the user was able to choose in a wise manner from the group of sellers presented by the search agent. This thinking is in line with previous findings. For example, Dhaliwal (1993) found that the use of feedback (reasoning trace) explanations had a positive effect on the accuracy of decision making. Mao (1995) found that increased use of deep explanations led novice subjects to make judgments that were similar to those of experts who contributed their knowledge to the development of the KBS.

Positive results on users' attitude are encouraging. In particular, the increased trust toward agent boosted by the explanation facility is certainly of practical significance. The importance of trust and how it may impact performance cannot be overemphasized (Zand, 1972). Given the e-commerce context, mechanisms that help increase trust in software agents are certainly welcome. User satisfaction was also found to be elevated by the

Table 2. ANOVA table for trust

Source	*Sum of Squares*	*df*	*Mean Square*	*F*	*p*
Power distance	0.10	1	0.10	0.47	0.50
Explanation facility	5.50	1	5.50	24.73	0.00
Power distance *Explanation	0.20	1	0.20	0.91	0.34

Table 3. ANOVA table for satisfaction

Source	*Sum of Squares*	*df*	*Mean Square*	*F*	*p*
Power distance	0.70	1	0.70	3.41	0.07
Explanation facility	9.76	1	9.76	47.42	0.00
Power distance *Explanation	1.85E-02	1	1.85E-02	0.09	0.77

provision of explanations. This finding is especially pertinent when viewed together with the corresponding effect on trust. Clearly, it is advantageous for system developers to provide due consideration to incorporating explanations into software agents.

Overall, the explanation facility has proven useful for the dependent variables of concern, i.e., performance, trust, and satisfaction, although its impact on performance needs to be qualified by power distance, an important dimension of culture. Whereas this study has examined two independent variables, the provision (or availability) of explanation facility is a fundamentally different type of independent variable from power distance, in that the former is artificial (man-made) and the latter natural (or a matter of predisposition). Because of this very nature, explanation facility should be exploited to the fullest for negotiation support.

This study has the following limitations. First, a specific tool was used; it is conceivable that other tools might produce other results. Second, it has tested only the buyer and not the seller. These have implications for future research, to be discussed next.

Future research should look into a closely related issue that is interface design, to see how best to present explanations, so that maximal benefits may be derived from this facility. Likewise, the moderating role of power distance should be greater explored in future extensions; such efforts should also examine those cultural dimensions not addressed in the current study. Along this line, more cultures can be studied. It should also be of interest to reverse experimental subjects' role from buyer to seller and study the possible differences in outcome measures. Further expansion can be made both in terms of the examination of explanation facilities and the use of other kinds of tools.

CONCLUSION

This study was conducted to address two questions: (1) What are the effects due to availability of explanation facility on negotiation outcomes? (2) How does power distance moderate the above relationships? Our findings have shown an almost penetrating effect of explanation facility, that its availability raised performance under certain circumstances, trust in agent, and satisfaction with outcome. Power distance, in fact, was the moderator on the relationship between the availability of explanation facility and performance, such that the benefits of explanations were visible in high power-distance context, but non-existent in low power-distance context.

REFERENCES

Anson, R.G. & Jelassi, M.T. (1990). A development framework for computer-supported conflict resolution. *European Journal of Operational Research*, *46*(2), 181-199.

Bazerman, M.H. & Neale, M.A. (1992). *Negotiating Rationally*. New York: The Free Press.

Baskerville, R.F. (2003). Hofstede never studied culture. *Accounting, Organizations and Society*, 28(1), 1-14.

Brett, J.M. & Okumura, T. (1998). Inter- and intra-cultural negotiation: U.S. and Japanese negotiators. *Academy of Management Journal*, 41, 495-510.

Bui, T.X. & Shakun, M.F. (1997). Negotiation processes, evolutionary systems design, and negotiator. In M.F. Shakun (Ed.), *Negotiation processes: Modeling frameworks and information technology*. Kluwer Academic Publishers (pp.39-55).

Chandrasekaran, B., Tanner, M.C., & Josephson, J.R. (1988). Explanation: The role of control strategies and deep models. In J.A. Hendler (Ed.), *Expert Systems: The User Interface* (pp. 219-247). Norwood, NJ: Ablex.

Chavez, A. & Maes, P. (1996). Kasbah: An agent marketplace for buying and selling goods. *Proceedings of the First International Conference on the Practical Application of Intelligent Agents and Multi-Agent Technology* (pp. 75-90).

Clancey, W.J. (1983). The epistemology of a rule-based expert system- A framework for explanations. *Artificial Intelligence*, *20*(5), 215-251.

Clancey, W.J. (1993). Notes on "Epistemology of a Rule-based Expert System." *Artificial Intelligence*, *59*(2), 197-204.

De Greef, H.P. & Neerincx, M.A. (1995). Cognitive support: Designing aiding to supplement human knowledge. *International Journal of Human-Computer Studies*, 42, 531-571.

Dhaliwal, J.S. (1993). *An Experimental Investigation of the Use of Explanations Provided by Knowledge-Based Systems*. Unpublished Doctoral Dissertation, University of British Columbia, Canada.

Dhaliwal, J.S. & Benbasat, I. (1996). The use and effects of knowledge-based system explanations: Theoretical foundations and a framework for empirical evaluation. *Information Systems Research*, *7*(3), 342-362.

Foroughi, A. & Jelassi, M.T. (1990). NSS solutions to major negotiation stumbling blocks. *Proceedings of the 23rd Hawaii International Conference on System Sciences* (Volume 4, pp. 2-11).

Gelfand, M. J. & Christakopoulou, S. (1999). Culture and negotiator cognition: judgment accuracy and negotiation processes in individualistic and collectivistic cultures. *Organizational Behavior and Human Decision Processes*, 79, 248-269.

Gilbert, N. (1989). Explanation and dialogue. *Knowledge Engineering Review*, *4*(3), 205-231.

Goldberg, D. (1989). *Genetic algorithms in search, optimization, and machine learning*. Reading, MA: Addison-Wesley.

Gregor, S. (1996). *Explanation from knowledge-based systems for human learning and problem solving*. Unpublished Doctoral Dissertation, University of Queensland, Brisbane, Australia.

Gregor, S. & Benbasat, I. (1999). Explanations from intelligent systems: Theoretical foundations and implications for practice. *MIS Quarterly*, *23*(4), 497-530.

Hall, E.T. (1976). *Beyond culture*. Garden City, NY: Doubleday Anchor Books.

Hall, E.T. & Hall, M.R. (1990). *Understanding cultural difference*. Yarmouth, ME: Intercultural Press.

Hayes-Roth, F. & Jacobstein, N. (1994). The state of knowledge-based systems. *Communications of the ACM*, *37*(3), 27-39.

Hill, C., Loch, K., Straub, D. & El-Sheshai, K. (1998). A qualitative assessment of Arab culture and information technology transfer. *Journal of Global Information Management*, *6*(3), 29-38.

Hofstede, G. (1980). *Culture's consequence: International difference in work-related values*. Beverly Hills/London: Sage Publications.

Hofstede, G. (1991). *Cultures and organizations: Software of the mind*. London: McGraw-Hill.

Hofstede, G. & Bond, M.H. (1988). The confucius connection: From cultural roots to economic growth. *Organizational Dynamics*, *16*(4), 5-21.

Holland, C.P. & Lockett, A.G. (1998). Business trust and the formation of virtual organizations. *Proceedings of the 31st Hawaii International Conference on System Sciences* (Volume 6, pp. 602-611).

Huang, W., Wei, K.K. & Lim, J. (2003). Using a GSS to support virtual team-building: A theoretical framework. *Journal of Global Information Management*, *11*(1), 72-89.

Jelassi, M.T. & Foroughi, A., (1989). Negotiation support systems: An overview of design issues and existing software. *Decision Support Systems*, *5*(2), 167-181.

Jones, B.H. (1988). *Analytical negotiation: An empirical of the effects of computer support for different levels of conflict in two-party bargaining*. Unpublished Doctoral Dissertation, Indiana University, USA.

Kahneman, D., Slovic, P. & Tversky, A. (Eds.). (1982). *Judgment under uncertainty: Heuristics and biases*. New York: Cambridge University Press.

Kaplan, M.F. & Miller, C.E. (1987). Group decision making and normative versus informational influence: Effects of type of issue and assigned decision rule. *Journal of Personality and Social Psychology*, *53*(2), 306-313.

Keen, P.G.W. (1999). *Electronic commerce relationships: Trust by design*. Englewood Cliffs, NJ: Prentice-Hall.

Keeney, R. & Raiffa, H. (1991). Structuring and analyzing values for multiple-issue negotiations. In H.P. Young (Ed.), *Negotiation analysis* (pp. 131-151). Ann Arbor: University of Michigan Press.

Kerr, N.L., MacCoun, R.J. & Kraemer, G.P. (1996). Bias in judgment: Comparing individuals and groups. *Psychological Review*, 103, 687-719.

Kramer, R.M. & Tyler, T.R. (1996). Whither Trust. In R.M. Kramer & T.R. Tyler (Eds.), *Trust in organizations: Frontiers of theory and research* (pp. 1-15). Thousand Oaks, CA: Sage Publications.

Lim, J. (1999). Multi-stage negotiation support: A conceptual framework. *Information and Software Technology*, 41, 249-255.

Luhmann, N. (1979). *Trust and power*. New York: John Wiley & Sons.

Mao, J. (1995). *An experimental study of the use and effects of hypertext based explanations in knowledge-based systems*. Unpublished Doctoral Dissertation, University of British Columbia, Canada.

Matwin, S., Szapiro, T. & Haigh, K. (1991). Genetic algorithms approach to a negotiation support system. *IEEE Transactions on Systems, Man, and Cybernetics, 21*(1), 102-114.

McKnight, D.H., Choudhury, V. & Kacmar, C. (2002). Developing and validating trust measures for e-commerce: An integrative typology. *Information Systems Research, 13*(3), 334-359.

Nunamaker, J.F. Jr., Dennis, A.R., Valacich, J.S., Vogel, D.R.& George, J.F. (1991). Electronic meeting systems to support group work. *Communications of the ACM, 34*(7), 40-61.

Oliver, J.R. (1997). A machine learning approach to automated negotiation andp prospects for electronic commerce. *Journal of Management Information Systems, 13*(3), 83-112.

Ortony, A. & Partridge, D. (1987). Surprisingness and expectation failure: What's the difference. *Proceedings of the 10th International Joint Conference on Artificial Intelligence* (Volume 1, pp. 626-635).

Rose, G. & Straub, D. (1998). Predicting general IT use: A study in Arab developing nations. *Journal of Global Information Management, 6*(3), 39-46.

Schank, R.C. (1986). Explanation: A first pass. In J.L. Kolodner & C.K. Riesbeck (Eds.), *Experience, memory, and reasoning* (pp. 139-165). Hillsdale, NJ: Lawrence Erlbaum Associates.

Simon, H. (1957). *Models of man: Social and rational*. New York: John Wiley & Sons.

Southwick, R.W. (1991). Explaining reasoning: An overview of explanation in knowledge-based system. *Knowledge Engineering Review*, 6, 1-19.

Straub, D., Loch, K., Evaristo, R., Karahanna, E. & Srite, M. (2002). Toward a theory-based measurement of culture. *Journal of Global Information Management, 10*(1), 13-23.

Straub, D., Loch, K. & Hill, C. (2001). Transfer of information technology to the Arab World: A test of cultural influence modeling. *Journal of Global Information Management, 9*(4), 6-28.

Tan, B.C.Y., Watson, R.T., Wei, K.K., Raman, K.S. & Keola, P.K. (1993). National culture and group support systems: Examining the situation where some people are more equal than others. *Proceedings of the 26th Hawaii International Conference on System Sciences* (Volume 4, pp. 132-141).

Terry, R.P., Terri, L.L., Susan, H., Michael, L. & Katia, S. (2000). Agent-based team critical task. *Proceedings of the 33rd Hawaii International Conference on System Sciences*.

Tinsley, C., Curhan, J. & Kwak, R.S. (1999). Adopting a dual lens approach for overcoming the dilemma of differences in international business negotiations. *International Negotiation*, 4, 1-18.

Toulmin, S., Rieke, R. & Janik, A. (1984). *An introduction to reasoning*. New York: Macmillan.

Tse, D.K., Francis, J. & Walls, J. (1994). Cultural differences in conducting intra- and inter-cultural negotiations: A Sino-Canadian comparison. *Journal of International Business Studies*, 25, 537-555.

Tsvetovatyy, M., Gini, M., Mobasher, B. & Wieckowski, Z. (1997). MAGMA: An agent-based virtual market for electronic commerce. *Applied Artificial Intelligence*, *11*(6), 501-524.

Watson, R.T., Ho, T.H. & Raman, K.S. (1994). Culture: A fourth dimension of group support systems. *Communications of the ACM*, *37*(10), 44-55.

Yan, Y., Yen, J. & Bui, T.X. (2000). A multi-agent based negotiation support system for distributed transmission cost allocation. *Proceedings of the 33rd Hawaii International Conference on System Sciences.*

Yen, J., Hu, J. & Bui, T.X. (2000). Intelligent Clearinghouse: Electronic marketplace with computer-mediated negotiation supports. *Proceedings of the 33rd Hawaii International Conference on System Sciences.*

Zand, D.E. (1972). Trust and managerial problem solving. *Administrative Science Quarterly*, *17*(2), 229-239.

ENDNOTES

1. It is noted that Hofstede's work is controversial (e.g., Baskerville, 2003) but remains a fairly standard approach for investigators.
2. A rival argument is applicable here, that is that low power-distance people are accustomed to thinking on their own and they would be more influenced by explanations with which they agree. It is apt for further research to look into intervening processes such as whether and to what extent people agree with explanations offered.

APPENDIX I: QUESTIONNAIRE ITEMS FOR TRUST AND SATISFACTION

(1) Trust

1. If I had it my way, I would not let the agent have any influence over issues that are important to my deal.
2. I would not be comfortable in giving the agent complete responsibility for completion of the task.
3. I really wish I had a good way to oversee the work of the agent on the task.
4. I would not be comfortable giving the software agent a task or problem that was critical to the task, even if I could monitor the agent.

(2) Satisfaction

1. I think the outcome of the negotiation was satisfactory for my company.
2. I am satisfied with my agent's negotiator performance.
3. I am satisfied with the number of points I earned.

APPENDIX II: CULTURE INCLINATION QUESTIONNAIRE

1. If you are the employee, how often will you be afraid to express disagreement with the managers in your company?
   1. Very frequently
   2. Frequently
   3. Sometimes
   4. Seldom
   5. Very seldom

The following explanations were given for the four different types of managers:

Manager A: Usually makes his/her decisions promptly and communicates them to his/her subordinates clearly and firmly. Expects them to carry out the decisions loyally and without raising difficulties.

Manager B: Usually makes his/her decisions promptly, but before going ahead, tries to explain them fully to his/her subordinates. Gives them the reasons for the decisions and answers whatever questions they may have.

Manager C: Usually consults with his/her subordinates before he/she reaches his/her decisions. Listens to their advice, considers it, and then announces his/her decision. He/she then expects all to work loyally to implement it whether or not it is in accordance with the advice they gave.

Manager D: Usually calls a meeting of his/her subordinates when there is an important decision to be made. Puts the problem before the group and tries to obtain consensus. If he/she obtains consensus, he/she accepts this as the decision. If consensus is impossible, he/she usually makes the decision him/herself.

2. For the four types of manager, please mark the one that you would prefer to work under.
   1. Manager A
   2. Manager B
   3. Manager C
   4. Manager D

3. To which one of the above four types of managers does your culture most closely corresponds to?
   1. Manager A
   2. Manager B

3. Manager C
4. Manager D
5. None of above

4. How often do you think an employee feels nervous or tense at work in your culture?
   1. Always feel this way
   2. Usually
   3. Sometimes
   4. Seldom
   5. Never feel this way

5. How long do you think an employee will continue working for this company?
   1. At most two years
   2. Two to five years
   3. More than five years
   4. Until he retires

6. Do you agree that in your culture, company rules should not be broken – even when the employee thinks it is in the company's best interests?
   1. Strongly agree
   2. Agree
   3. Undecided
   4. Disagree
   5. Strongly disagree

APPENDIX III: BACKGROUND QUESTIONNAIRE

1. What is your level of working experience?
2. What is your level of experience in making business decisions?
3. What is your level of experience in using computers?
4. What is your level of experience in using negotiation support systems?

Previously published in the Journal of Global Information Management, 12(3), 27-43, April - June 2004.

Chapter VI

How to Compete in a Global Education Market Effectively: A Conceptual Framework for Designing a New Generation eEducation System

Wayne Huang, Ohio University, USA

David C. Yen, Miami (Ohio) University, USA

Z.X. Lin, Texas Tech University, USA

J.H. Huang, Tsinghua University, China

ABSTRACT

Computer-aided instruction (CAI) has existed for decades. Its third generation system using Internet and Web technologies to deliver university education (so called virtual universities) has been a hot research topic in recent years. However, few such virtual universities have been as successful as expected. Why didn't eEducation systems turn out to be a silver bullet for virtual universities as expected? What are components and elements that have been missed in current eEducation systems? How can we learn from past experience and lessons so that a conceptual framework could be proposed to design a better next generation eEducation system that could help universities and corporations to gain competitive advantages in a global education market? This research paper tries to explore these important issues. Based upon a comprehensive literature review, a conceptual framework is proposed with the aim of guiding the design of a next generation eEducation system.

INTRODUCTION

Computer-aided instruction (CAI) has existed for decades. Its third generation systems using Internet and Web technologies to deliver education programs in global market has been a hot research topic in recent years (Al-Nuaimy, Zhang, and Noble, 2001; Bodendorf and Swain, 2001; Chandler, 2002; Chang, 2001; Jung, 2001; Kinshuk, 2002; Owston, 2000; Rovai, 2001). Online education and training (i.e., eEducation) is a multi-billion-dollar business (Bargeron et al., 2002; Snell, 1997) and also one of the important issues currently faced by global multi-national corporations and universities.

The American Bankers Association (ABA) predicts that spending by U.S. banks on all types of training, which is now approximately $2.3 billion a year, will increase 12 percent in 2003. The use of online courses (i.e., eEducational courses) in banks can dramatically lower costs of training (Wetzel, 2002) and easily deliver the educational programs across national geographical boundaries (Barker, 2001; Lee, 2001; Topper, 2002; Huang, Lienhart and Yen, 2003)

Many universities in the world are looking at the Internet as an effective way to help them compete in a global education market. In the U.S., there has been a 300% increase between 1995-1998 in the use of web pages for classes in public research universities and 500% increase during the same time period for private educational institutions. It is estimated that 55% of the United States' 2,215 four-year colleges and universities have courses available off-site, and over one million students are now attending virtual college classes (Huang, Lienhart and Yen, 2003). According to the management philosopher Peter Drucker: "Universities won't survive. The future is outside the traditional campus, outside the traditional classroom. Distance learning is coming on fast" (Forbes, 1997).

eEducation has a potential to decrease the cost of delivering education programs, which is especially important and attractive to educational programs that are marketed to international students in developing countries. For online students, the University of Phoenix in the U.S. charges $325 a credit for its B.S. program. With 102 course credits required for graduation, that is $33,150 for a degree program, just about one-third the cost of going to Yale University for four years. It costs the university $237 to provide one credit hour of education on-line, against $486 per hour for conventional education at Arizona State University (Huang, Lienhart and Yen, 2003). On average, the on-line students scored 5% to 10% higher than their traditionally educated peers and maintained that margin upon completing their coursework (Forbes, 1997).

Because of the potential substantial benefits and competitive advantages that may be provided by eEducation programs, many universities around the world invested substantially in building up eEducation systems to set up so-called virtual universities (Coppola, Hiltz and Rotter, 2002) that have no national geographical boundaries and could reach potential international students even in remote countries at much lower cost. As a result, it has been predicted that such virtual universities could at least partially substitute education programs offered by current physical universities. However, until now, few such virtual universities mainly supported by eEducational systems have been as successful as expected (e.g., Baumlin et al., 2000; Kearsley, 1998; Newman and Johnson, 1999; Zemke, 1998). Why didn't eEducation turn out to be a silver bullet for virtual universities? What are the elements that have been missed in current eEducation systems? How can we learn from past experience and lessons so that a conceptual framework could be proposed to design a better next generation eEducation system that could help universities and corporations gain competitive advantages in a global education market?

This paper tries to provide some answers to the above research questions. The next section of the paper will, based upon a comprehensive literature review on CAI and general education, present a holistic view of a university education system and examine what components are missed in current eEducation systems. We then propose a conceptual framework for designing a next generation eEducation system, followed by a discussion of the framework.

A HOLISTIC VIEW OF A UNIVERSITY EDUCATION SYSTEM

A physical university education system is very complex and includes many interrelated elements: teacher, student, classroom, teaching equipment, peer classmates, curriculum, course materials, course administration, teaching style and methods, university traditions and rituals, and a physical environment surrounding a university campus, etc. When eEducation systems are designed and developed to support virtual universities, many, if not all, of them are based on the assumption that adequate information from education curriculum and course materials plus learning collaboration should be enough for supporting learning in virtual universities (Zemke, 1998), ignoring many other important elements of a physical university education system.

Currently, the most widely used commercial eEducation systems are WebCT and Blackboard. Blackboard has a better user interface and is much easier to use to put course materials on a website; whereas WebCT provides more powerful yet complicated tools to facilitate instructors to conduct on-line teachings. Both systems are essentially aimed at providing good tools for putting enough information on a website for students to study as well as for supporting and enhancing collaborations between students and instructors. However, some important educational elements mentioned above, such as teaching style and methods, have been largely missed in these eEducation systems. Therefore, it is not convincing to many students and instructors that using such systems will be beneficial (Baumlin et al., 2000) and there exist many problems in current eEducation systems (Dillon, 2000). Sometimes eEducation technology could become a distraction to overall educational improvements and performance (Kearsley, 1998).

To design a viable and effective eEducation system, we first need to have a holistic view of a university education system so that all important components and elements of an education system will be considered for the design of a next generation eEducation system.

A comprehensive search on related literatures was conducted in following 14 databases/sources with a timeline ranging from 1995 to 2001: ABI, AEI, Computer Database, Compendex Plus, EiCompendex Plus, ERIC, Expanded, Academic, IEEE,

Figure 1. A holistic view of an education system

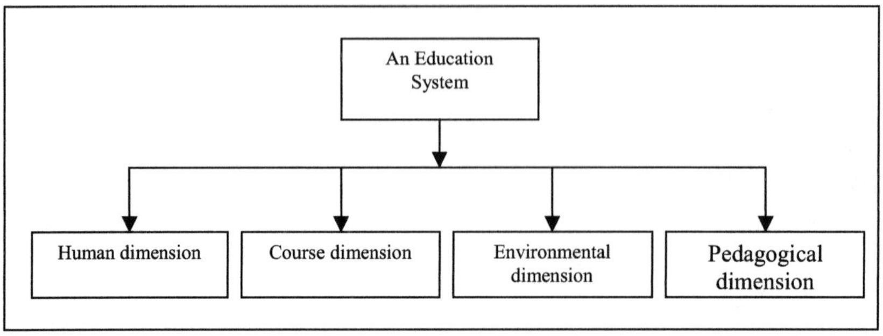

Inspect, NTIS, Science Direct, Science Citation Index, and Uncover. Over 7,700+ articles were searched from the 14 databases, which included a lot of duplicated articles from different databases. After screening to remove those duplicated articles, it was left with 1,500 unique entries. Further, two factors were used to filter out the entries. First, whether a paper is directly relevant to the current research theme; and second, whether a paper is published in an internationally quality journal. In this way, about 100 articles were short listed for detailed literature review.

The literature review indicates that a university educational system (physical or virtual) is generally composed of four main dimensions: human dimension, course dimension, environmental dimension, and pedagogical dimension (Figure 1), each of which consists of some important components as shown in Table 1.

Human Dimension

In Table 1, human dimension here refers to those people who use, develop or maintain an educational system, generally including components of Teacher, Student, Course Administrator, Peer, Tutor, Mentor, Faculty staff, Moderator, and External visitor.

Table 1. Dimensions and components of an education system

Dimensions	Components
Human dimension	Teacher, Student, Course Administrator, Peer, Tutor, Mentor, Faculty staff, Moderator, and External visitor.
Course dimension	Course Administration, Course materials, and Course Learning and Assessment
Environmental dimension	Easiness for access, Safety and security, integrity, Scalability, Flexibility, Compatibility/portability, and Easiness for setting up and maintaining; Multiple accesses, Multiple media, Instructor aids, Disability support, Virtual Reality, etc.
Pedagogical dimension	Pedagogical models and Instruction design

Teacher

The teacher is a course instructor who is one of the most important elements in an eEducation system. He/she, in most cases, is the most knowledgeable person in a course that he/she is about to teach. He/she is the person who decides when and what kind of materials or activities to be covered. As Davis (1996) suggested, the teacher acts like a content provider/filter providing the right level and amount of materials/activities for students. This is because students often lack the background knowledge of a course to guide them to the right path of learning. Sometimes the literature or knowledge of a particular course is so broad that it requires a teacher to break down complex materials into smaller chunks in order for students to understand in a systematic way (Kearsley, 1998). Moreover, he/she may have to take up roles of a course administrator as well as a tutor because of a limited budget and resources his/her educational institute allocates.

Because of the central role played by a teacher in an education system, more understanding of teacher's behaviors and characteristics would be very important to design and implement a successful eEducation system. For example, one of the problems that teachers often face is that they lack technology training and support from their faculty departments (Kearsley, 1998). As a result, they are often reluctant to adopt new eEducation systems. Further, Brown (1998) reports that it is time-consuming for a teacher to implement and maintain an online course both up-front and later. The workload on the teacher side alone may increase 40% to 50% more than a standard course of face-to-face type. As a result, even if an eEducation system is technically sound, teachers may not be ready to adopt and use the system. Therefore, the readiness of adopters for a new

Table 2. Teacher's characteristics that enhance the adoption of eEducation systems

Number	Characteristics
1	Serious, lifelong learners
2	Teachers favoring experimental and collaborative styles
3	Those who enjoy up-front conceptual work
4	Skilled group process facilitators
5	Teachers who make expectation explicit
6	Those who construct evaluation/assessment schemes
7	Those teachers willing to give feedback at frequent intervals
8	Those who provide detailed developmental feedback to students

Table 3. Student's characteristics that enhance the adoption of eEducation systems

Number	Characteristics
1	Independent, active learners
2	Those who enjoy working alone
3	Those who can structure and manage time well
4	Accomplished, busy professionals (adult students)
5	Students with superior verbal ability
6	Risk taking, creative problem solvers
7	Individuals committed to peers and group process
8	Those who are comfortable with asynchronous rhythms

Table 4. Top three benefits of using eEducation systems

Number	Benefits
1	Flexibility in learning pace and time
2	Guided access to useful resources on the web beyond those available in the conventional mode
3	The convenience of electronic contact with university staff

Table 5. Top three problems of using eEducation systems

Number	Problems
1	Loss of face-to-face contact with university staff and other students
2	Uncertainty about being able to study effectively in the unit using this new mode of learning
3	Concerns about access to the systems both off and on campus

technology (Chwelos, Benbasat, and Dexter, 2001) should be an important factor to be considered when designing and implementing an eEducation system.

Given the same level of training and support, perceptions of readiness in adopting a new eEducation system may still vary among different teachers. Brown (1998) reports that the teachers possessing the characteristics in Table 2 may be more ready to adopt and use an eEducation system in their daily teaching classes.

Student

Students are expected to learn and understand course materials and are able to meet the criteria set by teachers and retain the knowledge afterwards. Sometimes, students receive the knowledge from their teachers directly when teachers' pedagogy is instructor-centered; and sometimes, the knowledge is generated by themselves during the research of materials on their own when teachers' pedagogy is learner-centered.

Similarly, perceptions of the readiness of using an eEducation system could vary to different students.

Because an eEducation system is designed for frequent use of students and teachers, students' behaviours and attitudes towards the system should also be seriously considered in system design and implementation phases. The students possessing the characteristics listed in Table 3 are more ready to use an eEducation system and also more likely to benefit from the system (Brown, 1998).

Further, the survey carried out by Lund and Volet (1998) reports that students perceive the three top advantages of using an eEducation system (Table 4).

These students also mention three top problems in using an eEducation system (Table 5).

Course Administrator

A course administrator is dedicated to a particular course and is responsible for all kinds of course administration work such as student enrollment, class and tutorials scheduling and assignment submissions, etc. These works sometimes are taken by a teacher him/herself (Biggs, 1998).

Peers

Peers refer to those students who take the same course or participate in the same group discussion so that they are able to comment on and give suggestions to other students' works (Box, 1999; Phillips, 1998; Rada, 1998). Most current eEducation systems only provide simple group discussion forum function without more powerful and structured tools being provided. For example, if peers' comments on one peer student's work vary, there is no voting or ranking tool to facilitate them in reaching consensus.

Tutor

The role of a tutor is to answer any questions raised by students regarding course materials. Tutors are also to provide prompt feedback to students' assignments. Sometimes this role is taken by a teacher him/herself due to a shortage of manpower or other reasons (Craighead et al., 1998; Winn and Jackson, 1999). Many current eEducation systems normally do not provide separate supporting tools for tutors.

Mentor

The role of a mentor is to provide guidance to students on courses they are studying. A mentor may frequently check learning progress of students and lead them to the right path in solving difficult course questions. Sometimes a teacher takes up this role, but ex-course students can do this as well (Phillips, 1998). This role has been largely ignored by many current eEducation systems.

Faculty Staff

Faculty staff such as university administration clerk, secretary or librarian may need to use an eEducation system to facilitate teachings. Therefore, during the design phase of the educational system, their needs should be considered (Brown, 1998; Dringus and Terrell, 1999), which have been largely ignored by current eEducation systems.

Moderator

A moderator's role in an eEducation system is to control and maintain a clear and conscious discussion among a group of students. Functions provided by current eEducation systems include bulletin board threaded discussion, real time chat with text, voice and/or visual images (Brown 1998). However, powerful and effective tools facilitating a moderator to control and maintain an electronic discussion forum are inadequate in current eEducation systems. For example, an automatic alert function could be very useful to a moderator, so that he/she could be automatically alerted by possible slams or quarrels that could develop in the discussion forum, and then take actions to prevent them from happening.

External Visitors and/or Potential Students

eEducation systems may benefit other people outside a physical university. They may be used to promote courses to prospective students who may be interested in enrolling in the university (Auld and Pantelidis, 1999). The concepts of intranet and extranet could be suitable models for designing eEducation systems for such purpose.

Table 6. Main supporting components of the environment dimension

Educational Component	*Supporting Information Technologies*
Multiple/Universal accesses (Zhang, 1998; Zhao, 1998)	Web, Computer Terminals, Dial-In Connection (ISP), Telnet, FTP and Phone, etc.
Multiple media (Emerson and Mosteller, 1998; Levin et al., 1999)	Text, Graphics, Audio and Video, etc.
Multiple Security Mechanism (Firdyiwek,1999)	Account Password security, Different levels of File/Record access, Different levels of Application access and Firewall protection of intranet, etc.
Storage (Collis, 1999; Leh, 1999; Tinoco et al., 1997)	RAM, CDROM/CD-RW, Hard disk, Database, Tape drive, Zip drive and Network storage, etc.
Multiple distribution medium	Web/Internet, Intranet/LAN, CD-ROM, Email (Text. Audio, Video), Mail (Print, Audio, Video), Television, Radio, Phone, Satellite and Cable, etc.
Instructor aids (Hsu et al., 1998)	Centralized mouse control, Computer console, etc.
Course Administration Tools (Firdyiwek, 1999)	Student enrolment/registration, Course creation/allocation, Student account management and security management, etc.
Authoring Tools	Word processing tools, Webpage editor, Presentation tools and Interactive courseware tools, etc.

Environmental Dimension

An educational environment is where all the educational activities take place. These activities can either be carried out in a physical location or remotely in an online virtual environment. Through this environment, students can access educational materials and experience all kinds of interactions and discussions with their teachers and peers. For a virtual educational environment, it is the place where different hardware systems and software applications are integrated to create an environment for teachers to implement their courses, students to learn their courses, and all related people to collaborate and interact with each other.

Based on the literature review, an ideal eEducational environment should have the following characteristics: Easiness for access, Safety and security, integrity, Scalability, Flexibility, Compatibility/portability, and Ease of setting up and maintaining.

Ease of Access

One main concern from students with regards to eEducation systems is the access to the systems (Lund and Volet, 1998). There are generally two types of access problems. One problem refers to limited access or insufficient access to an eEducation system, especially for those students enrolled in an online course but who do not have Internet

Table 6. Main supporting components of the environment dimension (cont.)

Course Evaluation /Assessment Tools (Cleaver and Toole, 1999; Levin et al., 1999; Luck and Joy, 1995; Zhao, 1998)	Online tutorial, Online quiz, Assignment submission system and Text/exercise builder, etc.
Communication Tools (Alavi et al., 1995; Bandi and Nakatani, 1998; Bourne et al., 1997; Brown, 1998; Leidner and Jarvenpaa, 1995)	Hypertext (XML/HTML), Email (Text, Voice, Video), Phone, Voicemail, Fax, Mail, List-server, Electronic Bulletin Board, Internet chat room/IRC, Internet phone, Internet video, Web-enabled White Board, Telephone conference, Audio/Video conference, Newsgroup and Virtual Reality Environment, etc.
Student aids (Abowd et al., 1996; Bourne et al., 1997)	Electronic note taking pad, Play back, AV lecture clips
Collaboration Tools	Calendar, Application/File sharing, Memo, Collaborative writing tools and Configuration management tools, etc.
Modeling & Simulation (Noor, 1998)	Matlab, etc.
Disability support (Rankin, 2000)	Speech recognition software, Speech synthesis software, etc.
Adaptable educational system (Bonk & Cummings, 1998; Brusilovsky et al., 1998; Wang, 1997)	Active Server Page, Artificial Intelligent, Intelligently personalized educational functions, etc.
Virtual Environment (Crumpton and Harden, 1997; Winn and Jackson, 1999)	3-Dimensions Virtual Reality (VR) tools: Multi-user Object Oriented (MOO), Virtual Reality Markup Language (VRML) tool, etc.

connection at home (Lund and Volet, 1998). Another problem refers to the difficulties in accessing an eEducation system (Hill et al., 1998), even after students have required computing and networking facilities at home or in a university laboratory. Some students and/or teachers do not have enough computer knowledge and skills to use the system. Therefore, system designers should consider the access issue from an end user's perspective.

Safety and Security

An ideal educational system should provide a safe environment for teachers to teach and for students to learn online (Bonk and Cummings, 1998). The environment should provide communication channels allowing students to have all public, private and anonymous communications, and should provide different levels of securities such as restricted file access and enabling/disabling different features for different users (Firdyiwek, 1999).

Table 7. Course administration elements/features

No.	Course Administration Elements/Features	Fully Supported by Current eEducation Systems?
1	Course outline (Bourne et al., 1997; Cleaver and Toole, 1999)	Support
2	Enrollment criteria (Rankin, 2000)	Weak support
3	Honor code and process for Honor code violations (Rankin, 2000)	Weak support
4	A policy for the proper use of the web (Pan, 1998)	Support
5	Contacts of the lecturers/instructors (Rankin, 2000)	Support
6	Enrollment/Registration online	Weak support
7	Calendar/Schedule of classes	Support
	(Cleaver and Toole, 1999; Rankin, 2000)	
8	Announcements (Cleaver and Toole, 1999)	Support
9	Online Survey (Pre-course and Post-course)	Some support
10	Electronic Bulletin Board/Forum	Support
11	Security (Bonk and Cummings, 1998; Firdyiwek, 1999)	Some support
12	FAQ	Support

Integrity

A fully integrated environment should include an array of tools for both teachers and students to use, which provides teachers and learners with more flexibility and allows them to accomplish multiple tasks efficiently (Zhao, 1998). An integrated eEducation environment enables teachers and students to interact in multiple dimensions and/or communication channels (e.g., audio, video and textual messages), analyze the same data with various tools, and achieve multiple goals within a single application.

Scalability

Scalability is one of the necessary success factors for an eEducational environment (Bourne et al., 1997; Graziadei et al., 1997). A system should be easily scaled up or down when demands of the system change variably from time to time.

Flexibility

An eEducation system should have flexibility to allow frequent changes and reconfigurations to meet users' requirements (Graziadei et al., 1997). Multiplicity may be another word for the Flexibility (Levin et al., 1999). An ideal eEducation system should be able to provide maximum flexibility in terms of pedagogical perspectives and management styles (Zhao, 1998).

Compatibility/Portability

Another word describing compatibility and portability is Open Architecture, which means that an ideal eEducation system should be able to run on different network computing platforms (Graziadei et al., 1997).

Ease of Setting Up and Maintaining

The ease of setting up and maintaining an eEducation system is a long-term goal (Graziadei et al., 1997; Zhao, 1998). Features for the convenience of setting up and maintainability will make an eEducation system durable and effective. To build an ideal eEducation system environment, some necessary system components and supporting mechanisms/tools should be carefully considered during the system design phase which are summarized in Table 6.

Course Dimension

This important component consists of three main subcomponents: Course Administration, Course Materials, and Course Learning and Assessment.

Course Administration

An eEducation system with good course administration functions normally includes the elements and/or features listed in Table 7.

The Web page for course administration should have a course outline to state what this course is about and which topics it will cover. It should provide enough information to help students to know whether the course is suitable for them to take. Hence, enrollment criteria should list out any prior knowledge and prerequisites before they can be enrolled into this course.

Course Materials

An eEducation system with quality course materials function normally includes the elements and/or features listed in Table 8.

The course syllabus should give the students a more detailed description of what is going to be covered in a course. It tells students the expectation of a course that students should understand by the end of the course.

Besides the core lecture materials, some students may want to learn more about a particular topic they are interested in. Therefore, instructors should put up a list of extra readings for the topics being covered. Internet materials and information are frequently updated, and it is the instructors' responsibility to ensure the stability of materials (Harmon and Jones, 1999) and the accuracy and recentness of those materials and web sites (Branch et al., 1999; Sarapuu and Adojaan, 1998; Weston, 1999).

Table 8. Course material elements/features

No.	Course Material's Elements/Features	Fully Supported by Current eEducation Systems?
1	Syllabus (Bourne et al., 1997; Cleaver and Toole, 1999)	Support
2	Clear expectation of the course (Bonk and Cummings, 1998)	Support
3	Lectures materials online (Ulvund, 1998)	Support
4	AV clips of the lectures (Bourne et al., 1997)	Support
5	Class notes (Cleaver and Toole, 1999)	Support
6	A list of web resources (i.e., URLs)	Support
7	Extra Readings (Bourne et al., 1997)	Support
8	Digital Library (Tinoco et al., 1997)	Some support
9	Other resources such as Encyclopedia	Some support

Table 9. Elements/features of course learning and assessment

No.	Elements/Features of Course Learning and Assessment	Fully Supported by Current eEducation Systems?
1	Interactive learning modules (Bourne et al., 1997)	Weak support
2	Interactive tutorial (Cleaver and Toole, 1999)	Weak support
3	Structured Group Activities (Phillips, 1997 and 1998)	Weak support
4	Assignments details (Bonk and Cummings, 1998; Firdyiwek, 1999)	Some support
5	Online Quiz (Bandi and Nakatani,1998; Cleaver and Toole, 1999)	Some support
6	Homework/assignment submission (Bourne et al., 1997)	
	Luck and Joy (1995)	Support
7	Grade book (Cleaver and Toole, 1999)	Support

Course Learning and Assessment

An eEducation system with good course learning and assessment function normally includes the elements and/or features listed in Table 9.

Students can learn course materials themselves without the presence of an instructor through Interactive Learning modules of an eEducation system, which is designed to guide students in learning different topics at their own pace.

Table 10 summarizes the above discussion.

Pedagogical Dimension

The pedagogy component largely determines how to integrate the other three components of an eEducation system. It also largely differentiates teaching styles and performance even using the same eEducation system and course materials. However, this may be the weakest part in current eEducation systems because they do not have sufficient tools to support pedagogy effectively.

Pedagogical Models

There are four major pedagogical models commonly used by universities: the instructor-centered model, learner-centered model, learning-team-centered model (Noor, 1998), and problem-based learning model (Oliver and Omari, 1999).

Instructor-Centered Model

The instructor-centered pedagogical model has been commonly adopted in university education and current eEducation systems; it is based on the behavior educational theory (Firdyiwek, 1999) and/or objectivism educational theory (Leidner and Jarvenpaa, 1995). This model is often described as the model of one-way information transfer from an instructor to students (Noor, 1998).

Based on this model, a course web site is often used to display course information, make course announcements, and supply addition reading materials, etc. (Harmon and Jones, 1999). The reason for this model being so popular could be that an instructor has full control of course materials being learnted by students, fewer resources are needed,

Table 10. Three components and their main elements

Components	Main Elements	Fully Supported by Current eEducation Systems?
Course Administration	Contacts of the lecturers/instructors, Course outline, Calendar/Schedule of classes, Enrolment criteria, Enrolment/Registration online, Announcements, FAQ, Honor code and process for Honor code violations, A policy for the proper use of the web, Online Survey (Pre-course and Post-course), Electronic Bulletin Board/Forum, Other communication tools and Security.	Some support
Course materials	Syllabus, Clear expectation of the course, Lectures materials online, AV clips of the lectures, Class notes, A list of web resources, Extra Readings, Digital Library and Other resources such as Encyclopaedia.	Support
Course Learning and Assessment	Interactive learning modules, Interactive tutorial, Structured Group Activities, Assignments details, Online Quiz, Homework/assignment submission and Grade book.	Weak support

and the website is easier to develope and maintaine. Most, if not all, current eEducation systems seem to adopt this model in their system design.

Learner-Centered Model

The learner-centered model is opposite of the instructor-centered model. It posits that students learn most effectively through self-exploratory, self-discovery learning, and skills and knowledge acquisition through their own experiences. This model is based upon cognitive/constructivism theory, and it is implemented through interactive technologies such as virtual reality environment, simulation and modeling software, and interactive tutorial/quiz (Bonk and Cummings, 1998).

This model requires students to take a more proactive approach in their learning, which in turn requires a high level of self-discipline. In addition, students are encouraged to select their own learning path (Harbeck and Sherman, 1999; Smith and Northrop, 1998; Zhang, 1998). Courses equipped with interactive technologies to support this pedagogical model require a huge amount of time and resources for thorough planning, design, implementation, testing, and maintenance. Therefore, not many institutions or teachers have enough time and budget to use this pedagogical model. Many current eEducation systems are weak in supporting this model.

Learning-Team-Centered Model

This model emphasizes benefits of learning in groups, which posits that interactions and collaborations among students can improve learning performance (Noor, 1998). One involved in group activities is both a contributor and a beneficiary. The model requires different collaboration technologies such as group support environment (e.g., calendar, application/file sharing, and virtual environment) and synchronous and

Table 11. Main structured group activity methods

Group Method	Brief Description
Online Seminars	Similar to face-to-face seminars. Students read assigned materials, and then discuss, debate, and extrapolate on key issues. Other students may act as group leaders, initiating, and directing discussion.
Small Group Discussion	Students work in small groups to analyze a particular topic. This mode is suitable for active discussion in a large class, or situations where there are particular special interest groups.
Learning Partnerships	In this mode, students are grouped into pairs, or dyads, and work together on an assigned task. Dyads are useful early in a course because they provide peer support and act as icebreakers. Students realize that they have a contact with another person with similar problems and fears. After coming to groups with the dyad, students can progress to larger groups.
Learning Circles	This method is used for student collaboration on major tasks, such as a research project. The optimum size is 3 or 4, and this mode requires students to clearly define tasks, decision-making goals, roles and timelines.
Role-Plays	Online role-plays can be used to apply theoretical knowledge in a simulated environment, such as in management and business. They can also be used to discuss philosophical issues.
Debates	This method is related to role-plays but more formal. Students discuss contentious and controversial issues according to formal debating rules.

asynchronous communication technologies (e.g., e-mail, newsgroups, IRC, audio/video conference system, and other traditional technologies). This model receives some support from current eEducation systems (except the Structured Group Activity function as presented below).

Commonly used structured group activities with the aim to enhance group interactions that could in turn enhance group-learning performance are summarized in Table 11 (Phillips, 1997, 1998); these activities receive little support from many current eEducation systems.

Problem-Based Learning Model

Problem-based learning is a curriculum approach that helps a learner to frame experience as a series of problems to be solved, and where the process of learning unfolds through the application of knowledge and skills to the solution of real-world problems, often in the context of real practice (Bligh, 1995). Enhancing learning through problem-based activities requires learners to solve authentic problems, which reflect the way in which the learned information will be used outside classroom settings (Herrington and Oliver, 1997). This model has not been widely adopted in current eEducation systems.

Instruction Design

Having technologies as well as course materials online does not make learning a lot better. It is the pedagogy that plays a central role in integrating all of the other three educational components together and helps enhance learning performance. The pedagogy is mainly realized through instructional design. Table 12 summarizes some important issues for doing instruction design.

Interactivity

General ideas of interactivity are that it involves two-way communication, it enables learners to adjust the instruction to conform to their needs and capabilities, a learner becomes an active participant, rather than a passive observer, making significant decisions and encountering their consequences (Sims, 1998, 1999). Fenrich (1997) considers interactivity as embodied in instructional features that promote active learning.

Gilbert and Moore (1998) define two kinds of interactivity. The first is the instructional interactivity between teachers and students. The second interactivity is social interactivity between teachers and students. These interactivities often take forms like body language, greetings/socializing, and exchanging personal information, etc. Examples of technologies for social interactivity are face-to-face contact via audio and/or video, email, online chat, electronic bulletin boards and moderated discussion. Technologies, which are suitable for Instructional Interactivity, include shared whiteboard, computer application that allows information query and responding, file distribution, email, monitoring responses and providing correct answers as feedback, etc.

Motivation Factors

Motivation explains why someone chooses to put in effort to do one thing rather than another (Arnone and Small, 1999). Research on motivation factors focuses on how to enhance instructional design, improving classroom management, and meeting the needs of diverse student populations. Much research has been done to identify effective techniques and a motivation model to improve the teaching-learning environment. One popular model is the ARCS Motivation Model (Small, 1999), which identifies four essential strategic components for motivating instruction, as shown in Table 13.

A good instruction design should take the consideration of the above four motivational components and their subcomponents, which seems to be largely neglected by the design of current eEduction systems.

Psychological Factors

Psychology has played an important role in many areas of life. The American Psychological Association (APA) had written a report called 'The Learner-Centered Psychological Principles: A Framework for School Redesign and Reform" (Bonk and Cummings, 1998). The learner centered psychological principles provide a framework for developing and incorporating components of new designs for schooling. These prin-

Table 12. Important issues for instruction design

No.	Important Issues for Instruction Design	Fully Supported by Current eEducation Systems?
1	Interactivity	Support
2	Motivation factors	Very weak support
3	Psychological factors	Very weak support
4	Presentation and Language	Some support
5	Other issues	Weak support

Table 13. Four strategic components for motivating instruction

No.	Four Strategic Components for Motivating Instruction	Sub-components
1	Attention - strategies for arousing and sustaining curiosity and interest	Perceptual Arousal; Inquiry Arousal; Variability.
2	Relevance - strategies that link to learner's needs, interests and motives	Goal Orientation; Motive Matching; familiarity.
3	Confidence - strategies that help students develop a positive expectation for successful achievement	Learning Requirement; Success Opportunities; Personal Responsibility.
4	Satisfaction - strategies that provide extrinsic and intrinsic reinforcement for effort	Intrinsic Reinforcement; Extrinsic Rewards; Equity.

ciples emphasize the active and reflective nature of learning and learners. The 14 principles are classified into four categories (cognitive and metacognitive, motivational and affective, developmental and social, and individual difference) that influence learners and learning, which again seems to be largely neglected by the design of current eEduation systems.

Presentation and Language

The presentation and language factor concerns how to present course materials and use the right language to express thoughts (Weston, 1999). In general, the design of a course should be simple, clean and concrete (Harbeck and Sherman, 1999).

Other Issues

Instruction and eEducation systems designers should consider other important issues as well: social cultural issues (Kearsley, 1998), economical issues (Kearsley, 1998), political issues (Kearsley, 1998), time and distance (Noor, 1998), and group size for group interactions, which seems to be largely neglected by the design of current eEducation systems.

A CONCEPTUAL FRAMEWORK

Table 14 summarizes the above comprehensive literature review and discussion on the four dimensions and components of an education system, and how they are supported by current eEducation systems.

Many current eEducation systems have emphasized one or more educational dimensions. More specifically, the Course dimension is generally supported by current eEducation sytems, the Human and Environmental dimensions are weakly supported, and the Pedagogy dimension is little supported. Few current eEducation systems have had a holistic view of all the four dimensions in their system design, and many of the systems have not had specific tools and/or functions to effectively support the central dimension — pedagogy. In fact, supporting pedagogy is the weakest part of current

Table 14. Components of the four dimensions

Dimension	Sub-component	Fully supported by current eEducation Systems?
Human dimension	Teacher, Student, Course Administrator, Peer, Tutor, Mentor, Faculty staff, Moderator, and External visitor.	Weak support
Course dimension	Course Administration, Course materials, and Course Learning and Assessment	Some support
Environmental dimension	Easiness for access, Safety and security, integrity, Scalability, Flexibility, Compatibility/portability, and Easiness for setting up and maintaining; Multiple accesses, Multiple media, Instructor aids, Disability support, Virtual Reality, etc.	Weak support
Pedagogical dimension	Pedagogical models and Instruction design	Very weak support

eEducation systems. This could at least partially explain why virtual universities supported by current eEducation systems have so far failed to meet expectations (e.g., Baumlin et al., 2000; Kearsley, 1998). Therefore, supporting the pedagogical dimension should be the major concern of designing next generation eEducation systems.

Based upon the above comprehensive literature review, a conceptual framework is proposed that could be used to guide the design of next generation eEducation systems (see Figure 2).

In Figure 2, the pedagogical dimension is the central issue for designing a next generation eEducation system, which has been largely neglected by many current eEducation Systems. The arrow line "→" represents the relationship of controlling, namely, the pedagogical dimension controls how the other three dimensions should be included, formatted, presented, functioned, and interacted with each other. The dashed line "—" describes normal interaction relationships between two dimensions.

The human dimension represents users of an eEducation system mainly consisting of teachers and students. The course dimension covers course-related issues like teaching materials and course format, etc. The environment dimension consists of physical technological infrastructures, tools, and hardware, etc. How the three dimensions are configured and integrated is largely determined by the central dimension — pedagogy. The pedagogy dimension represents teaching methods that are adopted into courses and how an eEducation system should be configured and modified to meet specific educational missions and goals.

The proposed framework recognizes the important role of the pedagogy dimension, which should be the starting and ending point for designing a next generation eEducation system. For example, an adopted instructor-centered model (the PEDAGOGY dimension) may be able to cater to more students (the HUMAN dimension) and only requires relatively simply and commonly used technologies and limited other resources (the ENVIRONMENT dimension) to help distributing course materials to remote students.

Figure 2. A conceptual framework for designing a next generation educational system

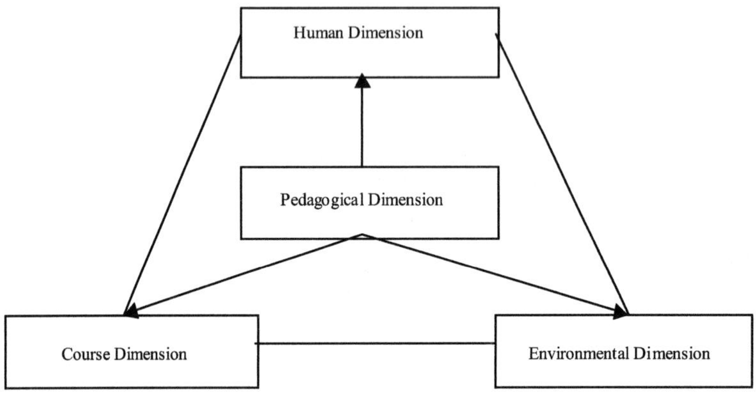

Course assessment materials and formats (the COURSE dimension) will be provided for students to work online individually. On the other hand, an adopted learner-team model (the PEDAGOGY dimension) will require technologies that facilitate group collaborations, group communications, and group projects (the COURSE dimension). Additional tutors (the HUMAN dimension) may be required to guide a number of student groups (the HUMAN dimension). All tutors will report to a teacher about students' progresses. Therefore, these two different teaching models (the PEDAGOGY dimension) demonstrate that different pedagogical models require different elements of the other three dimensions to be integrated into an effective eEducation system.

Surrounding the above four dimension sets are two layers of technical issues. They are eEducation services, which are network-based implementation and other relevant hardware/software systems, and information technologies, including commonly available network, software, hardware, and all information system development concepts and tools.

Although in this paper we emphasize the central role of pedagogical dimension in an eEducation system, we also need to take into account the impact of recent advances in IT/IS to eEducation system design and implementation. Without the Internet and web technology, there could have been no current generation of eEducation systems. Information technology has brought about revolutionary changes in the education sector, and its latest development may further push forward the revolution.

Overall, the four-layered conceptual framework captures the issues from inside to outside with increasing generality. It is obvious that the previously over-emphasized role of information technology in eEducation systems is not compliant with this framework. In a specific application field such as eEducation, domain-relevant issues always play dominant roles. The two layers of technical issues affect pedagogical dimension only via the three dimension sets: human dimensions, environmental dimensions and course dimensions.

FURTHER DISCUSSIONS

An Operational Scenario for the Conceptual Framework

Based upon the proposed conceptual framework, the following scenario describing a possible operational procedure may present a clearer guideline for designing a next generation eEducation system.

1) Given a course offered to a class in a university, the instructor(s) may choose a suitable pedagogical model from a pedagogical models database, based upon factors such as course characteristics and students' characteristics and background information.
2) After a specific pedagogical model is chosen, an electronic intelligent agent will come out to help identify components and/or sub-components of the other three dimensions that are saved in a system database (see Tables 6, 7, 8, 9, 11, 12, 13, and 14 for the specific components and/or sub-components), and integrate those identified components and/or sub-components to match the requirements of the chosen pedagogical model. In this way, an effective next generation eEducation system will be generated.
3) Based upon the feedback from students' learning and teachers' teaching during the course period (e.g., quiz result), the electronic intelligent agent will re-configure the identified components and/or sub-components to further improve the system's supporting functions and features and thus help improve learning/teaching performance.

Usefulness of the Conceptual Framework and Future Research

The proposed conceptual framework provides a holistic view of an education system in terms of key dimensions and components within it and their relationships. It can be used to assess current eEducation systems and helps identify weaknesses and strengths of current eEducation systems. As a result, it can be used to help universities to effectively choose suitable eEducation systems to meet their educational goals, and also to help guide the design of a next generation eEducation system. Specifically, the next generation eEducation systems should be a pedagogy-centered design that is different from many current eEducation systems that are relatively more technology-centered.

Table 14 presents a high level view of an assessment of current eEducation systems in terms of their support to the four educational dimensions and related components. The identification of weak support from current eEducation systems in the table provides some insightful and specific directions for eEducation systems developers to continuously improve their system design, and therefore, universities could compete better in the global education market in the future.

Further, the proposed framework helps identify some important research questions for future studies:

1) What are typical and effective educational methods for each of the four commonly used pedagogical models?
2) What are the important factors influencing the choice of a suitable pedagogical model? How can we effectively choose a suitable pedagogical model under given conditions?
3) For each of the four pedagogical models, what are the specific components and/or sub-components of the Course, Environment, and Human dimensions that can be integrated to match with the requirements of a specific pedagogical model for an effective eEducation system? (Tables 6, 7, 8, 9, 11, 12, 13, and 14 provide some clues to the answer but more studies are still needed to have a clearer understanding of the issue).
4) How can the latest Web technologies be used to support the pedagogy-centered next generation eEducation systems? For example, Web Service is a new Web technology being currently developed and upgraded by many big multi-national companies such as Microsoft .NET, Sun Open Net Environment (ONE), IBM Web Service Conceptual Architecture (WSCA), W3C Web Service Workshop, Oracle Web Service Broker, and Hewlett-Packard Web Service Platform (Myerson, 2002). The idea of Web Service is to allow application services and data resources being deployed in different geographical locations following the layered open system architecture, which may empower next generation eEducation systems that could support universities to more effectively compete in global education market. The Web Service application architecture has demonstrated its great potential value in e-commerce, enterprise information system integration, and public services (Myerson, 2002).

With answers to the above research questions being found by future studies, a next generation eEducation system designed under the guidance of the proposed conceptual framework is more likely to enhance online teaching and learning performance, and therefore helps universities and corporations to effectively compete in global education and/or training market.

REFERENCES

Abowd, G. D., Atkeson, C. G., Feinstein, A., Hmelo, C., Kooper, R., Long, S., Sawhney, N., & Tani, M. (1996). Teaching and learning as multimedia authoring: The classroom 2000. *Proceedings of the ACM International Multimedia Conference & Exhibition*, (pp. 187-198).

Alavi, M., Wheeler, B. C., & Valacich, J. S. (1995). Using IT to reengineer business education: An exploratory investigation of collaborative telelearning. *MIS Quarterly,* Sept, 293-312.

Al-Nuaimy, W, Zhang, J., & Noble, A. (2001). Web-based learning environment for a communications module. *Computer Applications In Engineering Education, 9*(2), 114-121.

American Society for Training & Development Inc. (1997). Distinctions and definitions of learning technologies. *Training, 51*(11), 48.

Arnone M. P., & Small, R.V. (1999). Evaluating the motivational effectiveness of children's Websites. *Educational Technology, 39*(2), 51-5.

Bandi, R., & Nakatani, K. (1998). Web-based distance learning: Choosing effective technologies for information systems courses. *Proceedings of the International Academy for Information Management (IAIM) Annual Conference,* Helsinki, Finland, December 11-13.

Bargeron, D., Grudin, J., Gupta, A., & Sanocki, E. (2002). Asynchronous collaboration around multimedia applied to on-demand education. *Journal of Management Information Systems, 18*(4), 117-145.

Barker, P. (2002). Innovation in open and distance learning – Successful development of online and web-based learning. *Innovations in Education and Training International, 39*(1), 82-85.

Bligh, J. (1995). Problem-based learning in medicine: An introduction. *Post-Graduate Medical Journal, 71*(8), 323-326.

Bonk, C.J., & Cummings, J. A. (1998). A dozen recommendations for placing the student at the center of Web-based learning. *Educational Media International, 35*(2), 82-89.

Bourne, J. R., McMaster, E., Reiger, J., & Campbell, J. O. (1997). Paradigms for on-line learning: A case study in the design and implementation of an asynchronous learning networks (ALN) course. *1997 Frontiers in Education Conference*, 245.

Branch, R. M., Kim, D., & Koenecke, L. (1999). Evaluating online educational materials for use in instruction. *ERIC Digest.* Report: EDO-IR-1999-07.

Brown, B. M. (1998). Digital classrooms: Some myths about developing new educational programs using the Internet. *THE Journal (Technological Horizons In Education), 26*(5), 56.

Brusilovsky, P., Eklund, J., & Schwarz, E. (1998). Web-based education for all: A tool for development adaptive courseware. *Computer Networks and ISDN Systems, 30*(1-7), 291-300.

Chandler, B. (2001). Innovation in open and distance learning: Successful development of online and web-based learning. *Studies in Higher Education, 27*(1), 122-123.

Chang, C.C. (2001). A study on the evaluation and effectiveness analysis of web-based learning portfolio (WBLP). *British Journal Of Educational Technology, 32*(4), 435-458.

Cleaver, T. G., & Toole, R. L. (1999). Design of a web-based education environment. *The 29th ASEE/IEEE Frontiers in Education Conference*, 12a3-1.

Collis, B. (1999). Applications of computer communications in education: An overview. *IEEE Communications Magazine, 37*(3), 82-86.

Coppola, N.W., Hiltz, S.R., & Rotter, N.G. (2002). Becoming a virtual professor: Pedagogical roles and asynchronous learning networks. *Journal of Management Information Systems, 18*(4), 169-189.

Crumpton, L. L., & Harden, E. L. (1997). Using virtual reality as a tool to enhance classroom instruction. *Computers & Industrial Engineering, 33*(1-2), 217-220.

Davis, B. H. (1996). Teacher of the future. *Journal of the American Society for Information Science, 47*(11), 849-853.

Emerson, J. D., & Mosteller, F. (1998). Interactive multimedia in college teaching (Part 1 - A ten-year review of reviews). *Educational Media and Technology Yearbook,* 23.

Fenrich, P. (1997). *Practical guidelines for creating instructional multimedia applications.* Fort Worth: The Dryden Press.

Firdyiwek, Y. (1999). Web-based courseware tools: Where is the pedagogy? *Educational Technology, 39*(1), 29-34.

Forbes magazine. *http://www.forbes.com/forbes/1997/0616/5912084a.html*

Gilbert, L., & Moore, D. R. (1998). Building interactivity into Web courses: Tools for social and instructional interaction. *Educational Technology, 38*(3), 29-35.

Graziadei, W. D. et al. (1997). Building asynchronous & synchronous teaching-learning environments: Exploring a course/classroom management system solution. *http://exon.itec.suny.edu/west/aslpaper.html*

Harbeck, J. D., & Sherman, T. M. (1999). Seven principles for designing developmentally appropriate Web sites for young children. *Educational Technology, 39*(4), 39-44.

Harmon, S. W., & Jones, M. G. (1999). The five levels of Web use in education: Factors to consider in planning online courses. *Educational Technology, 39*(6), 28-32.

Herrington, J., & Oliver, R. (1997). Avenues to understanding: A qualitative study into how students learn from multimedia. In T. Muldner & T. Reeves (Eds.), *Educational multimedia/hypermedia and telecommunications*, (pp. 473-478). Charlottesville, VA: AACE.

Hill, J. R., Rezabek, L. L., & Murry, B. (1998). Web-based delivery of instruction: Prospects and challenges. *The Annual Meeting of the American Educational Research Association*, San Diego, CA, April 13-17.

Hsu, S., Pandya, A., & Patel, R. (1998). Yet another virtual classroom over the Internet. *1998 FIE Conference*, 717.

Huang, W., Lienhart, T., & Yen, D. (2002). A preliminary study of eEducation using three case studies: Key factors for successful implementation. *Communications of the ICISA*, *5*(1), 1-13.

Ingram, A. L. (1996). Teaching with technology. *Association Management,* (June).

Jung, I.S. (2001). Building a theoretical framework of web-based instruction in the context of distance education. *British Journal of Educaitonal Technology*, *32*(5), 525-534.

Kearsley, G. (1998). Educational technology: A critique. *Educational Technology, 38*(2), 47-51.

Kinshuk, R. (2002). Web-based learning and teaching technologies: Opportunities and challenges. *Online Information Review*, *26*(1), 61-62.

Lee, M.G. (2001). Profiling students' adaptation styles in Web-based learning. *Computers & Education, 36*(2), 121-132.

Leh, A., & Sheng, C. (1999). Research on the characteristics of "the technology course." *SITE 99: Society for Information Technology & Teacher Education International Conference,* San Antonio, TX, February 28-March 4.

Leidner, D. E., & Jarvenpaa, S. L. (1995). The use of information technology to enhance management school education: A theoretical view. *MIS Quarterly,* September, 265-291.

Levin, J., Levin, S. R., & Waddoups, G. (1999). Multiplicity in learning and teaching: A framework for developing innovative online education. *Journal of Research in Computing in Education, 32*(2), 256-69.

Liu, M., Jones, C., & Hemstreet, S. (1998). Interactive multimedia design and production processes. *Journal of Research on Computing in Education, 30*(3), 254.

Luck, M., & Joy, M. (1995). Automatic submission in an evolutionary approach to computer science teaching. *Computers & Education, 25*(3), 105-111.

Lund, C., & Volet, S. (1998). Barriers to studying online for the first time: Students' perceptions. In C. McBeath and R. Atkinson (Eds.), *The Biennial Conference of the Australian Society for Educational Technology,* Perth, Western Australia, July 5-8, (pp. 123-136).

Myeson, J. (2002). Web service architectures – How they stack up. January 23: *http://www.web services architect.com/content/articles/myerson01.asp*

Noor, A. K. (1998). Future learning environment and impact of ISE. *Proceedings of the IEEE International Conference on Systems, Man and Cybernetics,* (Vol. 3, pp. 2692-2697).

Oliver, R., & Omari, A. (1999). Using online technologies to support problem based learning: Learners' responses and perceptions. *Australian Journal of Educational Technology, 15*(1), 58-79.

Owston, R.D. (2000). Evaluating Web-based learning environments: Strategies and insights. *Cyberpsychology & Behavior, 3*(1), 79-87.

Pan, A. C. (1998). Optimize the Web for better instruction. *The Proceedings of the Society for Information Technology & Teacher Education International Conference,* Washington, DC, March 10-14.

Pelton, J. N. (1996). Cyberlearning vs. the university: An irresistible force meets an immovable object. *The Futurist,* Nov-Dec.

Phillips, R. (1997). What is the most appropriate way to use the Internet for teaching and learning? In R. Pospisil & L. Willcoxson (Eds.), *Learning through teaching: The Proceedings of the 6th Annual Teaching Learning Forum,* Murdoch University, February, (pp. 276-278).

Phillips, R. (1998). What research says about learning on the Internet. In C. McBeath & R. Atkinson (Eds.), *The Biennial Conference of the Australian Society for Educational Technology,* Perth, Western Australia, July 5-8, (pp. 203-207).

Rankin, W. (2000). A survey of course Web sites and online syllabi. *Educational Technology, 40*(2), 38-43.

Rhodes, D. M., & Azbell, J. W. (1985). Designing interactive video instruction professionally. *Training and Development Journal*, Dec, 31-33.

Rovai, A.P. (2001). Building classroom community at a distance: A case study. *Educational Technology Research and Development, 49*(4), 33-48.

Sarapuu, T., & Adojaan, K. (1998). Evaluation scale of educational Web sites. *WebNet 98 World Conference of the WWW, Internet and Intranet Proceedings,* Orlando, FL, November 7-12.

Sims, R. (1998). Interactivity for effective educational communication and engagement during technology based and online learning. *The Biennial Conference of the Australian Society for Educational Technology,* Perth, Western Australia, July 5-8, (pp. 214-219).

Sims, R. (1999). Interactivity on stage: Strategies for learner-designer communication. *Australian Journal of Educational Technology, 15*(3), 257- 272.

Small, R. V. (1999). Motivation in instructional design. *Educational Technology Year Book,* 89.

Smith, K., & Northrop, K. (1998). The CLASS course design model for Web-based instruction. *Proceedings of the Annual Conference on Distance Teaching & Learning,* Madison, WI, August 5-7.

Snell, N. (1997). Why can't Johnny do client/serve? *Inside Technology Training*, July/August, 21-26.

Tinoco, L. C., Fox, E. A., & Barnette, N. D. (1997). Online evaluation in WWW-based courseware. *Proceedings of the 28th SIGCSE Technical Symposium on Computer Science Education,* San Jose, CA, February 27-March 1, (pp. 194-198).

Tolmie, A., & Boyle, J. (2000). Factors influencing the success of computer mediated communication (CMC) environments in university teaching: A review and case study. *Computers & Education, 34*(2), 119-140.

Topper, A. (2002). Web based learning and teaching technologies: Opportunities and challenges. *Teachers College Record, 104*(1), 151-154.

Ulvund, F. (1998). Hyperlectures - Teaching on demand. *UltiBASE In-Site* 1998 (October). *http://ultibase.rmit.edu.au/Articles/ulvund1.html*

Wang, H. (1997). LearnOOP: An active agent-gased educational system. *Expert Systems with Applications, 12*(2), 153-162.

Weston, C., Gandell, T., McAlpine, L., & Finkelstein, A. (1999). Designing instruction for the context of online learning. *The Internet and Higher Education, 2*(1), 35-44.

Wetzel, M. (n.d.). Banking on it. *Online Learning Magazine. http://www.online learningmag.com/new/mar01/feat2prt.htm*

Winn, W., & Jackson, R. (1999). Fourteen propositions about educational uses of virtual reality. *Educational Technology*, July-August, 5-14.

Zemke, R. (1998). Wake up! (And reclaim instructional design). *Training,* June.

Zhang, P. (1998). A case study on technology use in distance learning. *Journal of Research in Computing in Education, 30*(4), 398-419.

Zhao, Y. (1998). Design for adoption: The development of an integrated Web-based education environment. *Journal of Research on Computing in Education, 30*(3), 307-328.

Previously published in the Journal of Global Information Management, 12(2), 84-107, April - June 2004.

Chapter VII

Adaptive Strategies of Firms in High-Velocity Environments: The Case of B2B Electronic Marketplaces

Jai Ganesh, Infosys Technologies Limited, India

T.R. Madanmohan, Indian Institute of Management, Bangalore, India

P.D. Jose, Indian Institute of Management, Bangalore, India

Sudhi Seshadri, Indian Institute of Management, Bangalore, India

ABSTRACT

The adaptation-evolution strategies of firms have been a major area of interest. Internet based businesses operate in highly unstable environments witnessing shakeouts and changes in the industry structure. In this paper, we focus on the adaptive strategies and paths of adaptation of independent B2B marketplaces. The independent B2B marketplaces have undergone tremendous change with regard to their business models and products/services. Since their inception as pure market makers, these marketplaces have morphed into integration service providers with market making being one of the functions. What makes the adaptive strategies of B2B marketplaces interesting is the extremely short time period within which these changes happened. The research uses the case study methodology to identify the adaptive strategies of B2B marketplaces in terms of their product/service class evolution. The results based on the case studies suggest a three-stage adaptation model wherein the marketplaces progress through stages such as aggregation, transactions and integration.

INTRODUCTION

The Internet is having a profound influence on the nature of interorganisational commerce. According to Forrester Research, business-to-business (B2B) electronic commerce (e-commerce) in the US now stands at US $2.4 trillion. Inspite of the failure of hundreds of B2B exchanges, B2B e-commerce is expected to reach US $3.9 trillion in 2003 (*Business Week*, 2003). eMarketer predicts that worldwide business-to-business ecommerce would reach US $1.41 trillion by the end of 2003. According to eMarketer, the worldwide e-commerce revenues will touch US $2.4 trillion by 2004. B2B e-commerce refers to transactions wherein both the buyer and seller are organizations. Business-to-business e-commerce is not a new concept and we can trace it back to EDI (Electronic Data Interchange). EDI allows businesses and their partners to exchange information through a standard set of transactions over value-added-networks (VANs). EDI facilitates transactions such as purchase orders, invoices, shipping notices and a multitude of other documents. However, the high costs associated with EDI can be afforded only by the large companies (Khazanchi, 1995, 1999). The common forms of B2B e-commerce include electronic procurement, electronic catalog management systems, B2B marketplaces, etc. The broader business adoption of the Internet laid the foundation upon which B2B marketplaces evolved. A number of business-to-business marketplaces emerged to facilitate efficient commerce among organizations. These marketplaces facilitate efficient search and transactions by offering services such as buyer/supplier and product/services searching, transactions such as procurement and asset disposal. In addition to their market making function, the marketplaces also offer integration services such as supply chain and ERP integration and have entered into alliances with various firms to offer value-added services such as vendor rating, logistics, payment processing, etc.

The Internet reduces search costs of buyers, resulting in price competition among sellers Bakos (1991, 1997). Bakos (1997) explores the implications of reducing buyers' search cost in electronic marketplaces, the benefits for buyer and seller, the social welfare and the incentives to create such a marketplace. Garicano and Kaplan (2000) provide empirical evidence that the benefits of electronic marketplaces may come from price advantage as well as process improvement. B2B marketplaces differ from the traditional marketplaces, as they offer increased personalization and customization of product offerings, and aggregation and disaggregation of information-based product components to match customer needs. They are able to overcome some of the problems related to richness vs. reach of information, as they are able to facilitate real-time transactions. They also enable new types of price discovery to be employed in different markets. B2B marketplaces also improve information sharing between buyers and sellers, helping lower the cost of logistics and promoting quick, just-in-time deliveries and reduced inventories.

B2B marketplaces can be broadly classified into horizontal vs. vertical marketplaces, buy-side vs. sell-side marketplaces, etc. For this research, we classify them into

Figure 1. Sciquest

Year	About the company	Revenue model
1995-1999	**Online marketplace for Scientific Products targeted at** researchers, scientists, and purchasing managers	Advertising revenues
1999 – 2000	Web-based, interactive marketplace for scientific and laboratory products	Advertising revenues Revenue from e-commerce transactions Advertising revenues Commissions received from auctions
2001	Technology and solutions company providing integrated e- commerce and asset management solutions for research enterprises and their supplier partners worldwide	Advertising revenues Revenue from e-commerce transactions through the Sciquest website Advertising revenues Commissions received from auctions
2002 – 2003	Software, information and services company that provides solutions to enhance research operations for pharmaceutical, biotechnology and other research-based organizations.	License fee Subscription fee Maintenance fees Fees from professional services Advertising revenue

public/independent, private and consortia marketplaces. A public/independent marketplace, also known as neutral marketplace or third-party marketplace, brings together buyers and sellers within a particular industry for the purpose of commerce. It provides content, value-added services and transaction capabilities e.g., Freemarkets. Private marketplaces are owned and operated by one company to transact with a select group of suppliers. They have the potential to provide high performance and tight integration with current suppliers, a facility provided by the traditional EDI model. Examples of enterprises that have embraced this model include: Wal-Mart, Dell, Sun Microsystems, Amtrak, and Cisco. Consortia marketplaces are jointly owned by several large enterprises that deploy applications and infrastructure to facilitate collaboration and conduct business among trading partners. They are highly customized and integrated with the process of its founders, and hence require a large investment and have longer implementation schedules. Examples include Covisint, E2Open, etc.

Between mid-1998 and mid-2000, nearly 1,900 public e-marketplaces (1,500 independent and 287 industry-sponsored) were formed (Deloitte Research, 2001). Growth in the number of B2B marketplaces has been stupendous. Very often, the business models of many of the B2B marketplaces have undergone rapid changes. To illustrate the rapid changes in the business models of the B2B marketplaces, we take the example of two prominent marketplaces, Verticalnet and Sciquest. Both the marketplaces were founded during the year 1995. Tables 1 and 2 list the details about these marketplaces. We chronologically list the company description as well as their revenue model as per their filings with the SEC (Securities and Exchange Commission). Sciquest a prominent B2B

Table 2. Verticalnet

Year	About the company	Revenue model
1995-2000	Creator and operator of Vertical trade communities that acted as industry-specific comprehensive sources of information, interaction and electronic commerce	Advertising revenues Web site development fees Sponsored newsletters Virtual trade shows Auctions
2001	Operates industry-specific e-marketplaces The firm offers software solutions to industry alliances, independent net market makers and enterprises	Software licensing Professional services Advertising revenues Web site development fees Auctions
2002 - 2003	Provider of Collaborative Supply Chain solutions allowing organizations to communicate, collaborate and conduct commerce more effectively across the extended supply chain	License fee Fees from professional services Advertising revenue

player acted as an online marketplace for scientific products until 1999. The firm generated revenues from advertising fees. During 1999-2001, the firm acted as a marketplace for scientific and laboratory products, and generated revenues from transactions and advertising. During 2001, Sciquest acted as a technology and solutions company providing integrated e-commerce and asset management solutions for research enterprises. Currently, the firm is a software, information and services company that generates revenues from license fee, subscription fee and fee from services. Similarly, Verticalnet acted as a creator and operator of vertical trade communities until 2000. Sources of revenue were advertising, web site development fees and auction fees. During 2001, the firm started offering software solutions thereby generating revenues from software licenses and professional services. During 2002, the firm morphed into a provider of Collaborative Supply Chain solutions with the main sources of revenue being license fee and fee from professional services.

Both these firms started as independent B2B marketplaces, and after continuous adaptation over the years, Verticalnet has morphed into a Collaborative Supply Chain solutions provider. Sciquest has emerged into a software, information and services company. A similar trend is witnessed in the case of many of the B2B marketplaces. Either the marketplaces successfully adapted into new forms or they closed down. After the fall of the NASDAQ in April 2000, the marketplaces were faced with one of three options: adapt and survive on their own revenues, sell out, or go bankrupt. An increasing number of marketplaces have since adapted themselves beyond market making, merged with other marketplaces, have been acquired or exited the market. According to industry sources (mySupplyChain, 2001), about 400 marketplaces have closed down or been acquired by others. The closure of many independent B2B marketplaces brings into focus the viability of their business models. In the case of B2B marketplaces, the dynamic nature of the environments in which they operate makes transformation a key research issue. Thus, analysis of the transformation paths chosen by the marketplaces would offer us

Table 3. Brief description of the cases

Firm	Business description	Launched in	Target industry segment	Current Ownership	Products/services
Firm 1	Business-to-business (b2b) auction site	August 1999	Across industries	Fully owned subsidiary of leading two-wheeler manufacturer in India	Forward and reverse auctions. Comprehensive listing and matchmaking facility.
Firm 2	E-Procurement service provider	September 2000	Across industries	Owned and operated by a leading Indian software service provider	Forward/reverse auctions, supplier/price discovery, catalogues, online negotiations, e-procurement consulting.
Firm 3	Value-added hosted services provider	August 1999	Across industries	Owned by a leading engineering and manufacturing firm	Procurement and disposal of assets, supply chain management, dynamic catalogue management services

useful insights. Studying the adaptive strategies of firms that operated in such environments would improve our understanding of other current/future firms operating in similar environments. The paper aims to understand the adaptive strategies of B2B marketplaces in terms of their products/service class evolution. Specifically the objectives of the study are:

1. Understand the adaptive strategies of B2B marketplaces and identify the stages in the evolution
2. Analyse the product/service class evolution of B2B marketplaces
3. Understand the survival strategies of B2B marketplaces

The paper is organized into four sections. In the first section, we describe the theoretical bases we use to analyse firm adaptation. In the second section, we explain the research design and methodology adopted by us. The third section will describe the three case studies. The last section draws some conclusions and presents guidelines for further research.

THEORETICAL PERSPECTIVES

Firms operating in high-velocity environments exhibit behavior that differs significantly from those operating in static environments (Thomas, 1996). In high velocity environments, firm strategies are often more concerned with change (Eisenhardt and Brown, 1997, 1999), speed (Eisenhardt, 1989; Eisenhardt and Tabrizi, 1995) and flexibility (D'Aveni, 1994). However we have very limited understanding of adaptive strategies of Internet-based businesses that operated in such environments. This paper is an attempt to bridge this gap in literature.

Eisenhardt and Bourgeois (1988) define high-velocity environments as those exhibiting rapid and discontinuous change in four domains. The domains are demand, competition, technology, and regulation. B2B marketplaces have experienced rapid

Table 4. List of personnel interviewed in the three firms

Firm	CEO	COO	CIO	CFO	Vice President	Assistant Vice President	Technical Head	Managers
Firm 1	X	X	X	X	X	X	X	X
Firm 2	X	X	X	X	X	X	X	-
Firm 3	X	X	X	X	X	X	X	-

changes in all four categories and hence they can be classified as operating in high velocity environments. The tenets of high velocity environments are similar to the concept of hypercompetition (D'Aveni, 1994). Firms competing in hyper-competitive environments exhibit behavior that differs significantly from behavior in more static environments (Thomas, 1996). According to D'Aveni the key drivers for a shift towards hypercompetition are consumer demand, the knowledge base of firms and associated workers, declining entry barriers, and the increasing frequency of alliances among firms.

Organizations operating in highly competitive environments rely on strategies that are more adaptive (Davis et al., 1991; Jennings and Seaman, 1994). The success of a business is a function of its appropriate and timely adaptation over time. If we take adaptation of organizations along a continuum, we can see that at one end organizations can maintain an external focus, with an accompanying ability to adapt to market change, but at a significant cost. At the other end, organizations can focus on a narrowly defined product-market. These organizations can focus internally, but with an accompanying risk of failing to adapt when market changes occur. Adaptation involves changes in strategic behavior to improve competitive posture and achieve better fit between the organization and its environment. This may involve comprehensive, continuous changes in products, services, resources, capabilities, and external linkages through which firms seek to regenerate competitive advantage. Adaptation can be conceptualized in terms of a number of dimensions affecting the business, such as products or services offered, customer profile, marketing, distribution, personnel, financial and physical facility requirements (Morris et al, 1999). Further, it typically entails incremental, continued change as opposed to dramatic or discontinuous change, leading to major transformation of an organization.

Research on organization adaptation has tended to emphasize the adaptive capacity of firms, or the extent to which particular strategies reflect more or less ability to adapt. Adaptation strategy concerns specific ways in which the firm makes adjustments, as it seeks to survive and capitalize on external circumstances. Such adjustments can be made in a variety of product, market and resource management areas (Ginsberg, 1988; Snow and Hrebiniak, 1980) such as a broader product mix, new product development, exploration of new markets and market segments, speed of response, outsourcing and resource leveraging, formation of strategic alliances, etc. Chakravarthy (1982) argues that organizations survive by pursuing high, medium or low levels of adaptation, depending on their defined market niche. Determining the total amount of adaptation in a firm would require that one combines the amount of change that occurs in the various product, market and resource management activities of the firm over time (Morris and Zahra, 2000).

Figure 1. Stages of transformation

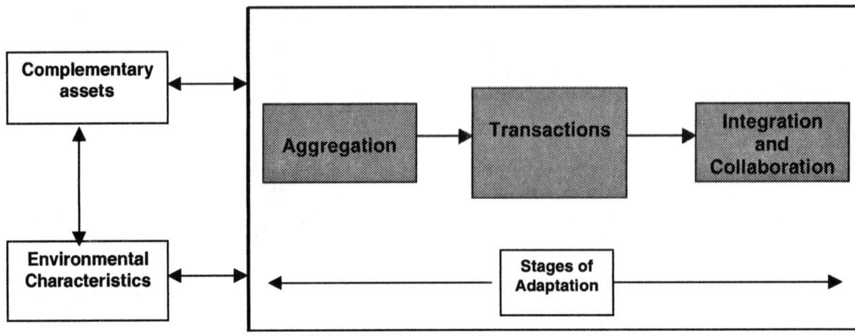

Table 5. Products/ services offered by the B2B marketplaces during the first stage

	Firm 1	Firm 2	Firm 3
Aggregation	Product Search, Catalogue, Directories	Firm 2 skipped the first stage as the firm was formed during late 2000	Catalogue, Directories

Key characteristics of the Internet, such as interconnectivity and network externalities, make it necessary for Internet based firms to adopt inherently dynamic strategies, including product versioning, rapid product development, direct relationships with users and frequent partnering (Shapiro and Varian, 1999). To succeed over the long haul, firms have to periodically reorient themselves by adopting new strategies and structures that are necessary to accommodate changing environmental conditions.

RESEARCH DESIGN AND METHODOLOGY

The industry-level cross-sectional studies are not very helpful in capturing the evolution of Internet-based businesses, as they would fail to capture how over time they have adapted and evolved. To understand this, we need to look inside firms to understand evolution at the organizational level. The case methodology was chosen, as the purpose of this work was exploratory. We chose purposeful sampling, that is, selecting a sample from which the maximum can be learned. According to Yin (1994), sample selection should be dictated by replication logic instead of a statistical one. Each case was considered as an experiment in itself, subsequent sites being used to either confirm or refute previous findings. According to Eisenhardt, cases may be chosen to replicate previous cases or extend emergent theory, or they may be chosen to fill theoretical categories and provide examples of polar types (1989). A group of potential B2B marketplaces was first identified.

The following criteria were used to choose firms for case study:

- Firm is a market maker
- Operate in the business-to-business marketplace
- Use the Internet as a key mode of delivery of their services

Firms were picked one at a time, starting with those that seemed to best meet the selection criteria. Based on the above criteria, three Indian firms were chosen for the case study research (see Table 3 for a brief description of the firms). These firms were some of the leading business-to-business marketplaces in India. In order to keep the project's cost within reasonable limits, we selected B2B marketplaces in India. The multi-case design was adopted to allow for replication logic. The cases were treated as experiments, each case serving to confirm or reject the inferences drawn from the previous ones (Yin, 1984).

Data was collected through semi-structured interviews with the CEOs, CIOs, CFOs, Vice Presidents, Assistant Vice Presidents, Technical heads as well as the Managers of the firms. We also conducted interviews with independent industry experts. Interviews lasted from one to four hours. We conducted 22 unstructured and semi-structured interviews during 2001-2002. The interview guideline was used to ensure that the focus areas of our research were discussed during the interviews. A similar number of interviews were conducted with personnel with same/equivalent designations to ensure uniformity of data sources. Table 4 lists the personnel interviewed in the three firms. Our interviews focused on how the B2B marketplaces have adapted and evolved. The case interpretations are based on the interviews, secondary data sources such as press reports, company releases, business plan and analyst reports.

An interview guide was used to avoid losing focus and to ensure that all relevant questions were asked. The interview guideline had 15 questions and the questions were both closed and open-ended. Respondents were thus given the opportunity to express their thoughts on the topic of interest as freely as possible. The interview guideline concerned the history and implementation of eCommerce in the respective organizations, strategies for its development, expectations concerning its impact, perceived changes in the business, business models and organization, perceived success and critical success factors, and future plans.

Table 6. Products/services offered by the B2B marketplaces during the second stage

	Firm 1	Firm 2	Firm 3
Transactions	Product Search, RFP/RFQ, Catalogue, Auction (Public), Auction (Private), Exchange	Catalogue Auction (Public) Auction (Private) Exchange	Catalogue Auction (Public) Auction (Private) Exchange
Value Added Services (VAS)	Logistics, Escrow, Appraisal, Valuation, Industry news, Market Analysis and reports	Payment, Industry News, Market analysis and reports	Logistics, Escrow, Appraisal, Valuation, Industry news, Market Analysis and reports

The following are some sample questions from the interview guide:

1. What are the key products/services offered by you across the years? What are the reasons for offering these products/services?
2. Is your firm in the aggregation stage? What are the aggregation services on offer and how have they changed over the years and why?
3. What caused the firm to adapt through the various stages? Are they internal factors or was the adaptation forced due to the market characteristics?
4. Are these variables a good depiction of firm adaptation?

The first question was aimed at understanding the product/service class evolution of B2B marketplaces. The objective of the second question was to understand the stages of evolution. Questions three and four aimed to understand the adaptive strategies of B2B marketplaces. In all cases, the respondents allowed the tape recording of the interview. Notes were reviewed the same day or the day after the interview, and the within-case analysis was performed as soon as possible, while the information concerning the case was still fresh in the mind of the researcher. Even when the interviews were taped, the investigator tried to do the individual case analysis shortly after meeting with the respondents. Tapes were listened to and notes were taken.

To enhance the quality of our research design, several of the methods recommended by Yin (1994) for case studies were used. We collected data from more than one member of each respondent firm, giving us greater confidence in the measures of the constructs. We also obtained information about the firms from consultants, analyst reports, business magazines and promotional material, thereby achieving triangulation of sources and methods (Patton, 1999). Internal validity checks were conducted using the pattern-matching mode of analysis after having performed the cross-case search for patterns. As a form of theoretical validation, the emerging framework (empirically based pattern) was compared to the initial theoretical framework (predicted pattern). We also used the explanation-building mode of analysis. External validity was ensured by iteratively comparing and contrasting the firms for similarities (literal replication) or differences (theoretical replication). Following such replication logic both strengthens and broadens analytical generalizations.

We started data analysis with an in-depth study of each individual site, and this first step is called within-case analysis. This is helpful to develop an in-depth understanding of each case before moving on to the next level of analysis. This entails sifting through

Table 7. Products/services offered by the B2B marketplaces during the third stage

	Firm 1	Firm 2	Firm 3
Integration and Collaboration	Private exchange	Catalogue integration, private exchange, e-procurement consulting, services	Dynamic catalogue management services, e-sourcing, Supply chain management

all the data, discarding whatever is irrelevant and bringing together what seems most important. The idea is to allow the most significant observations to emerge from all data gathered in the field, while reducing the volume of data. To facilitate the cross-case analyses, all three cases were written in the same format: a brief introduction describing the organization and its business environment; a detailed description of the adaptation and evolution of the firm since inception, analysis of each of the product/service classes, external networks and the effect of industry dynamism on the variables. One section was kept open for any variables that may emerge from the field. The second step of the analysis consisted of a cross-case search for patterns. The cross-case search for patterns was executed along the same dimensions as mentioned in the within-case analysis. A third level of analysis was used to explain the phenomenon by stipulating a set of possible causal links about it (Yin, 1994). Yin (1994) suggests beginning such an analysis by taking the data collected from a first case to build a logical sequence of events explaining the case outcomes. The hypothesized set of events was then verified in a second case and if it is confirmed, we proceed with the third case. If at any point in the process the hypothesized explanation did not hold, an alternative explanation was developed and verified again until one holds good for all the cases.

RESULTS

The case analysis reveals that B2B marketplaces transformed and evolved in response to a highly dynamic environment. Based on the type of products/services offered by the B2B marketplaces across the three stages, we identified three stages of evolution: aggregation stage, transactions stage and integration and collaboration stage (see Figure 1 for the stages of transformation). The aggregation stage is the formative stage of the firm. The transaction stage corresponds to the expansion stage, as it involves expanding the firm's operations, revenues, as well as the number of transactions. The integration stage corresponds to the consolidation stage, as it requires attaining maturity in both the nature and the process of service delivery undertaken by the firm.

Aggregation

In the formative period, the B2B marketplaces were acting as aggregators. They were offering catalogues detailing the product listings. Table 5 lists the products/services offered by the three B2B marketplaces during this stage. This stage witnessed intense competition among the various B2B marketplaces, as the entry barrier was very low. Further, all the players were offering similar services such as static product listings and directory services. Thus there was very little differentiation among the various B2B marketplaces. At this stage, the main purpose of B2B marketplaces was to reduce transaction costs, bring together buyers and sellers as well as facilitate price discovery.

All three firms evolved in a similar direction, providing similar services with no dynamic transactions. The firms were targeting a particular industry segment. Firm 1 focused on engineering items, Firm 2 targeted the IT industry and Firm 3 focused on MRO (Maintenance, Repair and Operating supplies) items. The similarities in the services offered by the B2B marketplaces in this stage, such as static product listings and directory services, resulted in these services being perceived as a commodity. As all the

Table 8. Evolution of firm characteristics

	Aggregation (1999-2000)	Dynamic Transactions and VAS (2000-2001)	Integration and Collaboration (2002 onwards)
Products/Services	Static listings, directories	Auctions, reverse auctions, logistics, escrow, etc.	Supply chain integration, collaborative planning, forecasting
Strategic Focus	Match making	Match making, reducing transaction costs	Interorganisational processes
Revenue model	Listing fee	Listing fee, transaction fee, fee for VAS	License fee, listing fee, transaction fee, fee for VAS
Competitors	Online and physical directories, other B2B marketplaces	Brick and mortar auction houses, other B2B marketplaces	Software service providers, consulting firms, other B2B marketplaces
Sources of Competitive advantage	First mover advantage, network externalities	First mover advantage, network externalities, trust, brand, reputation, complementarity among internal and external resources, low transaction and search costs	Interorganisational processes, lock-in, network externalities, trust, brand, reputation

B2B marketplaces were offering identical services, there was very little differentiation among them. They also faced problems in terms of their inability to achieve the critical mass of transactions (value of transaction required to achieve break-even). Hence, the B2B marketplaces had to find new sources of competitive advantage by re-examining their services. This resulted in them building resources and capabilities to offer transactions and move to the next stage of their evolution.

Transactions

B2B marketplaces were exploring a new way of delivering the services, and steps were taken to enter into alliances to access complementary assets. The three B2B marketplaces continued to invest substantial amounts in marketing and promotional activities. At this stage Firm 1 spent close to 60% of its total expenses towards marketing and promotional activities. The B2B marketplaces focused on building their domain name as a brand and used print advertisements, hoardings as well as advertisements in various Internet search engines. All three firms were similar in their approach towards brand building.

The most important characteristics of this stage were low transaction volume and the inability of the B2B marketplaces to achieve critical mass of transactions. This was caused by low industry participation in the B2B marketplaces. To increase the number of transactions, Firm 3 even divested a minority stake to four engineering industry players thereby ensuring more industry participation. During this stage, the B2B marketplaces started offering a number of Dynamic Transaction services. Firm 1 started offering a bundle of services which comprised dynamic forward and reverse auctions, logistics, appraisal, valuation services, RFQ (Request for Quotation) and industry news. Table 6 lists the products/services offered by the three B2B marketplaces during this stage. The evidence presented in Table 6 indicates that all the three B2B marketplaces evolved in similar directions by adding new and innovative transactions and services.

Firm 2 was offering dynamic forward and reverse auctions, payment services, valuation services, industry news and market analysis and reports. Firm 3 was offering services such as auctions, appraisal, valuation and market analysis and reports. Firm 3 also introduced innovative services such as the Global TIO auction services. However, the most drastic shift in focus was witnessed in the case of Firm 2, which evolved from an online marketplace to an eProcurement service provider, providing reverse and forward auction services, eProcurement consultancy, setting up virtual private marketplaces and licensing of their indigenously developed procurement software.

The B2B marketplaces entered into alliances with partner firms to offer value-added services (VAS). Value-added services are defined as services, which supplement the actual transaction, cataloguing and search capability. These include financial services, logistics services, analytics, inspection and settlement of disputes. The B2B marketplaces considered logistics services and financial settlement as two key areas of differentiation. This can be seen in the case of Firm 2, which entered into alliances with banks to offer financial settlement services, and with rating service providers to offer rating and verification services. Firm 2 also entered into alliances with various content providers. Firm 1 had alliances with logistics firms, consulting and appraisal service providers. Firm 3 had alliances with verification and dispute settlement agencies, rating service providers, logistics service providers and content providers. Firm 3 had also entered into an alliance with another B2B marketplace.

However, similar to the previous stage, there was very little differentiation among the B2B marketplaces, as most of them started offering transactions and VAS. The industry continued to witness a large number of new entrants. As all the B2B marketplaces were offering identical services, there was very little differentiation among them. For some B2B marketplaces such as Firm 1 and Firm 3, the problems were accentuated by their inability to achieve the critical mass of transactions. All these conditions resulted in the B2B marketplaces losing their competitive advantage. This made it imperative for the B2B marketplaces to attract long-term industry participation, which was not possible only through dynamic transactions, as in many cases the average transaction fee ranged between 2-5% of the value of the transaction. As compared to transaction fee based business models, business models based on license fee became more attractive to the B2B marketplaces as it ensured minimum revenue. Moreover, the customers were demanding integration with the entire supply chain rather than just procurement or disposal of assets, which was possible only by integrating the solutions of the B2B marketplace with the processes of their customers. This triggered the next stage in the evolution of B2B marketplaces wherein they started offering integration and collaboration services.

Integration and Collaboration

Between Stages 2 and 3, a major punctuated shift happened in the form of the crash of B2B company stocks. This was accompanied by the non-availability of venture capital funding. As most of the B2B marketplaces at this stage were not generating enough revenues to sustain their operations, they were putting on hold their expansion plans and were restructuring, downsizing and retrenching. In all the companies, a number of people were laid off, and in many cases the firms changed their products/services, revenue model, mission statements etc. For example, Firm 2 exited its e-selling business and started to focus only on e-procurement. In order to improve the margins from e-Procurement services such as auctions, reverse auctions, catalogues, etc., the firm started offering consultancy services. This included consultancy in strategic sourcing, which is the step before e-procurement and supplier relationship management (SRM), which is the step after procurement. Firm 3 evolved from a marketplace to a value-added application service provider (ASP) and was offering an integrated e-business solution suite for the manufacturing sector comprising of e-procurement, supply chain management, etc.

Table 7 lists the products/services offered by the three B2B marketplaces during this stage. From Table 7, we can infer that the three B2B marketplaces were integrating their products/services with those of their customers. In Stages 2 and 3, long-term contracting and service delivery became salient, in contrast to the previous focus on sales and marketing. The knowledge base in the B2B marketplaces went through many changes due to the layoffs, as most of the firms lost about a quarter of their employees. At his stage, the license fee was the primary source of revenue for B2B marketplaces, and secondary sources of revenue were transaction fee as well as non-transactional usage fees such as tracking fee, listing fee, etc.

Integration of their products and services with their customers assured the B2B marketplaces of regular revenues in the form of license fees. Integration with corporate systems is achieved through enterprise application integration methods. Typically, a common metadata repository maintains process and data maps for translating the data, and a message broker routes the data to the right application. Integration also involves modeling, automating, and integrating business processes and trading relationships between partners. Integration of information refers to the sharing of information among participants of the B2B marketplace such as buyers and suppliers. This includes data such as inventory data, demand data, capacity plans, production schedules, promotion plans, and shipment schedules. B2B marketplaces at this stage also facilitated collaborative activities such as joint design and execution plans for product introduction, demand forecasting and replenishment. They also facilitated workflow coordination, which refers to streamlined and automated workflow activities between supply chain partners.

To facilitate integration and collaboration, the B2B marketplaces were building their technological capabilities such as data mapping repositories, including XML (Extended Markup Language) document exchange formats and trading partner agreements, which allowed the customers to switch from one market to the other at any time. This was important as the customers may not find the relevant buyers or sellers in one particular marketplace. Therefore, B2B marketplaces such as Firm 3 entered into partnerships with B2B marketplaces from other countries such as GoIndustry. Firm 3 was building its core strengths such as its inventory of modular functional components, domain specialists and technical resources. This was supplemented by managerial capabilities such as end-to-end consulting. What distinguishes the integration stage of B2B marketplaces is their ability to achieve sustained competitive advantage by exploiting stage 1 and stage 2 activities. The stage 2 activities involved the ability to offer stand-alone dynamic transactions and leveraging the complementary assets of partners. However, stage 3 activities involved the ability to create inter-organizational processes.

From the results, we can see that firms rely on continuous adaptation to regenerate competitive advantage under conditions of rapid change (see Table 8 for the evolution of firm characteristics across stages). As firms adapt and evolve into new forms, they need to regenerate competitive advantage relative to the new competitors they encounter in these domains. This triggers the adaptation process again. Adaptation regenerates competitive advantage through continuous redefinition of the products and services, and changes in the resources and capabilities. In the aggregation stage the focus of the B2B marketplaces was to bring together buyers and sellers over a common platform. In the transactions stage, the focus of the B2B marketplaces was to reduce search and transaction costs as well as offer VAS. In the integration and collaboration stage, the focus of the B2B marketplaces was to design and implement interorganisational processes. Porter (1985) hypothesized a tendency for firms to unbundle complementary products over the course of the industry life cycle. In the case of B2B marketplaces, we find that the reverse is true. In B2B marketplaces, as we move along their industry life cycle, there is an increasing tendency to bundle their products/services along with complementary products/services.

Discussion and Implications

This research involved an empirical examination of the evolution of B2B marketplaces in terms of their product and service evolution. Our study makes several important contributions to the literature on B2B marketplaces in general and evolutionary strategies in emergent stages of the industry in particular. Our results allow for a complete understanding of the stages of evolution of B2B marketplaces. The focus of B2B marketplaces across the world was to achieve a critical mass of transactions by targeting small and medium-sized buyers and sellers. They aimed to facilitate this by providing a platform for displaying product offerings and enabling price competition. B2B marketplaces across the world failed mainly due to the lack of supplier and buyer participation. This occurred because of the following reasons:

- **Supplier enablement and participation**–Supplier enablement was a key issue and often involved complex procedures as well as high upfront costs. The suppliers were required to make investments in software and/or hardware to get started. Suppliers did not want to join a number of B2B marketplaces unless there is a clear value proposition. Moreover, the price savings provide benefits for buyers, but there are no similar benefits for suppliers.
- **Path dependency**–The brick and mortar companies had existing supplier contracts and had spent many years developing these partnerships. Hence, they were not willing to break away from their existing network of relationships.
- **Asset specificity**–The brick and mortar companies had made considerable investments in building their network of contracts. Hence, they were not ready to forego these investments.
- **Partnerships and relationships**– They were also not open to the cultural changes necessary to compete solely based on price within an open marketplace, abandon important relationships, and adopt new technology.
- **Privacy issues**–There were privacy concerns as competitors were able to access the pricelists thereby escalating price competition resulting in reduced margins.
- **Technology adoption**–The suppliers were reluctant to adopt the new technology, as participation in marketplaces involved issues related to the electronic formatting of inventory records, learning new technology and making process changes.

- **Price competition and commoditization**–Suppliers were not comfortable with pure price competition and were wary of the fact that their products/services would be commoditised if they compete solely based on price.

All the above factors resulted in negating the network effects and lesser value for the customers. There were too many exchanges targeting too few participants. The key reasons for the failure of the B2B marketplaces can be attributed to supplier related issues such as enablement and participation, path dependency, asset specificity, technology adoption, etc. The target customers of B2B marketplaces were not ready to embrace the new way of doing transactions. Thus the B2B marketplaces were adapting and evolving their business models to meet the requirements of the market. In other words, the B2B marketplaces were operating in a high velocity environment, whereas their customers were not moving fast and were slow to respond to the changes.

Managers could gain from these results as it would help them to identify the type of products/services to be offered. The study also proposes that B2B marketplaces have survived by building stable interorganisational processes. This builds a case for managers to invest in interorganisational information systems to hedge some of the risks posed by high velocity environments. Managers could use the results to understand the role of complementary assets in firm adaptation. This paper thus offers valuable insights about the influence of complementary assets in the evolutionary stages of firms. Increased focus on integration required firms to develop products/services that were compatible with multiple standards and solutions. Thus, system integration rather than isolated processes became critical. The multi-stage model of firm adaptation could act as a useful tool by managers. The model could help them identify the type of resources and capabilities, which needs to be built in high velocity environments.

Limitations

Studying the adaptation of firms in high velocity environments poses a number of challenges. The case studies, used to identify and understand the variables were limited to firms in India. Markus and Soh (2002) argue that structural conditions play a key role in business-to-business activity. These include physical, social and economic arrangements that shape e-commerce business models and influence individual and organizational use of the Internet. They posit that structural conditions differ from country to country and even from location to location within a country. Therefore, valid explanations of geographical differences in e-commerce activity require a careful assessment of relevant structural factors. Gottschalk and Khandelwal (2002) emphasise that with the increasing globalization of business, management of IT is becoming more complex. They posit that there are significant differences in the IT management needs in different geographic regions. In this paper, we have not examined the framework in terms of multiple contexts. Examining the influence of structural conditions on the adaptation of firms in high-velocity environments is a possible area for future research.

The study does not link adaptation of firms with their financial performance, as we were not able to access the financial performance measures of these firms. By limiting our research to three case studies, we are not able to generalize our findings. Despite these limitations, our study develops key constructs that can advance understanding and stimulate research on the evolution of firms in highly dynamic environments.

CONCLUSION

This research involved an empirical examination of the adaptation of B2B marketplaces. The research argues that we need to have a comprehensive understanding of the factors that could influence the adaptation of firms, including complementary resources at the emergent stage of industries. This research contributes to a growing body of research, seeking to understand the determinants of organizational ability to adapt and gain competitive advantage in new competitive and high-velocity environments. The framework aims to capture the evolution and adaptation of B2B marketplaces. The link between an attribute and its adaptive implications can be studied by analyzing the organization's history and the effect of the attribute. As adaptation is dynamic and cumulative, the effect of an attribute or a combination of attributes on adaptation would be felt in the future. These aspects suggest that summing up the history of the organization offers a means of studying the same. The model presented in this paper captures the paths of adaptation of B2B marketplaces in terms of its products/services as well as the industry dynamism. The research indicates that B2B marketplaces pass through three stages, which correspond to the three stages in their products/service class evolution: aggregation, transactions and integration.

A state of adaptation for a business organization is one in which it can survive the conditions of the environment. The higher the environmental complexity that can be handled by the firm, the better are the chances of its long-term survival and thus higher the level of its adaptation. It is not necessary that all B2B marketplaces must pass through all stages or must pass through the stages one at a time. Many firms do not reach stage 3 or even stage 2, and most firms contain activities that are at more than one stage. In order to survive in the long run, B2B marketplaces have to reach the integration stage. Although the B2B marketplaces in the transaction stage have marginal competitive advantage, they can rarely demand premium for their services. On the other hand, B2B marketplaces in the integration stage minimize environmental uncertainty by creating interorganisational processes. Thus, the evolutionary process for B2B marketplaces involves them starting as pure aggregators to transaction facilitators and finally evolving into integrators.

REFERENCES

Bailey, J., & Bakos, Y. (1997). An exploratory study of the emerging role of electronic intermediaries. *International Journal of Electronic Commerce, 1*(3), 7-20.

Bakos, Y. (1991). A strategic analysis of electronic marketplaces. *MIS Quarterly, 15*(3), 295-310.

Bakos, Y. (1997). Reducing buyer search costs: Implications for electronic marketplaces. *Management Science, 43*(12), 1676-1692.

Business Week (2003). The e-biz surprise. (May 12), 44-52.

Chakravarthy, B. S. (1982). Adaptation: A promising metaphor for strategic management. *Academy of Management Review, 7*, 35-44.

Davis, D., Morris, M., & Allen, J. (1991). Perceived environmental turbulence and its effect on selected entrepreneurship, marketing, and organizational characteristics in industrial firms. *Journal of the Academy of Marketing Science, 19*, 43-51.

D'Aveni, R. A. (1994). *Hypercompetition: Managing the dynamics of strategic maneuvering*. New York: Free Press.

Eisenhardt, K., & Schoonhoven, C. B. (1990). Organizational growth: Linking founding team strategy, environment, and growth among U.S. semiconductor ventures, 1978-1988. *Administrative Science Quarterly, 35*(3), 504-526.

Eisenhardt, K. M. (1989). Making fast strategic decisions in high-velocity environments. *Academy of Management Journal, 32*(3), 543-576.

Eisenhardt, K. M., & Bourgeois, L.J. III. (1988). Politics of strategic decision making in high-velocity environments. *Academy of Management Journal, 31*(4), 737-770.

Eisenhardt, K. M., & Brown, S. L. (1997). The art of continuous change: Linking complexity theory and time-paced evolution in relentlessly shifting organizations. *Administrative Science Quarterly, 42*(1), 1-34.

Eisenhardt, K. M., & Brown, S. L. (1998). Time pacing: Competing in markets that won't stand still. *Harvard Business Review, 76*(2), 59-69.

Eisenhardt, K. M., & Brown, S. L. (1999). Patching: Restitching business portfolios in dynamic markets. *Harvard Business Review, 77*(3), 72-82.

Eisenhardt, K. M., & Tabrizi, B. N. (1995). Accelerating adaptive processes: Product innovation in the global computer industry. *Administrative Science Quarterly, 40*(1), 84-110.

eMarketer (2002). *E-Commerce Trade and B2B Exchanges.*

Faber, M., & Proops, J. L. R. (Eds.). (1991). *Evolution in Biology, Physics and Economics: A conceptual analysis*: Harwood Academic Publishers.

Forrester Research (2001). *eBusiness Propels Productivity.*

Galbraith, J. (1982). New venture planning—The stages of growth. *Journal of Business Strategy, 3*(1), 70-79.

Garicano, L. & Kaplan S.N.(2000). *The effects of business-to-business e-commerce on transaction costs.* NBER Working Paper No. W8017.

Ginsberg, A. (1988). Measuring and modeling changes in strategy: Theoretical foundations and empirical directions. *Strategic Management Journal, 9*, 559-575.

Gottschalk, P., & Khandelwal, V. K. (2002). Global comparison of stages of growth based on critical success factors. *Journal of Global Information Management, 10*(2), 2002, 40-49.

Greiner, L. E. (1972). Evolution and revolution as organizations grow. *Harvard Business Review, 50*(4), 37-46.

Hanks, S. H., Watson, C. J., Jansen, E., & Chandler, G. N. (1993). Tightening the life-cycle construct: A taxonomic study of growth stage configurations in high-technology organizations. *Entrepreneurship Theory and Practice*, 5-29.

Hirshleifer, J. (1982). Evolutionary models in economics and law: Cooperation versus conflict strategies. *Research in Law and Economics, 4*, 1-60.

Jennings, D. F., & Seaman, S. L. (1994). High and low levels of organizational adaptation: An empirical analysis of strategy, structure, and performance. *Strategic Management Journal, 15*, 459-475.

Khazanchi, D. (1995). Spoke enterprises: A preliminary assessment of expectations and performance of EDI. *Proceedings of the 25th Annual Decision Sciences Institute Meeting*, Boston.

Khazanchi D. (1999). The nature and structure of impediments to EDI adoption and integration: A survey of small- and medium-sized enterprises. *Failure & Lessons in Information Technology Management,* 3, 101-110

Markus, L.M., & Soh, C. (2002). Structural influences on global e-commerce activity. *Journal of Global Information Management, 10*(1), 5-12.

Metcalfe, J. S., & Saviotti, P. P. (1991). *Evolutionary theories of economic and technological change.* Harwood Academic Publishers.

Miller, D., & Friesen, P. H. (1984). A longitudinal study of the corporate life cycle. *Management Science, 10,* 1161-1183.

Morris, M. H., & Zahra, S. (2000). Adaptation of the business concept as a factor in entrepreneurship: The case of historically disadvantaged South African entrepreneurs. *Journal of Small Business Management, 38*(1), 92-100.

Patton, M. Q. (1999). Enhancing the quality and credibility of qualitative analysis. *Health Services Research, 34*(5), 1189-1208.

Porter, M. E. (1998). *Competitive advantage: Creating and sustaining superior performance.* New York: The Free Press.

Quinn, R., & Cameron, K. (1983). Organizational lifecycles and shifting criteria of effectiveness: Some preliminary evidence. *Management Science, 29*(1), 33-51.

Riggins, F. (1998). *A framework for identifying Web-based electronic commerce opportunities.* Working Paper, Dupree School of Management, Georgia Institute of Technology.

Shan, W. (1990). An empirical analysis of organizational strategies by entrepreneurial high-technology firms. *Strategic Management Journal, 11*(2), 129-139.

Shapiro, C., & Varian, H. R. (1998). *Information rules: A strategic guide to the network economy.* Boston: Harvard Business School Press.

Snow, C. C., & Hrebiniak, L. G. (1980). Strategy, distinctive competence, and organizational performance. *Administrative Science Quarterly, 25,* 142-193.

Spulber, D. F. (1998). *The market makers: How leading companies create and win markets.* New York: McGraw-Hill.

Thomas, L. G., III. (1996). The two faces of competition: Dynamic resourcefulness and the hyper-competitive shift. *Organization Science, 7*(3), 221-242.

Yin, R. K. (1984). *Case study research: Design and methods.* Beverly Hills, CA: Sage Publications.

Yin, R. K. (1994). *Case study research: Design and methods* (2nd ed.). Thousand Oaks, CA: Sage.

Previously published in the Journal of Global Information Management, 12(1), 41-50, January - March 2004.

Chapter VIII

Development Attributes of Quality for E-Supply Chain System

Wing S. Chow, Hong Kong Baptist University, Hong Kong

ABSTRACT

E-supply chain system is an enabler/system that enriches the information service quality in supply chains. This chapter uses factor analysis to determine four extranet success factors: system quality, information quality, service quality, and work performance quality. A critical analysis of areas using discriminant analyses that require improvement is also conducted.

INTRODUCTION

Traditionally, the focus of a supply chain was on connection, transaction, and delivery within an organization. In today's faster-paced markets, the focus has shifted to rapidly, correctly, and profitably meeting the market demands on an organization. With materials flowing downstream from suppliers, manufacturers, warehouses, and stores to customers, and information flowing in both directions, a supply chain must maintain and sustain technology-based and quality-driven capabilities to minimize system-wide costs, reduce lead time and transit time, and improve customer service levels. To this end, the supply chain must be managed effectively. Thus, the supply chain literature reveals that the focus has now shifted from the traditional company-centered setting to a complete supply chain system with the implementation of e-commerce; that is, e-supply chain system (Kuei et al., 2002).

The implementation of e-supply chain system (e-SCS) helps organizations to effectively manage their global competitive strategy, which includes strategic source materials and components worldwide, select global locations for key depots and distribution centers, and transfer exiting logistics technologies to new markets (Coyle et al., 2003). The e-SCS practice has played a strategic role in the fast-changing business environment in the Asia Pacific region. In Hong Kong, the economical structure has merged gradually with mainland China; the local government has propelled the idea of setting a logistics center, which includes the development of a supercomputer platform that aids fast information exchange among trading partners. An e-SCS is an effective information system that enables supply chain partners to keep track of their work performance and the customer services that they offer to their clients electronically.

The modeling and study of the success elements of supply chain practices are well documented in the literature. For example, Narasimhan and Kim (2001) proposed a structured model for developing a strategy that could integrate all relevant organizational components among supply chain firms. Kuei and Madu (2001) identified a set of success factors in exercising quality management in a supply chain. Gallupe and Tan (1999) stated that IT adoption in a supply chain is very much dependent on national culture. Studies have reported different practices in Korea (Kim, 1996), Hong Kong (Wong et al., 1999), the United States (Tan, 2002), and other Asia-Pacific countries (McMullan, 1996). However, there have been relatively few studies on the performance of information services such as e-SCS. Further, information services that are designed for e-SCS are very different from the information system that is internally adopted by each supply chain partner. Hence, the prime objective of this chapter is to identify a set of success factors encompassing key e-SCS system features and practices in a supply chain.

In brief, this chapter identifies a set of e-SCS features and practices from the relevant literature. Exploratory factor analysis is used to extract the e-SCS success factors from these system features and practices. In addition, areas for improvement in e-SCS features and practices are identified by using the discriminant analysis. The service industry is the focus here because most firms are now using the services of outsourcing logistic firms in their non-core businesses so that they can concentrate on their core business and re-engineering (Voon & Ho, 2001). The following sections will review e-SCS features and practices, discuss the study design and measures, and present the findings before concluding the chapter.

E-SUPPLY CHAIN SYSTEM FEATURES AND PRACTICES

Since existing literature on e-SCS features and practices are relatively sparse, we borrow a set of relevant system features and practices from the intranet literature. Twenty-four system features and practices were initially identified in the intranet literature, but these were reduced to 18 after the verification of applicability in personal interviews with five local extranet providers and five supply chain practitioners. Table 1 outlines the 18 remaining system features and practices with supporting references.

STUDY DESIGN AND MEASURES

Study Subjects

Two groups of respondents were involved in this study: extranet providers and e-SCS users. The e-SCS providers were either from the IT departments of supply chain firms or from outsourcing logistic firms. The observations gathered from these providers were treated as expected values of e-SCS performance, and the scores were later used to determine the critical success factors. Observations collected from extranet users were considered as the actual performance of e-SCS services and were used to measure the performance deviation as described in the SERVQUAL method (Kettinger & Choong, 1997).

Measures

The 18 extranet system features and practices were used to develop two questionnaires. The first questionnaire asked e-SCS providers to evaluate expected performance on a measurement scale of 1 to 5, where a value of 5 represented a high expectation and a value of 1 represented a low expectation. The second questionnaire was directed at extranet users, who were asked to rate the actual performance of e-SCS services on a measurement scale of 1 to 5, where a value of 5 represented strong agreement and a value of 1 represented strong disagreement. In evaluating the overall satisfaction of extranet services, three questions were adopted with modification from Chow and Lui (2001). E-SCS users were asked to rate these three questions on a measurement scale of 1 to 5, with a value of 5 representing excellent performance and a value of 1 representing poor performance. The average scores of the three questions were used to determine the significance of extranet service quality. The respondents were asked to treat the term "the system" in the questionnaire as "the e-supply chain system" and to consider their e-fulfillment system as an example.

Data Collection Procedure

Structured questionnaires with cover letters were used to collect data through a mail survey. Three hundred and eighty potential respondents were randomly selected from http://www.tdctrade.com, which provided a comprehensive list of e-SCS adopting companies. Telephone calls were made to confirm adoption.

The two questionnaires were sent either to the e-SCS providers, who were asked to fill out the first questionnaire and pass the second to their users, or to supply chain firms, who were asked to complete the second questionnaire and direct the first to their e-SCS provider. A separate return envelope was attached to each questionnaire so that they could be completed and returned separately. Each set of questionnaires was labeled with a code for matching when returned. One hundred and forty-six replies were received from e-SCS providers and 110 replies were received from e-SCS users. However, only 103 complete sets could be formed, which constituted a response rate of 27.6%. The average experience of e-SCS providers was 2.95 years, with a standard deviation of 0.51 year. The e-SCS providers had been with the present company for an average of 5.46 years, which indicated that they were conversant about their system. The e-SCS users were from delivery services (including air cargo services), freight, transportation, wholesale,

Table 1. e-SCS system features and practices

Code	Extranet System Features and Practices	Sample References
(a)	The system is consistently and regularly maintained	Cupito (1997)
(b)	The system provides user-menu/instructions	Misic & Hill (1994)
(c)	The system enhances collaboration among trading partners	Lynch (1997)
(d)	The system produces accurate search results	Foo & Lim (1997)
(e)	The system provides a standardized display format (screen layout)	Bhanager et al. (1999)
(f)	The system adopts data security/privacy	Fuller & Pagan (1997)
(g)	The system improves personal productivity	Sridhar (1998)
(h)	The organization provides competent technical support	Huff & Munro (1985)
(i)	The system improves business communications	Cupito (1997)
(j)	The system enhances fast response time	Tabor et al. (1997)
(k)	The system provides a standardized retrieval procedure	Sridhar (1998)
(l)	The organization provides minimal system training	Cronan & Douglas (1990)
(m)	The system provides up-to-date information	Baines (1996)
(n)	The system provides a standardized search procedure	Foo & Lim (1997)
(o)	The system is easy to use	Tabor et al. (1997)
(p)	The system improves the quality of decision-making	Sridhar (1998)
(q)	The system aids faster decision-making	Sridhar (1998)
(r)	The system allows a faster exchange of information	Tabor et al. (1997)

trading, and logistics firms. About 76% of respondents were middle-level managers, and 5% were top-level managers (Table 2). Of the 103 returned questionnaires, 92 came from outsourcing logistics firms.

Data Analysis

Factor analysis with varimax rotation was used to determine e-SCS success factors. A factor loading of ≥ 0.5 was considered as a significant contribution to a factor cluster. The internal consistency method was adopted to check for homogeneous data, with $\alpha \geq 6$ as a sufficient condition for an exploratory study. KMO (Kaiser-Meyer-Olkin) values were used to measure the adequacy of the samples, with KMO ≥ 0.5 as an acceptable condition. The perceptual scores from the SERVQUAL method proposed by Kettinger and Choong (1997) were adopted to study e-SCS service quality. Perceptual scores were computed by subtracting the expected scores of e-SCS providers from the actual scores of e-SCS users, and then the gap correlations were computed by studying the effects of perceptual scores on the overall satisfaction.

RESULTS

Exploratory Factor Analysis

Table 3 shows the e-SCS success factor results and their corresponding gap correlations. The four extracted factors are labeled as system quality, information quality, service quality, and work performance quality. The α and KMO values are shown in the first column of Table 3, and these values meet the standards outlined earlier. All expected

Table 2. Summary of background information of extranet users

		Number	Percentage (%)
Industry Type	Delivery Services	11	11
	Transportation	14	14
	Freight	18	17
	Wholesalers	20	19
	Trading	18	17
	Logistic Firms	22	21
Position	Top Management	5	5
	Middle Management	78	76
	Operational	20	19

Table 3. Results of factor loadings and gap correlation

	Code	Factor Loadings	Expected Performance		Actual Performance		Perceptual Scores		Gap Correlation
			ξ	σ	ξ	σ	ξ	σ	
System	(f)	0.771	4.08	1.09	3.30	0.89	-0.78	1.12	0.133
Quality	(k)	0.754	3.72	1.02	3.17	0.97	-0.54	1.17	0.287*
	(n)	0.716	3.93	0.94	3.18	1.06	-0.75	1.27	0.482*
α =0.76	(j)	0.642	4.33	0.83	3.24	1.02	-1.09	1.25	0.305*
KMO=0.76	(o)	0.642	4.19	0.88	3.57	1.02	-0.62	1.00	0.259*
	(b)	0.590	3.60	1.30	2.50	1.24	-1.10	1.55	0.097
Information	(m)	0.904	4.33	0.94	3.41	0.94	-0.92	1.22	0.063
Quality	(d)	0.880	4.19	0.99	3.30	1.04	-0.89	1.07	0.194**
α =0.71	(e)	0.607	3.48	1.06	3.09	1.03	-0.39	1.21	0.267*
KMO=0.72									
Service	(a)	0.835	4.03	1.09	3.11	0.88	-0.92	1.28	0.092
Quality	(h)	0.771	4.21	0.85	3.21	0.96	-1.00	1.26	0.219**
α =0.66	(l)	0.729	3.77	1.05	2.56	0.97	-1.20	1.40	0.117
KMO=0.63									
Work	(i)	0.813	4.10	0.94	3.35	1.08	-0.75	1.25	0.201**
Performance	(q)	0.791	3.74	0.96	2.86	1.08	-0.87	1.27	0.267*
Quality	(p)	0.776	3.60	1.10	2.81	0.94	-0.80	1.29	0.173
	(c)	0.758	3.90	0.94	2.95	0.99	-0.95	1.21	0.138
α =0.85	(g)	0.708	3.95	0.97	2.95	0.86	-1.00	1.16	0.142
KMO=0.76	(r)	0.696	4.11	0.96	3.46	1.06	-0.65	1.09	0.371*

*Where * $\rho \leq 0.05$ and ** $\rho \leq 0.01$*
Mean value for the overall satisfaction = 3.18

values of performance are greater than 3.0, which indicates that all proposed system features and practices are important. The perceptual scores are all reported as negative values, which implies that the performance of each e-SCS practice falls short of user expectations. In the last column, values with "*" or "**" indicate that service quality improvement is needed. The 10 areas that needed improvement correspond to the following e-SCS features and practices: (1) the system enhances fast response time; (2) the system is easy to use; (3) the system provides a standardized search procedure; (4)

the system provides a standardized retrieval procedure; (5) the system provides a standardized display format (screen layout); (6) the system produces accurate search results; (7) the organization provides competent technical support; (8) the system allows a faster exchange of information; (9) the system improves business communication; and (10) the system helps to provide faster decision-making.

Discriminant Analysis

The four e-SCS success factors have reported a moderate to strong positive correction with the e-SCS over user satisfaction (r values range from 0.42 to 0.83, p-value <0.001). This result implies that e-SCS success factors have a positive relationship with its performance. This chapter separated respondents into a high satisfaction (HS) group and a low satisfaction (LS) group. Respondents with the mean score of overall performance higher than a value of 3 were grouped into the HS group, and respondents with the mean score lower than a value of 3 were grouped into the LS group. In this study, respondents with the mean score equal to a value of 3 were discarded because they were considered as impartially different. In total, HS had 88 respondents, and LS had 34 respondents.

This study adopted two-factor discriminant analyses to discriminate e-SCS practices between the HS and LS groups. The stepwise discriminant procedure of the SPSS program was deployed for the data analysis. This study used the maximum chance criterion to check for the adequacy of samples, and the cutting score was reported as 69.4%.

Table 4 summarizes the results of stepwise discriminant analysis. It reveals that two e-SCS success factors are significantly discriminated between HS and LS groups: system quality and work performance quality. The e-SCS success factor of system quality has the highest discriminating power (r=0.853), and the work performance quality is second (r=0.650). The discriminant function has a canonical squared correlation of 0.818, and it is significant with Wilks Lambda=0.330, ρ<0.000.

Table 4. Discriminant analysis of e-SCS success factors

Success Factors	Discriminant Loadings
System Quality	0.853*
Work Performance Quality	0.650*
Information Quality	0.427
Services Quality	0.397
Centroid for HS[a]	0.927
Centroid for LS[b]	-2.1
Wilks' Lambda	0.330
P-Value	0.000
Canonical Squared Correlation	0.818
Hit Ratio	86.7

* *Factor entered into the stepwise model,* [a] *High satisfaction group,* [b] *Low satisfaction group*

The discriminating model has a hit ratio of 86.7%, much higher than the pre-defined cutting score. Hence, the proposed model is an effective model in clustering the HS and LS groups (Chow & Lui, 2001). The two significant success factors have positive discriminant values, which implies that the HS group pays more attention to these two success factors than the LS group. There is no negative significant loadings reported; therefore, no observations could be drawn for the practice of the LS group. The other two insignificant success factors have discriminant loadings ranging from 0.397 to 0.427.

DISCUSSION

Although many supply chain practitioners recognize the importance of e-SCS adoption, they lacked knowledge about the critical dimensions of its success factors. Based on a survey of extranet provider perceptions, this study suggests four success factors: information quality, system quality, service quality, and work performance quality. The achievement of these four factors involves the design, development, and implementation of extranet system features and practices. E-SCS providers can examine the identified e-SCS features and practices closely so that the identification and implementation of the relevant system functions and features can take place. To further improve the satisfaction levels of users, this study identifies 10 areas in which e-SCS features and practices should be improved. E-SCS providers should examine these areas and work with their users closely so that the system features and practices can be enhanced. It is desirable to develop a strategic e-SCS plan that is cost justifiable. The success factors identified herein can serve as a foundation for developing such a strategy. Furthermore, the results of this study can serve as a basis for a strategy of supply chain quality management (SCQM), as proposed by Kuei and Madu (2001). However, modifications will be needed for formulating SCQM because the Kuei and Madu model deals with supply chain management, whereas the present findings concentrate on supply chain information systems.

This study also contributes to the understanding of the e-SCS performance. Two e-SCS success factors are singled out as highly practiced by the high satisfaction group: system quality and work performance quality.

CONCLUSION

The globalization of business ventures has changed the pattern of how firms use their information systems. In supply chains, success depends on collaboration among trading partners. The effectiveness of e-based supply chain system management is also a critical issue in the success of supply chain practices. This chapter identified four success factors and 10 areas in which to improve the service quality of extranet adoption.

REFERENCES

Baines, A. (1996). Intranets. *Work Study, 45*(5), 5-7.

Bhatnagar, R., Sohal, A.S., & Millen, R. (1999). Third party logistics services: A Singapore perspective. *International Journal of Physical Distribution & Logistics Management, 29*(9), 569-587.

Chow, W.S., & Lui, K.H. (2001). Discriminating factors of information systems function performance in Hong Kong firms practicing TQM. *International Journal of Operations and Production Management, 21*(5/6), 749-771.

Coyle, J.J., Bardi, E.J. & Langley, C.J. (2003). *The management of business logistics: A supply chain perspective* (7th ed.). Quebec, Canada: South-Western.

Cronan, T.P., & Douglas, D.E. (1990). End-user training and computing effectiveness in public agencies: An empirical study. *Journal of Management Information Systems, 6*(4), 21-39.

Cupito, M.C. (1997). Intranets: Communication for the internal universe. *Health Management Technology, 6*, 20-58.

Foo, S., & Lim, E.P. (1997). Managing World Wide Web publications. *Information Management & Computer Security, 5*(1), 11-17.

Fuller, S., & Pagan, K. (1997). *Intranets firewalls*. USA: Ventana.

Gallupe, R.B., & Tan, F.B. (1999). A research manifesto for global information management. *Journal of Global Information Management, 7*(3), 5-18.

Huff, S.L., & Munro, M.C. (1985). Information technology assessment and adoption: A field study. *MIS Quarterly, 9*(4), 327-340.

Kettinger, W.J., & Choong, C.C. (1997). Pragmatic perspectives on the measurement of information systems service quality. *MIS Quarterly, 21*(2), 223-240.

Kim, J.I. (1996). Logistics in Korea: Current state and future directions. *International Journal of Physical Distribution & Logistics Management, 26*(10), 6-21.

Kuei, C., & Madu, C.N. (2002). Identifying critical success factors for supply chain quality management. *Asia Pacific Management Review, 6*(4), 409-423.

Kuei, C.H., Madu, C.N., Lin, C.H., & Chow, W.S. (2002). Developing supply chain strategies based on the survey of supply chain quality and technology management. *International Journal of Quality and Reliability Management, 19*(7), 889-901.

Lynch, G. (1997). Intranets—Just another bandwagon? *Industrial Management & Data Systems, 97*(4), 150-152.

McMullan, A. (1996). Supply chain management practices in Asia pacify today. *International Journal of Physical Distribution & Logistics Management, 26*(10), 79-95.

Misic, M.M., & Hill, J.A. (1994). Keys to success with the Internet. *Journal of Systems Management, 11*, 6-10.

Narasimhan, R., & Kim, S.W. (2001). Information system utilization strategy for supply chain integration. *Journal of Business Logistics, 22*(2), 51-75.

Sridhar, S. (1998). Decision support using the intranet. *Decision Support Systems, 23*, 19-28.

Tabor, S.W., Pryor, A.N., & Gutierrez, C.F. (1997, Fall). Improving corporate communications with intranets. *Information Strategy: The Executive's Journal*, 7-12.

Tan, K.C. (2002, Winter). Supply chain management: Practices, concerns, and performance issues. *The Journal of Supply Chain Management*, 42-53.

Voon, T.J., & Ho, K.S. (2001). *Economic impact of logistics infrastructure development: The case of Hong Kong*. Huntington, California: Nova Science Publishers.

Wong, A., Tjosvold, D., Wong, W.Y.L., & Liu, C.K. (1999). Relationship for quality improvement in the Hong Kong-China supply chain. *The International Journal of Quality and Reliability Management, 16*(1), 24-41.

Chapter IX

Information Technology Professional Career Development: A Progression of Skills

Seokha Koh, Choong Buk National University, Korea

Sooun Lee, Miami University (Ohio), USA

David C. Yen, Miami University (Ohio), USA

Douglas Havelka, Miami University (Ohio), USA

ABSTRACT

No industry seems to change as much or as quickly as the information technology (IT) marketplace. The skills necessary to be a successful professional in this industry change as the technology changes and as one advances through a career. The results of prior research are used to develop a field study of IT professionals' skill sets throughout their careers. Our findings suggest that IT professionals need to have different technical skills as their careers progress, and that; in fact, they do have different technical skills at different stages in their careers.

INTRODUCTION

Continuous professional development is an indispensable component of any job in today's business environment. This is especially true for IT professionals in the Web-based, Internet-driven e-commerce era. Changes in the market, technology, and work processes reduce and may even eliminate the need for old skills and demand the development of new ones. Performance standards are becoming more complex and demanding (Ilgen & Pulakos, 1997). To remain competitive and advance in their careers, IT professionals should identify skills gaps, recognize areas to improve current performance, keep up with advances in their profession, and try to anticipate how changes throughout their organizations and their industries affect work demands and skills requirements (Carnevale & Desrochers, 1999; London & Smither, 1999). In fact, some research has suggested that IT professionals who aspire to the CIO (chief information officer) position must have more than technical and interpersonal expertise to be effective (Kwak, 2001).

It is becoming more difficult to specify and prepare for the skill demands of professional positions prior to beginning work activities (Barley & Orr, 1997; Mirvis & Hall, 1996). Web-based applications that support e-commerce throughout the value chain require different technical skills than those that some experienced IT professionals possess. Not only are these new skills required, but also the rate of skill and technology change is continuously increasing. Also, the number of factors influencing these changes is constantly increasing (Howard, 1995; Weick, 1990). Given these dynamic work environments, contemporary work roles require a significant amount of activity that is contingent and hard to predict (Darrah, 1994). As a result, skill deficiency in IT professionals has been found in various areas (Lee et al., 1995; Nelson, 1991; Trauth et al., 1993). To address this deficiency, IT professionals are continuously adding, replacing, and retrofitting their skill set to ensure an adequate stock of knowledge and expertise to perform their jobs (Alder, 1992; Carnevale et al., 1994). It should be noted that most of these issues and phenomena were observed prior to the push for Internet and Web-based applications. These technologies and enterprise systems, mobile communication, new security concerns, and the emphasis on global management and outsourcing only add to the changes and diversity in technical skills required by IT professionals. Combine these ever-shifting technical requirements with the notion that most professions require different skills sets as individuals advance in their careers, and you have a very demanding learning requirement for IT professionals to be successful.

This chapter presents an investigation into the technical skills *required* by organizations for the IT professionals they employ as these professionals move through different stages of their careers. Also, the technical skills that these IT professionals actually *possess* in these career stages are investigated. Obviously, as IT professionals progress through their careers, the jobs and responsibilities allocated to them change. It is to be expected that the roles and responsibilities of IT professionals at later stages in their careers demand different technical skills than those who are at the entry-level or in the middle stage of their careers. It also seems reasonable that as technology and the general business environment changes in response to the e-commerce era, so will the demands on IT professionals and their skill sets. The study undertaken provides some new knowledge with regard to the technical skills required and those currently possessed by IT professionals.

BACKGROUND

Technical Skill Development

A great deal of this learning takes place without any connection to either trainers or training programs (Marsick & Watkins, 1990; Van der Krogt, 1996). Marsick and Watkins (1990) estimate that up to 83% of workplace learning occurs informally[1] or incidentally while on the job. Additionally, learning results from training and development activities provided by employers. A few outstanding organizations sustain competitive advantage through training and development activities (Olian et al., 1998). The average organization, however, fails to provide adequate training and is improving its training investments and practices at a slow rate (Bassie & Van Buren, 1999).

Computer skill training consumes a major part of training and development activities in organizations and is known to constrain training of other skills (Carnevale & Desrochers, 1999). This situation is only exacerbated by the new demands on IT professionals and other business professionals to develop skills necessary to support e-commerce applications. Recent research has suggested that an "information systems approach" for developing computer skills is more effective than the traditional methods used for this training (Naqvi, 2002).

Organizations sometimes choose to buy desired skills through hiring. Todd et al. (1995) found that the relative frequency and proportion of stated technical skill requirements in recruiting advertisements for IT professionals have increased dramatically, while the relative frequency of requirements for business and systems knowledge has actually decreased slightly. This suggests that new IT-related skills are imported from outside by hiring, whereas development of business knowledge and other skills originates mainly from inside the organization. Regardless, the burden is clearly on the individuals to maintain and develop the skills necessary to successfully perform their jobs and for career advancement.

The emphasis on human resource management and learning systems in the workplace is shifting from training to learning (Van der Krogt, 1998). Schools and universities should take this shift from training to learning in the workplace into explicit consideration to optimize their students' lifelong learning process. As it is not economical for organizations to train their personnel in areas where the personnel can learn the materials efficiently in an informal fashion in the workplace, it is also not economical for schools to instruct students on topics the students can learn more efficiently and effectively after graduation. In such areas, it is desirable to equip students with the basis upon which to learn (Couger, 1995). This argument is especially relevant when the skills and knowledge are not needed imminently, but are required for roles and positions attained during the later stages of their careers.

Technical and Other Skill Sets

A recent study by Liu et al. (2003) identified the types of skills required by IT professionals in the e-commerce era. Their results indicate a shift in the demand for traditional programming languages, operating systems skills, and networking skills to more contemporary programming languages and Web-development skills. Additionally, they suggest that students should focus on developing a wide diversity of job skills and tools. Their results emphasize the speed in which technical skill demands change.

Ashenhurst (1972) generated 37 skills and abilities that IT graduates or, equivalently, IT professionals should possess. He classified them into six categories: people, models, systems, computers, organizations, and society. Various classifications of the abilities, knowledge, or skills that an IT graduate must possess have been suggested since Ashenhurst (Couger et al., 1995; Lee et al., 1995; Nelson, 1991; Todd, 1995; Young & Lee, 1996). Despite the absence of a universally accepted classification scheme, the literature generally agrees that people regard general knowledge such as interpersonal skills and business knowledge as more important than IS-related skills (Garner, 1998; Lee et al., 1995; Leitheiser 1992; Nelson, 1991; Trauth, 1993). This is true even for entry-level IT personnel (Young & Lee, 1996).

Young and Lee (1996) found that the level of technical skill often determines an IT major's access to an internship, which may directly result in permanent hiring after the internship. Todd et al. (1990) also found that recruiters put more emphasis on technical skill than on business or systems knowledge when they recruit IT personnel. Recognizing that their finding contradicts the literature concerning skill requirements for IT personnel, Todd et al. (1995) concluded that there is a mismatch between what is achieved and what is desired. Therefore, the recruitment process does not respond to the true organizational requirements and is misdirecting its attention towards technical skills.

However, another interpretation of these results is plausible: recruiters really do desire technical skill, as opposed to other skills, when they recruit IT personnel. Almost three fourths (75%) of firms place new IT hires into technical training programs, while less than one-fourth (25%) of firms place new IT employees into management training (Young & Lee, 1996). These findings imply that organizations assign their newly hired, technically-focused IT personnel to technically-oriented jobs. And after they have acquired technical job experience, organizations assign them to an increasingly greater proportion of non-technical responsibilities.

Other research has suggested that the changing IT work environment will force IT professionals to acquire new and different skills to be successful in the post-dot-com bust and the e-commerce-driven business world (Prager, 1998). The importance of IT technical skills in the design and development of Web-based, e-commerce systems has also been emphasized (Henneman, 1999). Additionally, recent research has suggested that virtual learning environments for IT skills will have a significant impact on computer self-efficacy, a significant negative impact on satisfaction with the learning process, and little or no difference in performance when compared to traditional classroom instruction (Piccoli, Ahmad, & Ives, 2001). This indicates that not only will the specific skills needed in the future be different, but how IT professionals learn these new skills will also change.

In a study of the technical and organizational skills of webmasters, it was found that employers seek technical skills over organizational skills, but that webmasters felt that organizational skills were more important to performing their job (Wade & Parent, 2002). Additionally, the study found that a deficiency in either technical or organizational skills leads to lower job performance, but the negative effect of an organizational deficiency was larger. Another survey of employers found that employers rated non-technical skills higher than technical skills, and that gaps between expected performance and actual performance were greatest for non-technical skills (Cappel, 2002). In a similar study of IT students' perceptions of the importance of technical versus non-technical skills to be a successful IT professional, the results indicate that students believe that non-technical skills such as communication, analytical, and managerial skills are at least as important as technical skills for a successful IT professional (Medlin, Dave, & Vannoy, 2001).

Another study focused on computer skills for business professionals through the year 2005 found that 12 hardware and software skills and 13 Internet/Web-centric communication skills are and will be important for all business professionals (Zhao, 2002). Additionally, it was reported that programming skills remain important for IT professionals and that discipline-specific applications are important for other business professionals. Finally, this study found that the discipline-specific applications were considered integrated components of the Web-enabled enterprise; this also could imply the need for a change in the IT professionals' skill set.

RESEARCH METHOD AND ANALYSIS

Given the findings of this prior research and the lack of research on how IT professionals' skill sets match the organizations' needs, a study was undertaken to investigate how well IT professionals' technical skills match the organizations' requirements relative to the stage of the IT professional's career. It is reasonable to presume that less experienced IT professionals possess different kinds and levels of technical skills compared with more experienced IT professionals whose skills may meet the organization's needs to varying degrees. To investigate these observations, the following hypotheses are postulated for testing in this study:

H1: The technical skills *required* by IT professionals will be significantly different in different stages of their IT career.

H2: The technical skills *possessed* by IT professionals will be significantly different in different stages of their IT career.

As part of a larger research project that investigated general skill requirements for IT professionals (Koh, Lee, Yen, & Havelka, 2001), this study focuses on the evolution of particular technical skills as IT professionals move through their careers in the e-commerce era. A list of technical skills widely used was assembled and consolidated arbitrarily into 29 items (see Table 4, 5, & Appendix A). This set of 29 technical skills was used to test hypotheses 1 and 2 stated above. For each item of the survey, IT professionals were asked to assess the proficiency level that is required for that item (H1) and also the proficiency level that they possess for that item (H2). Every variable is measured by using a 1-5 rating scale (1 = lowest level, 5 = highest level). Two pilot tests of the questionnaire were conducted before the administration of the survey questionnaire.

The questionnaire contains two parts: (1) a section for demographic data and (2) a section to measure the technical skill levels of IT professionals in industry. Organizations from several industries were identified that had a history of employing IT professionals. A total of 470 survey questionnaires were mailed, and a follow-up mailing was made to non-respondents. A total of 92 questionnaires were returned from the 470 survey questionnaires mailed. After excluding seven invalid questionnaires because of incomplete responses, a total of 85 records were used for the analysis giving a response rate of 19.57%. Table 1 shows the demographics of the sample. The respondents represent employment in a variety of industries and a diverse range of positions (see the demographic results in Table 1). Recipients were assured that all data were confidential and that only aggregate results were to be reported.

Table 2 summarizes the results of a preliminary analysis followed by a multivariate analysis. As would be expected, IT professionals indicate that they possess the technical skills required to perform their jobs successfully. The overall technical skill level possessed is equivalent to or higher than the technical skill level required in almost every area. It can be concluded safely that IT professionals feel virtually no deficiency in the level of IT technical skills that they *possess* when compared to what they consider to be *required* to perform their jobs.

Table 1. Demographics of subject organizations and IT professionals responding

Number of employees	**Percentage**
Under 100	1.3%
100-499	8.8%
500-999	7.5%
1,000-4,999	21.3%
5,000-9,999	5.0%
10,000 & over	56.3%
Gross Revenue ($ millions)	**Percentage**
Under 100	9.3%
100-249	5.3%
250-499	6.7%
500-999	10.7%
1,000-1,999	9.3%
2,000 and over	58.7%
Age	**Percentage**
Under 20	1.2%
20-24	10.7%
25-29	23.8%
30-39	35.7%
40-49	22.6%
50-59	6.0%
60 & over	0.0%
Types of Jobs (multiple entries allowed)	**Percentage**
AI/expert systems specialists	1.2%
Application programmer	25.9%
Data communication specialist	1.2%
Database specialist	4.9%
End user support	6.2%
IT manager	34.6%
Operator	0.0%
Systems programmer	4.9%
Systems analyst	18.5%
Other IT specialists	28.4%
Non-IT specialists	2.5%

Table 2. Results of preliminary analysis

(a) Statistics of the variable sets

Variable Set	Minimum	Quartiles			Maximum
		1st	2nd	3rd	
Technical Skills: Required	1.1	2.2	2.4	2.9	3.9
Technical Skills: Possessed	1.6	2.3	2.7	3.1	3.9

*(b) Variables with significant difference between the required and possessed scores (*significant at α = .1, **significant at α = .05, and ***significant at α = .01)*

Assembly language (.5)***
Presentation graphic tools (.5)***
Spreadsheet tools (.5)***
Fourth generation language (.4)***
Internet/navigation browser (.4)***
Statistics tools (.4)***
Word processing tools (.4)***
High-level procedural language (.3)***
PC-based DB tools (.3)**
Client-server based DB tools (.2)**
Decision support systems (.2)*
PC operating systems (.2)*
Web page production tools (.2)*
Query language (.1)*
Enterprise resource planning tools (-.1)**
Simulation/optimization tools (-1)**

To analyze the influence of work experience, we divided the respondents into three groups representing different career stages for IT professionals: those with less than three years experience (Group 1); those with three to 10 years of experience (Group 2); and those with more than 10 years[2] experience (Group 3). The multivariate analysis of variance (MANOVA) procedure in SAS V6.12 is used to test the hypotheses. The procedure uses only the records with no missing values for analysis. Because of the large number of variables, there are missing values in the data, making it inappropriate to test the hypotheses related to the technical skill at the variable level. Instead, the MANOVA is applied against the technical variables set in total to test the hypotheses. Table 3 summarizes the MANOVA results. The MANOVA procedure in SAS V6.12 uses four statistics to calculate the p-value. The results on all four of the statistics are presented.

In addition to the MANOVA, the univariate analysis of variance (ANOVA) is conducted for each variable to determine if significant differences exist among the different IT experience levels for specific technical skills. The results of the analysis are presented and discussed below for each hypothesis proposed.

H1: The technical skills *required* by IT professionals will be significantly different in different stages of their IT careers.

Table 3. Results of MANOVA

	P-VALUE			
Variable Set	Wilks' Lambda	Pillai's Trace	Hotelling-Lawley Trace	Roy's Greatest Root
Technical skills set: Required	.007	.014	.004	.001
Technical skills sets: Possessed	.000	.000	.000	.003

The MANOVA results (Table 3) indicate that Hypothesis 1 is accepted (using all four statistics). This result indicates strong support for the proposition that different technical skills are required at different stages of an IT professional's career. Using ANOVA to gain additional insight reveals six specific items that have significant differences in required technical skill among the groups (see Table 4). In Table 4, the variables with a p-value greater than .10 are omitted for simplicity. All the variables with a p-value greater than .10 are presented in Appendix A.

Clear trends exist in the scores of two of the required technical skills (see Table 4), CASE/Structured Programming Tools and Object-oriented Languages. In both cases, the trend is a decrease in the required level of skill as the IT professional attains more experience. Several explanations for this pattern are plausible. These trends may be explained by the fact that the availability of CASE tools and the use of object-oriented languages are more recent phenomena, and, therefore, the younger, less experienced IT professionals in the beginning of their careers are required to possess these skills for entry level positions as organizations attempt to use these new technologies. It also may be the case that these skills have been taught in college and training programs relatively recently and that as the individuals that were exposed to these skills move into more senior positions, the skills are applied to projects that they lead, thereby making these skills required for the more junior IT professionals that follow. Alternatively, it may be that both of these skills relate to the low-level, "grunt" work involved with constructing technical and, therefore, is required primarily by the personnel performing this work (i.e., the more junior personnel) and the personnel immediately supervising (i.e., the intermediate level of experience).

It is also interesting to note that for both Enterprise Resource Planning tools and Transaction Processing Systems, the highest level of technical skills is required by the

Table 4. ANOVA results: Experience level vs. technical skill required (Variables with p-value < 0.1)

	CAREER STAGE			
VARIABLE	G1	G2	G3	P-VALUE
Expert systems/shells	1.9	1.3	1.6	.031
Transaction processing systems	2.1	3.2	2.5	.042
CASE/structured programming tools	2.9	2.3	1.9	.001
Enterprise resource planning	2.3	3.0	1.8	.026
Object-oriented languages	3.5	2.6	2.4	.090
PC-based database tools	3.4	2.2	2.7	.014

middle group (G2: three to 10 years of experience). This may be explained by the nature of these technologies. Both ERP and customized transaction processing systems may require some degree of sophistication to understand and to work with; therefore, only those personnel with proven track records are assigned to these tasks. And as the IT personnel gain even more experience, they move into more technical areas as technical experts or into more managerial roles.

The other two areas of technical skills that demonstrate a significant difference across IT personnel experience levels are Expert Systems/Shells and PC-based Database Tools. Both of these technical skills demonstrate a pattern where the least experienced and the most experienced are required to have higher levels of competence in these areas than the intermediate IT personnel. For expert systems technical skills, this may be explained by the need for more experienced IT personnel to act as technical experts and the less experienced to bring in the latest technical skills from their educational programs. This would be consistent with the notion presented above that mid-level IT personnel will have a tendency to become primarily technical experts or to take on more managerial roles as their careers progress further. In the case of PC-based Database Tools, it appears that the significant differences found may be due primarily to the large difference between the level required by the least experienced and both the mid-level and more experienced IT personnel. This seems logical as the newer IT personnel most likely would be assigned to tasks involved with supporting user requests to get information from the existing database.

H2: The technical skills *possessed* by IT professionals will be significantly different in different stages of their IT careers.

Hypothesis 2 is also supported by all four statistics of the MANOVA results (see Table 3). A more detailed analysis using ANOVA results in 10 technical skill variables whose p-values are less than 0.1 (see Table 5) indicate differences in the level of technical skills possessed among the groups.

Of the significant variables in the table, the senior IT group (G3) did not score the highest mean value in any variable. In fact, for eight of the 10 significant variables, there is a trend in the reported scores from the junior IT professionals (G1) to the senior IT professionals (G3) going from the highest to the lowest. Moreover, the senior IT group scored the highest mean value in only two variables out of the 19 variables whose p-values are greater than 0.1 (see Table 6). These results are consistent with the notion presented above: as IT personnel advance in their careers, they take on more managerial types of responsibilities and their technical skills become less important.

Similar to the trends observed for the skills required, the skills possessed shows a clear trend in the scores of the CASE/Structured Programming Tools and Object-Oriented Languages from a high level possessed in the early career stage to lower levels possessed in the later career stage. Also, the same result is seen in the skills possessed level for the Enterprise Resource Planning tools that occurred for the skills required. In this case, the intermediate level IT personnel exhibit the highest level of skill.

It is very interesting to note that the personal productivity skills (i.e., electronic mail tools, word processing, Internet/Web browsers, presentation graphics, and spreadsheets) were rated as the highest technical skills among all the skills possessed for both

Table 5. ANOVA results IT career stage vs. technical skills possessed (Variables with p-value < 0.1)

VARIABLE	CAREER STAGE G1	G2	G3	P-VALUE
CASE/structured programming tools	2.9	2.6	2.0	.019
Enterprise resource planning	2.1	2.8	1.7	.006
Object-oriented languages	3.5	2.9	2.3	.017
PC-based database tools	3.8	2.8	2.7	.024
Electronic mail tools	4.6	3.8	3.8	.045
Internet/navigation browser	4.3	4.1	3.5	.010
Presentation graphics tools	4.1	3.8	3.4	.031
Spreadsheet tools	4.1	3.9	3.4	.023
Web page production tools	3.5	2.8	2.4	.028
Word processing tools	4.4	4.0	3.7	.049

junior and senior IT groups, although those in the early career stage group are consistently higher than those in the later career stage groups. This makes sense on its face, due to the fact that nearly all IT personnel should be expected to be "expert" users of these productivity tools, regardless of their level of experience. In addition, the newest hires would certainly be expected to be most up-to-date on the latest tools due to their more recent formal education.

Overall, there appears to be a tendency that less experienced IT professionals possess and are required to possess higher levels of proficiency in most technical skills. The difference between the later career stage group and the earlier career stage groups is more prominent in the skills-possessed scores than in the skills-required scores. Again, this is consistent with the observation that most IT personnel advance from a more technical position when they get right out of college to more management-oriented roles as they gain knowledge and experience. The one caveat to this pattern may be the IT personnel that choose to become technical experts. And even these specialists appear to view their positions more as management or business consultants than they would as technical specialists.

DISCUSSION AND IMPLICATIONS

The most obvious implication of the results of this study is that IT professionals must be dedicated to lifelong learning. Given that the nature of the profession is based on a continuously changing and evolving knowledge base, the IT professional must diligently monitor the technological environment and take action to learn and apply new technical skills to remain a productive and valuable employee (or consultant). This study adds to the evidence in support of the notion that as IT professionals move through their careers, the skills required to perform their job duties will change. This might be considered an "IT career stage theory" with related hypotheses. Some of these hypotheses can be deduced and tested by further research.

First, this theory implies that only those individuals that have the ability and the motivation to learn multiple types of skills will advance into higher levels of IT management. Technical skills are required to get the first job and to succeed at it. As IT professionals gain experience, they are given more managerial responsibilities, perhaps as a technical team leader and then as a project manager of a technical project. These roles demand technical expertise as well as planning, coordinating, and communication skills. If successful in these positions, the IT professional is then expected to take on responsibilities that may include a budget and hiring, again demanding new skills. These new responsibilities require less technical ability and more managerial skills. Attaining the highest levels of IT management (chief information officer or partner) requires different skills again (i.e., selling to customers, strategic thinking, risk-taking). Only those individuals able and willing not only to learn, but to master, different types of skills can take this path.

Second, the career stage theory would suggest that organizations require different types of technical skills from the IT professionals in different stages of their careers. New hires are expected to possess skills related to the newest technologies. This is not to say that they possess high levels of technical skill, but that they should have been exposed to the most recent advances that have become accepted in the field. Mid-level IT professionals, like mid-level professionals in all areas, are the true workhorses. They are expected to have high levels of technical expertise for the technologies that the organization uses, and they are expected to have high levels of understanding of business processes and industry-specific nuances. In addition, this is where the individuals that move to the next level are discovered. Individuals that have strong technical skills and demonstrate the ability to plan, lead, and coordinate are tapped for the management path. Last, in the later stages of IT professionals' careers, they are expected to be technical experts or managers.

A third hypothesis of the IT career stage theory would be that the skills required by IT professionals will change, based on the changing technological environment. As new technologies replace older ones, the specific technical skills required by new hires will change. This may have a ripple effect on the skills required at each stage of the IT professional's career, as each generation moves through the career stages. Or this could have onetime "shocks" to the organization that impact IT professionals at every career stage, requiring additional training and learning as these technologies are adopted.

These three hypotheses, based on an IT career stage theory, are suggested by this research study, but further research is needed to test these propositions. In the study, IT professionals were asked to give their perceptions of the technical skills required to perform their jobs and the technical skills that they possess. Overall, the IT professionals indicated that they possess the skills required for them to perform their jobs. Additional insight about the level of technical skills required and the level of skills that the IT professionals possess may be obtained by querying system end-users or managers in the same organizations or departments about their perceptions of the technical skills required by the IT staff and the level of technical skills possessed by the IT staff. It would seem reasonable that these stakeholders would hold a different opinion than the IT professionals and may provide a more objective assessment of the current stock of IT skills, at least in terms of the stakeholders' different perspectives.

This study has several limitations. Given the number of skill variables that were being investigated, more definitive analyses could be performed if a larger sample were taken. Also, the skills that were included in the study were taken from prior research and may not include the most recent skills needed and possessed by IT professionals for e-commerce applications. The study also considered IT professionals in three stages. More detailed research into these stages may be useful to design appropriate training and learning mechanisms for IT professionals. It may also be useful to make further distinctions among the skill sets required by different subsets of IT professionals (i.e., programmers versus project managers).

The findings of this study have implications for both academia and industry—primarily, the importance of technical skill transition and lifelong learning in the IT professional's career development. For academia, the study implies the importance of giving undergraduate students the technical skills they need to get that first job and strategies to learn quickly. For organizations, the study shows the importance of designing their continuing education and training programs to enable the e-commerce era workforce.

CONCLUSION

As new demands are placed on IT professionals to design, develop, and support Web-based e-commerce applications, the technical skills required by these individuals are expected to change. This study reports statistically significant relationships between an IT professional's career stage, as measured by their work experience, and technical skills required and technical skills possessed. It appears that IT professionals (1) require different technical skills at different stages in their careers; (2) possess different levels of technical skills in their careers; (3) require more technical skills in early stages of their career; and (4) possess more technical skills in early stages of their career.

REFERENCES

Adler, P.S. (1992). *Technology and the future of work.* New York: Oxford University Press.

Ashenhurst, R.R. (ed.) (1972). Curriculum recommendations for graduate professional programs in information systems. *Communications of the ACM, 15*(5), 364-384.

Barley, S.R., & Orr, J.E. (1997). *Between craft and science: Technical work in U.S. settings.* Ithaca, NY: ILR Press.

Bassie, L.J., & Van Buren, M.E. (1999). Sharpening the leading edge. *Training & Development,* January, 23-33.

Cappel, J.J. (2001-2002). Entry-level IS job skills: A survey of employers. *Journal of Computer Information Systems*, Winter, 76-82.

Carnevale, AP., & Desrochers, D. (1999). Training in the dibert economy. *Training & Development,* December, 32-36.

Carnevale, W.A., Gainer, L., & Shultz, E. (1994). *Training the technical workforce.* San Francisco: Jossey-Bass.

Couger, J.D. et al. (1995, September). IS'95: Guideline for undergraduate IS curriculum. *MIS Quarterly, 19*(3), 341-359.

Darrah, C. (1994). Skill requirements at work. *Work and Occupation, 21*(1), 64-84.

Garner, R. (1998, September 28). IT leadership: Are you the right fit? *Computerworld*, 28-82.

Henneman, R.L. (1999). Design for usability: Process, skills, and tools. *Information Knowledge Systems Management, 1*(2), 133-145.

Howard, E. (Ed.) (1995). *The changing nature of work.* San Francisco: Jossey-Bass.

Ilgen, D.R., & Pulakos, E.D. (1997). *The changing nature of work performance: Implications for staffing, personal actions, and development.* San Francisco: Jossey-Bass.

Koh, S., Lee, S., Yen, D., & Havelka, D. (2001). Evolution of IS professionals' competency: An exploratory study. *Journal of Computer Information Systems*, Summer, 21-30.

Kwak, M. (2001). Technical skills, people skills: It's not either/or. *MIT Sloan Management Review*, Spring, 16.

Lee, D.M.S., Trauth, E.M., & Farwell, D. (1995, September). Critical skills and knowledge requirement of IS professionals: A joint academic/industry investigation. *MIS Quarterly, 19*(3), 313-340.

Leitheiser, R.L. (1992, Summer). MIS skills for the 1990s: A survey of MIS managers' perceptions. *Journal of Management Information Systems, 9*(1), 69-91.

Liu, X., Liu, L.C., Koong, K.S., & Lu, J. (2003). An examination of job skills posted on Internet databases: Implications for information systems degree programs. *Journal of Education for Business*, March/April, 191-196.

London, M., & Smither, J.W. (1999). Empowered self-development and continuous learning. *Human Resource Management, 38*(1), 3-15.

Marsick, V.J., & Watkins, K.E. (1990). *Informal and incidental learning in the workplace.* New York: Routledge.

Medlin, B.D., Dave, D.S., & Vannoy, S.A. (2001). Students' views of the importance of technical and non-technical skills for successful IT professionals. *Journal of Computer Information Systems, 42*(1), 65-70.

Mirvis, H.P., & Hall, D.T. (1996). New organizational forms and the new career. In D.T. Hall and Associates (Eds.). *The career is dead: Long live the career* (pp. 72-101). San Francisco: Jossey-Bass.

Naqvi, S.J. (2002, Fall). Practising information systems approach for developing information technology skills used in business. *Journal of Research on Technology in Education, 35*(1), 97-106.

Nelson, R.R. (1991, December). Educational needs as perceived by IS and end-user personnel: A survey of knowledge and skill requirements. *MIS Quarterly, 15*(4), 503-525.

Olian, J.D., Durham, C.C., Kristof, A.L., Brown, K.G., Pierce, R.M., & Kunder, L. (1998). Designing management training and development for competitive advantage: Lessons from the best. *Human resource planning, 21*(1), 20-31.

Piccoli, G., Ahmad, R., & Ives, B. (2001). Web-based virtual learning environments: A research framework and a preliminary assessment of effectiveness in basic IT skills training. *MIS Quarterly, 25*(4), 401-427.

Prager, K. P. (1998). Assessing career goals and skills. *Information Systems Management, 15*(2), 73-83.

Todd, P.A., McKeen, J.D., & Gallupe, R.B. (1995, March). The evolution of IS job skills: A content analysis of IS job advertisements from 1970 to 1990. *MIS Quarterly, 19*(1), 1-27.

Trauth, E.M., Farwell, D.W., & Lee, D. (1993, September). The IS expectation gap: Industry expectation versus academic preparation. *MIS Quarterly, 17*(3), 293-303.

Van der Krogt, F.J. (1998). Learning network theory: The tension between learning systems and work systems in organizations. *Human Resource Development Quarterly, 9*(2), 157-177.

Wade, M.R., & Parent, M. (2002). Relationships between job skills and performance: A study of webmasters. *Journal of Management Information Systems, 18*(3), 71-96.

Weick, K.E. (1990). Technology as equivoque: Sensemaking in new technologies. In P.S. Goodman, & L.S. Sproull and Associates (Eds.), *Technology and organizations.* San Francisco: Jossey-Bass.

Young, D., & Lee, S. (1996). The relative importance of technical and interpersonal skills for new information systems personnel. *Journal of Computer Information Systems*, Summer, 66-71.

Zhao, J.J. (2002, Spring). Computer end-user skills important for business professionals now and toward 2005. *Journal of Computer Information Systems*, 31-37.

ENDNOTES

1 Learning may be classified as formal, non-formal, or informal according to the mode of learning [13]. Informal learning generally denotes the learning from daily experience and exposure to the environment. Non-formal learning and formal learning denote the learning associated with deliberate instructions and programs. Job training is frequently classified as formal or non-formal according to whether the trainee leaves his or her workplace or not. We will denote "learning in the workplace" as any learning during a professional's career.

2 This classification is arbitrary. For some tests, the p-value is reduced if we classify the groups differently. The results of the analysis, however, do not change much whichever classification is used. We used this classification for all tests for consistency. This classification provides a better balance among the groups with respect to the sample size.

APPENDIX A

Technical Skills Variables with p-value > 0.1 for both technical skills required and technical skills possessed

Variable Item	Required/ Possessed	Experience Short	 Middle	 Long
Assembly language	Required	1.2	1.1	1.1
	Possessed	1.6	1.4	1.8
Client-server-based DB tools	Required	3.2	2.6	2.6
	Possessed	3.5	2.9	2.8
Data warehouse/mart tools	Required	1.8	1.7	2.0
	Possessed	1.9	1.8	1.9
DSS	Required	2.2	2.2	2.3
	Possessed	2.8	2.3	2.4
EDI tools	Required	2.7	2.7	2.2
	Possessed	2.6	2.6	2.0
E-mail tools	Required	4.5	3.6	3.8
Expert systems/shells	Possessed	2.0	1.3	1.7
4th generation languages	Required	2.5	2.4	2.2
	Possessed	2.9	3.0	2.4
High level procedural languages	Required	2.5	2.6	2.4
	Possessed	2.7	3.0	2.7
Internet/navigation browser	Required	3.5	3.3	3.3
LAN tools	Required	2.7	3.0	2.9
	Possessed	3.1	3.3	2.7
Mini/mainframe OS	Required	2.3	2.4	2.4
	Possessed	2.4	2.7	2.5
Multimedia production tools	Required	1.5	1.3	1.7
	Possessed	1.5	1.3	1.7
PC OS	Required	2.5	3.0	3.1
	Possessed	3.2	3.4	3.0
Presentation graphics tools	Required	3.1	3.0	3.2
Project management tools	Required	2.9	3.5	3.2
	Possessed	3.0	3.4	3.0
Query languages	Required	3.4	3.2	3.0
	Possessed	3.4	3.8	3.0
Simulation/optimization tools	Required	1.6	1.3	1.5
	Possessed	1.8	1.6	1.6
Spreadsheet tools	Required	2.9	3.0	3.3
Statistics tools	Required	1.8	1.4	1.8
	Possessed	2.7	2.0	2.0
Telecommunication tools	Required	2.5	2.2	2.1
	Possessed	2.6	1.8	2.3
Tele/video-conference tools	Required	1.9	1.7	1.8
	Possessed	2.1	1.6	1.8
Transaction processing systems	Possessed	2.5	3.0	2.6
Web page production tools	Required	2.6	2.3	2.5
Word processing tools	Required	3.5	3.3	3.5

An earlier version of this chapter was published in the Journal of Global Information Management, 12(1), 68-82, January - March 2004.

Section II:

Regional Themes

Chapter X

Perceptions of IS Risks and Failure Types: A Comparison of Designers from the United States, Japan and Korea

Dane K. Peterson, Southwest Missouri State University, USA

Chung Kim, Southwest Missouri State University, USA

ABSTRACT

Information systems (IS) designers from the United States, Japan, and Korea were surveyed to explore potential similarities and differences in their views on two IS risk factors, various types of IS failure and the overall failure rate on IS projects. While there were only a few differences between the U.S. and Japan, there were a number of differences in the views of designers from the U.S. and Korea. The results revealed that a lack of user involvement and a lack of experienced IS personnel were perceived as greater risk factors in Korea than in the U.S. and Japan. The data also revealed that unmet project goals and missed deadlines were perceived by designers from Korea as more likely to contribute to IS failure than did the designers from Japan and the U.S. Finally, the designers from Korea perceived a higher overall failure rate on IS projects than did the designers from the U.S. The findings were discussed in terms of the importance of national differences in technology development and national culture.

INTRODUCTION

For a number of years, researchers have investigated potential methods of improving the rate of success on Information Systems (IS) projects. Some of the earliest attempts were based on the critical success factors approach. This approach assumed that the success of an IS project was dependent on a small number of key factors. Thus, investigations were conducted to identify factors that were capable of statistically distinguishing between successful and unsuccessful projects (e.g., Ginzberg, 1981; McDoniel, Palko, & Cronan, 1993). While these studies provided many useful insights with regards to the successful development of IS projects, they typically were only concerned with the distinction between successful and unsuccessful projects. Thus, these studies generally did not take into consideration the types of IS failure associated with unsuccessful projects.

Recent studies have demonstrated that there are a variety of ways in which IS projects may fail. For example, studies have reported that up to 90 percent of all IS projects fail to meet their goals (Clegg et al., 1997). In addition, investigations have estimated that between 40 and 80 percent of IS projects exhibit some degree of budget escalation (Clegg et al., 1997; Keil, Mann, & Rai, 2000). Empirical evidence also reveals that between 30 and 70 percent of IS projects are designated as sometimes or usually late (Phan, Vogel, & Nunamaker, 1988; van Genuchten, 1991). Thus, it is perhaps not surprising that three-fourths of all large-scale IS developed in the U.S. are considered to be "operational failures" (Gibbs, 1994).

While a number of studies have investigated the prevalence of various types of IS failures, most of the research has been conducted in the U.S. or other countries with a Western culture (Keil et al., 2000; Schmidt, Lyytinen, Keil, & Cule, 2001). However, the recent growth in multinational businesses has resulted in a significant increase in IS applications which cross national boundaries and span diverse cultures. It has been suggested that the IS development process in global environments is likely to differ from the IS development process in the U.S. due to national differences in government regulations, worker expectations, information sharing, competitive strategies, and cultural factors (Duval, 1995; Hunter & Palvia, 1996; Shore & Venkatachalam, 1996). Thus, the types of IS failure most common in global environments may differ from what has been reported in studies conducted in the U.S. National differences in the types of IS failure that are most prevalent and the factors contributing to these differences may require managers of global IS to alter their approach to IS development in order to cope with the most frequent problems in the host country (Katz & Townsend, 2000). Therefore in an era of corporate multi-nationalism and globalization of markets, cross-national research on possible differences in the types of IS failure may prove to be extremely valuable to managers of global IS.

Software Risk Management

An approach to improving IS success that may be particularly conducive for taking into consideration the various types of IS failure is the concept of software risk management (Boehm, 1991; Hall, 1998; Karolak, 1996). This approach attempts to reduce the possibility of IS failure by identifying and analyzing threats to IS success and then developing strategies to overcome potential risks. The risk factors are defined as any condition that can present a serious threat to the completion of an IS project (Schmidt et al., 2001). From this perspective, it may be possible to relate specific types of risk factors directly with certain types of IS failure (Jones, 1994).

The initial step in the project risk management approach is the identification of common risk factors and the determination of which factors are likely to represent the greatest risks (Schmidt et al., 2001). Identifying the common risk factors may present a major obstacle due to the possibility that the relevance and importance of potential risk factors may depend on a number of variables. For instance, Jones (1994, p. 28) noted that the major risk factors are not the same across all software domains. His research suggests that management information systems projects, systems software projects, commercially marketed software projects, as well as other types of projects are likely to differ in terms of the most relevant risk factors.

Another possibility is that there may be international differences with respect to the relevant risk factors. Evidence to support the possibility of national differences in the importance of IS risk factors was illustrated in a recent international study using a Delphi survey procedure (Schmidt et al., 2001). This study examined the views of a panel of IS experts from three culturally diverse countries. The results revealed a number of national differences in terms of the risk factors identified. In addition, while there was some general agreement concerning the relative importance of the 11 risk factors that were identified by the experts from all three countries, there were also some large differences regarding the opinions on the relative importance of some of the identified risk factors. For example, the experts from all three countries generally agreed that a lack of top management support was one of the most critical risk factors. However, there was considerable disagreement about the relative importance of other risk factors such as the lack of user involvement and a lack of required expertise on the part of the IS personnel.

Scope of the Study

The present study investigated the possibility that there may be national differences with respect to the most crucial IS risk factors which could result in differences in the types of IS failure most prevalent in a given country. To examine this possibility, the present study compared the views of IS designers from the U.S., Japan, and Korea. The decision to compare the U.S. with Japan and Korea was based primarily on the fact that there are vast cultural differences between Western nations like the U.S. and Eastern nations like Japan and Korea (Hofstede, 1991). These vast cultural differences are generally assumed to contribute to the observed differences in management styles between the countries and the reported national differences in IS development procedures (Grover, Segars, & Durand, 1994; Guimaraes, Sato, & Kitanaka, 1999; Shore 1998). Thus a comparison between the three countries may provide some insights into the differences between the Western and Eastern cultures with respect to IS risks and failure types.

While Japan, like the U.S., is considered to be a highly technologically advanced nation, many of the countries in Eastern Asia, including Korea, may be more appropriately classified as newly industrialized nations (Palvia, Palvia, & Whitworth, 2002). Thus, Korea may provide a better representation of IS development in most East Asian countries than Japan. Therefore, this study compared both Japan and Korea with the U.S.

To investigate potential national differences in IS risks and their consequences, this study examined the views of IS designers. Ideally all stakeholders, including management, users, and designers, would have access to all the relevant information concerning IS projects and share similar views. However, this is frequently not the case. For a number of reasons, management, users, and designers often have varying opinions on IS-related issues (Dos Santos & Hawk, 1988; Lyytinen, 1988). Since all three groups play a crucial role and have an interest in the successful development of IS projects, the opinions of all stakeholders are worthy of investigation. However, this study chose to focus on the views of IS designers since they are often responsible for making many of the important decisions in the IS development process. Based on their perceptions of IS risks, designers have to make a series of important judgments regarding the design and implementation of a system during the complex and unstructured process of IS development (Kumar & Bjorn-Andersen, 1990). The designers often have to define, interpret, and execute IS development strategies to overcome perceived risks, and in so doing they often must rely on their own judgments. Thus, although the views of designers may differ from those of the users and management, designers' views are highly important because their perceptions are likely to influence many critical decisions related to project management and resource allocation (Kumar & Bjorn-Andersen, 1990).

To narrow the scope of the investigation, the present study focused on two IS risk factors, namely lack of user involvement and lack of experienced IS personnel. These two risk factors were chosen largely based on evidence suggesting that there are likely to be national differences in the importance of these risk factors. As previously noted, an international Delphi study on IS risk factors found considerable difference among the opinions of experts from the various nations regarding the importance of user involvement and IS personnel experience. In addition, there are a number of reasons to suspect that the three nations investigated in this study may have different views on the severity of the risks associated with lack of user involvement and lack of experienced IS personnel. These issues along with the rationale for the proposed hypotheses are discussed in the following sections. The variables investigated in the present study and the expected relationships among the variables are illustrated in Figure 1.

Figure 1. Proposed national differences and relationships among variables

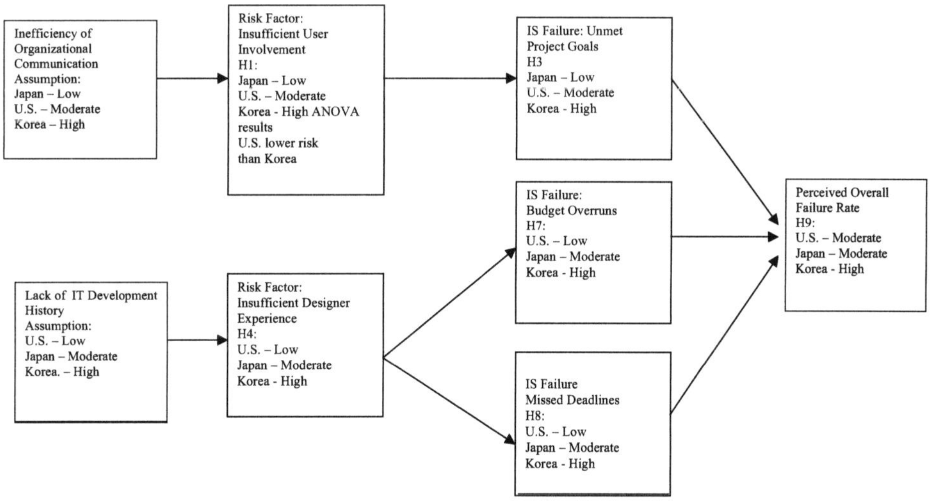

Lack of User Involvement

A number of investigations conducted in the U.S. have documented the importance of user involvement (Cleland & King, 1983; Ginzberg, 1981; Jiang, Klein, & Balloun, 1996). However, there are reasons to suspect that a lack of user involvement may represent a more serious problem in some nations than in others. Successfully involving users in the systems development process often necessitates effective organizational communication and the sharing of information between functional areas (Lyytinen & Robey, 1999; Phan et al., 1988). A high degree of departmentatilization and inefficient organizational communication can severely hamper attempts to involve users in the IS development process (Lyytinen & Robey, 1999). Since prior studies suggest that the effectiveness of organizational communication varies greatly among nations (Chen, 1995), it may be logical to expect national differences in the extent to which IS designers perceive lack of user involvement as a critical IS risk factor.

Of the three countries investigated in this study, the literature clearly suggests that organizational communication and information sharing is most likely to be a problem in Korean organizations (Chung, Lee, & Jung, 1997, p. 14). The communication process in Korean firms is almost entirely one from top-down. The top-down form of communication stifles information flow within Korean firms. It has often been noted that the poor horizontal communication between departments is a major barrier of many Korean companies to efficient organizational performance (DeMente, 1991, p. 94). Thus, based on established research documenting the high degree of departmentalization and poor organizational communication in most Korean firms (Chung et al., 1997) and the assumption that these factors contribute to a lack of user involvement, it was anticipated that designers from Korea would perceive lack of user involvement as a more serious risk factor than designers from the U.S.

Organizational communication and information sharing is generally considered more efficient in Japanese firms than in organizations located in the U.S. Organizations in Japan stress the importance of organizational communication, which is necessary for the group consensus style of decision making that is predominant in Japanese firms (Chen, 1995). Organizations in Japan are typically characterized by an efficient top-down and bottom-up communication process. In addition, vertical information sharing is also considered a strong point in Japanese firms (Chen, 1995, p. 295). Thus organizational communication should not be a major barrier to effective involvement of the users in the IS development process. Therefore, it might be expected that designers from Japan would be less likely to perceive problems with insufficient user involvement than would designers from the U.S. Based on the above analysis, the following hypothesis was proposed.

H1: IS designers from Japan will be least likely to view insufficient user involvement as a serious risk factor in the IS development process and IS designers from Korea will be most likely to view insufficient user involvement as a serious risk factor in the IS development process.

User participation is considered to be necessary for the assessment of the user's needs and thus enhances the likelihood that the system will meet the user's requirements (Ginzberg, 1983; Jiang et al., 1996; Pinto & Prescott, 1990). Hence, a lack of user involvement is often assumed to be associated with unsuccessful projects due to failure to achieve the project's goals (Jones, 1994). Although it is generally assumed that a lack of user involvement is related to failure to successfully achieve the project's goals, much of the evidence supporting this relationship is based on case studies or anecdotal reports. Thus, this study examined the relationship between designers' perceptions on lack of user involvement and failure to achieve project goals. The second hypothesis tested was as follows.

H2: There will be a positive relationship between perceptions of insufficient user involvement and perceptions of unmet project goals.

If the above hypotheses prove to be valid, then it would seem to follow that designers from Japan should be least likely to view unmet goals as a cause for IS failure. That is, because lack of user involvement is not assumed to represent a substantial risk in a country known for efficient organizational communication, unmet project goals were expected to be viewed as less likely to be a cause of IS failure in Japan than in the U.S. Conversely, since Korean firms are characterized by poor organizational communication which is expected to be associated with inadequate user involvement, it was anticipated that IS designers from Korea would view unmet goals as a likely cause of IS failure. Therefore the third hypothesis tested was as follows.

H3: IS designers from Japan will be least likely to view unmet goals as a contributor to unsuccessful IS projects, and IS designers from Korea will be most likely to view unmet goals as a contributor to unsuccessful IS projects.

LACK OF EXPERIENCED IS PERSONNEL

A number of studies have noted the necessity of an experienced IS personnel in order to achieve IS success (Baker, Murphy, & Fisher, 1983; Ewusi-Mensah, 1997; Jiang et al., 1996; Liebowitz, 1999). While maintaining an experienced IS staff is often difficult even for highly technological advanced nations, the scarcity of experienced IS personnel represents one of the primary risks to IS success in many of the less technologically advanced nations (Hassan, 1994; Odedra, 1993). Therefore, it would seem logical to expect that a lack of experienced IS personnel would represent a major risk factor in Korea, since Korea is generally classified as a newly industrialized nation. This assumption is supported by recent articles (Palvia et al., 2002; Simon & Middleton, 1998). Thus it was anticipated that designers from Korea would have less years of IS experience than designers from the U.S.

It has also been reported that a lack of experienced IS personnel is apt to represent a more serious risk factor in Japan than in the U.S. (Duvall, 1995; Friedland, 1994; Lecht, 1989). Some of the reasons given for why Japan may have less experienced IS personnel than the U.S. include the rapidly growing rate of the software industry in Japan and the perceptions among Japanese students that IS personnel are paid less, work longer hours, and have less status than other professionals (Bensaou & Earl, 1998; Humphrey, Kitson, & Gale, 1991). Based on this evidence, it might be expected that the IS designers from Japan would also have less years of experience in the IS field than designers from the U.S. Therefore, the following hypothesis was proposed.

H4: IS designers from the U.S. will have more years of experience than IS designers from Japan, while designers from Japan will have more years of experience than designers from Korea.

Studies have suggested that inexperienced designers have a tendency to provide overly optimistic projections for budgets and completion times on IS projects (Jones, 1994; Phan, Vogel, & Nunamaker; 1988; van Genuchten, 1991). It is also assumed that inexperienced designers are likely to take longer to develop systems and are prone to make more design mistakes leading to budget overruns and missed deadlines than experienced designers. Although it seems reasonable to assume that a lack of experience may be at least partially responsible for budget overruns and missed deadlines, there appears to be little empirical evidence demonstrating such a relationship (Jiang & Klein, 1999). Thus, the current study examined the relationship between the level of IS experience and perceptions of IS failure due to budget overruns and missed deadlines. The following represents the fifth and sixth hypotheses tested.

H5: There will be an inverse relationship between years of experience and the frequency with which budget overruns are perceived to be responsible for unsuccessful IS projects.

H6: There will be an inverse relationship between years of experience and the frequency with which missed deadlines are perceived to be responsible for unsuccessful IS projects.

If Hypotheses 4 through 6 prove to be true, then it should follow that designers from the U.S. would be less likely to view budget overruns and missed deadlines as a cause of IS failure. That is, due to the presumed more experienced IS personnel in the U.S., it was anticipated that missed deadlines and budget overruns would be viewed as less likely to contribute to IS failure in the U.S. than in Japan and Korea. The following represents the seventh and eighth hypotheses tested.

H7: IS designers from the U.S. will view budget overruns as less likely to be a contributor to unsuccessful projects than IS designers from Japan, and designers from Japan will view budget overruns as less likely to be a contributor to unsuccessful projects than designers from Korea.

H8: IS designers from the U.S. will view missed deadlines as less likely to be a contributor to unsuccessful projects than IS designers from Japan, and designers from Japan will view missed deadlines as less likely to be a contributor to unsuccessful projects than IS designers from Korea.

National Differences in Perceived IS Failure Rate

This study also investigated the designers' perceptions on the overall failure rate of IS projects. Although designers' views on IS success and failure may not correspond with the views of users and managers, the perceptions of designers are likely to affect many important IS development decisions. For instance, it is unlikely designers will alter their development strategies if they perceive that their past projects have generally been successful. Based on the previously stated hypotheses suggesting that budget overruns, missed deadlines, and unmet goals would represent greater problems in Korea than in the U.S. and Japan, the following hypothesis was proposed for the estimated failure rate on IS projects.

H9: The IS designers from the U.S. and Japan will perceive a lower failure rate on IS projects than will IS designers from Korea.

RESEARCH METHODOLOGY

The Sample

A total of 310 IS designers from 26 companies participated in the study. The designers were selected primarily because of previous contacts with the companies through personnel recruitment with a host university in each of the three countries. The host universities included a large state university in the mid-western U.S, Kyonggi University in Korea, and Tokyo Metropolitan University in Japan. The role of the host universities in Japan and Korea were primarily for data collection only. Questionnaires were mailed from the host university to an appointed liaison officer at each company with instructions on how to distribute them confidentially to IS designers using a systematic random sampling process. Based on information provided by the contact person, all of the respondents who received the questionnaire supplied data. While the selection of organizations was not random, the use of this procedure avoided the usual problems associated with non-response bias. A supervisor from each company was asked to complete a survey regarding company information.

The data for the U.S. sample were obtained from eight organizations, including three in manufacturing and one each in transportation, technology, wholesale/retail, public utility, and government. For the total number of employees in the companies, three had between 500 and 1,000 employees, three had between 1,000 and 5,000 employees, and the remaining two had over 5,000 employees. The number of IS employees ranged from 50 to over 150. The IS budget was between \$5.1 and 10 million for two companies, between \$10.1 and 100 million for three companies, over \$500 million for two companies, and one company left this item blank.

The data for the Japanese sample were obtained from eight organizations, including three in financial services/banking/insurance, three in manufacturing, and two in wholesale/retail. For company size, four had less than 1,000 employees, two had between 1,000 and 5,000 employees, and the remaining two had over 5,000 employees. The number of IS employees ranged from 9 to over 170. The IS budget was between \$5.1 and 10 million for two companies, between 10.1 and 100 million for two companies, over \$500 million for two companies, and two companies left this item blank.

Data for the Korean sample were obtained from 10 organizations. Five were consulting-computer service companies, four were in manufacturing, and one company was in the oil industry. Three companies had less than 500 employees, two had between 500 and 1,000 employees; two had between 1,000 and 5,000 employees, and the remaining three companies had over 5,000 employees. The number of IS employees ranged from 20 to over 250. The IS budget was under one million for one company, between $1.1 and 5 million for six companies, between $5.1 and 10 million for one company, and over $10 million for two companies.

The Questionnaire

The survey used in this investigation was first pre-tested by distributing it to 21 CIS faculty members in the U.S., of which several provided written and oral feedback on the questionnaire. The questionnaire was then pilot-tested with IS designers from two local companies in the mid-western United States. For the Japanese and Korean sample, the questionnaire was translated by a faculty member at the university in the host country. The translated version of the questionnaire was pre-tested by several graduate students and faculty members working at the same institution using the same pre-test methods used for the English language version. A back translation was performed on the Korean version of the questionnaire. Although the Japanese version was inadvertently distributed before a back translation was performed, a back translation on the Japanese version was later conducted and there was no evidence to suggest a need to revise the Japanese version of the survey used in this study.

The survey items of primary relevance for this study are presented in the Appendix. When possible, items were worded in manner that required a numerical response (or a percent value). It was assumed that quantitative responses would provide more detailed information and reduce possible comparison problems due to language translations. As in other studies of this nature, perceptions of IS success were subjectively measured based on a single item (Clegg et al., 1997). Designers were asked to provide an estimate of IS success based on their own perceptions of success. No attempt was made to define success for the respondents, as the objective was to measure the designers' views on success. Therefore, it was intentionally left for the designers to define what constitutes a successful project. Respondents were asked to assess the percentage of IS projects that they considered to be a success and the percentage that were unsuccessful. The purpose of asking both questions was to encourage respondents to think in terms of their experience with both successful and unsuccessful projects. Since the estimated rate for successful projects was always the complement of the estimated failure rate, only the estimated failure rates were used in the analysis.

The respondents also provided estimated percentages for each of the three types of IS failure in terms of how frequently they were considered a major reason a project was viewed as unsuccessful. Due to the difficulty of quantifying perceptions on lack of user involvement, a Likert-scale item was included to assess this risk factor. Respondents indicated their views regarding the degree to which insufficient user involvement represented an obstacle to successful IS development on a scale from 1 - Not at all to 7 - A great extent. Finally, the respondents were also asked to provide some biographical information, including their years of experience in the IS field.

Some of the relevant biographical information is presented in Table 1. As can be seen in Table 1, the average age of Korean designers was slightly lower than the average age of designers from the U.S. and Japan. The lack of older IS personnel in Korean firms is likely due to its later entry into the technology industry and indicative of Korea's status as a newly industrialized nation. Table 1 also reveals a larger percentage of female designers in the U.S. than in Japan and Korea which is likely due to cultural differences

Table 1. Profile of respondents

		United States	Japan	Korea
Sample Size		137	46	127
Age				
	Mean	35.45	38.09	32.24
	Standard Deviation	8.42	8.21	4.42
Gender				
	Males	70.1%	89.1%	90.6%
	Females	29.9%	10.9%	9.4%
Education				
	Graduate Degree	27.0%	4.3%	10.2%
	College Degree	56.9%	80.4%	73.2%
	No College Degree	16.1%	15.3%	16.6%
Major				
	Computer Science	31.2%	10.9%	42.5%
	MIS/CIS	29.8%	4.3%	4.7%
	Business	9.1%	15.2%	3.9%
	Other	29.9%	69.6%	48.9%
Job Title				
	Project Leader	27.0%	23.9%	13.4%
	Programmer	8.8%	17.4%	43.3%
	Analyst	7.3%	17.4&	2.4%
	Programmer/Analyst	41.6%	15.2%	31.5%
	Other	15.3%	26.1%	12.6%

regarding the acceptable roles of females in the three nations (Chen, 1995). All three nations were about equal in terms of the number of designers completing a college degree. However more designers in the U.S. had completed a graduate degree. This difference can be attributed to the fact that designers in Asian countries are more likely to participate in company-sponsored, in-house training programs rather than attend graduate school (Duvall, 1995). Table 1 also illustrates the differences in the three countries in terms of college major and job title. While many of the IS designers in the U.S. have a business-related degree and are classified as programmers or analysts, it is common for designers in Asian countries to receive a degree outside the business school (e.g., engineering) and to be classified as technicians rather than programmers and analysts (Bensaou & Earl, 1998).

RESULTS

Table 2 presents the means for both IS risk factors, the three types of IS failure, and the overall estimated mean failure rate on IS projects. As can be seen in Table 2, for each country the estimated percentages for the three types of IS failure summed to less than

Table 2. Means for IS risk factors and types of IS risks by nationality

	United States	Japan	Korea	Grand Mean
IS Risk Factors				
Insufficient User Involvement	5.29[b]	5.49	5.87	5.64
IS Experience	10.17[b]	13.37[a]	6.26	8.94
IS Failure Types				
Unmet Project Goals	.215[a]	.288[b]	.395	.299
Budget Overruns	.159	.186	.186	.174
Missed Deadlines	.318[b]	.301[b]	.416	.355
Estimated IS Failure Rate	.200[b]	.250	.397	.247

*(a) indicates the mean is significantly superior to the means for the other two countries. (b) indicates the mean is significantly superior to the country with the least desirable mean value. * Some missing values created discrepancies in the grand means compared to the weighted average based on the means for the individual countries.*

100 percent. This would seem reasonable since there are a number of factors that may contribute to unsuccessful projects that were not included on this survey. However, some of the designers in the study provided estimated percentages for the three types of IS failure that summed to more than 100 percent. Again this would seem feasible, since an unsuccessful project could have multiple causes.

The results for each hypothesis are summarized in Figure 2. To test Hypothesis 1, a completely randomized ANOVA was performed to examine possible national differences in perceptions on insufficient user involvement. The difference between the three countries was significant ($F=4.18, p=.017$). Pairwise comparison tests were conducted with Tukey's procedure. The results of the procedure indicated that IS designers from the U.S viewed insufficient user involvement as a less serious risk factor than did the designers from Korea. The difference between the U.S. and Japan as well as the difference between Japan and Korea was not significant.

Hypothesis 2 was tested by computing the correlation between perceptions on insufficient user involvement and the designers' views on unmet project goals based on the data obtained from all three countries combined. The results revealed a significant relationship between the two variables ($r=.189, p<.05$). Thus designers who perceived that users were not sufficiently involved in the IS development process also rated unmet goals as a frequent cause for unsuccessful IS projects.

To test Hypothesis 3, a completely randomized ANOVA was conducted on the estimated percentage of project failures due to unmet project goals. The results revealed a significant difference between the three countries ($F = 6.30, P < .001$). The results of Tukey's procedure indicated that unmet goals was significantly less likely to be viewed as a cause of IS failure in the U.S. than in Japan and Korea. The results also demonstrated that designers from Japan viewed unmet goals as less likely to contribute to project failure than did the designers from Korea. Thus, the results are consistent with those obtained for lack of user involvement. That is, Korean designers were most likely to view lack of user involvement as a problem in IS development and also most likely to perceive unmet goals as a cause for project failure.

To examine Hypothesis 4, a completely randomized ANOVA was used to analyze differences between the three countries in terms of the designer's years of IS experience. The ANOVA results were significant ($F = 26.13$, $p < .01$). Tukey's procedure revealed that the differences between all three countries were significant. As shown in Table 2, the designers from Japan had significantly more years of experience than the designers from both the U.S. and Korea. In addition, the designers from the U.S. had significantly more years of IS experience than the designers from Korea.

Hypothesis 5 was tested by computing the correlation between years of experience in the IS field with budget overruns based on the data obtained from all three countries combined. The results demonstrated that IS experience was not related to budget overruns ($r = -.03$). The same analysis conducted on missed deadlines resulted in a significant relationship ($r = -.135$). Thus the designers with less years of experience were more likely to view missed deadlines as a cause for unsuccessful IS projects providing support for Hypothesis 6.

To test Hypothesis 7, a completely randomized ANOVA was conducted on the estimated percentage of project failures due to budget overruns. The results revealed that the difference between the three countries was not significant ($F = 0.29$).

Table 2 illustrates that missed deadlines was viewed as the most likely cause of unsuccessful IS projects by designers from all three countries. A completely randomized ANOVA on the data for missed deadlines yielded significant differences between the three countries providing support for Hypothesis 8 ($F = 3.35$, $p < .05$). The results of Tukey's tests revealed that designers from the U.S. and Japan were less likely to view missed deadlines as a cause for unsuccessful IS projects than were designers from Korea. This finding is consistent with the results on IS staff experience which indicated that designers from Korea also had less experience than designers from the U.S. and Japan. To further establish the relationship between the variables, an ANOVA was conducted to examine if the differences between the three countries in terms of missed deadlines would be eliminated once years of user experience was included as a covariate. The ANOVA revealed there were no significant differences between the three countries with respect to missed deadlines once level of IS experience was included as a covariate ($F = 2.56$). This result, along with the significant correlation between the two variables, is consistent with the assumption that a lack of experienced IS personnel is likely to contribute to problems associated with missed deadlines.

As shown in Figure 2, the correlations between each of the three failure types and the perceived overall IS project failure rates were almost identical ($r = .138, .134$, and $.138$, for unmet goals, missed deadlines, and budget overruns, respectively). To examine Hypothesis 9, a completely randomized ANOVA was used to test for differences between the three countries in terms of perceived overall failure rates on IS projects. The results indicated that the differences between the countries were statistically significant ($F = 5.73$, $p < .01$). Using Tukey's tests to make pairwise comparisons revealed that the only significant difference was between the U.S. and Korea. As shown in Table 2, the designers from the U.S. perceived a lower failure rate on IS projects than did the designers from Korea.

It is conceivable that some of the observed differences between the countries could be due to differences in the types of organizations investigated in each country. To examine the potential influence of organizational differences, a number of statistical tests were conducted to examine possible differences between the organizations. The following describes the results of some of the tests conducted. To compare the various organizations from the U.S., separate one-way ANOVA tests were conducted on each of the three types of IS failure. The results revealed a significant difference between the eight organizations only for missed deadlines ($F = 4.19$, $p < .05$). The same tests conducted on the data collected from the eight companies from Japan revealed a significant difference between the organizations only for unmet project goals ($F = 3.75$, $p < .05$). For the Korean data, a t-test was conducted comparing the five consulting computer service companies with the remaining five companies. No significant differences were observed on any of the three types of IS failure. Comparisons between the organizations based on other organizational characteristics (e.g., organizational size) also revealed no significant differences between the organizations. These results would seem to suggest that the differences between the countries are probably not due to differences in the types of organizations surveyed.

DISCUSSION

The results revealed a number of differences between the views of designers from the U.S. and Korea. Designers from Korea perceived that a lack of user involvement was a greater risk factor than did the designers from the U.S. This result could be interpreted as being consistent with the poor organizational communication and information sharing that is characteristic of many Korean organizations. Perceptions on lack of user involvement were also related to the perceived frequency of unmet project goals supporting the assumption that a lack of user involvement increases the likelihood of failure to achieve a system's requirements. Consistent with these findings, Korean designers also viewed unmet goals as more likely to contribute to unsuccessful projects than the designers from the U.S.

The implications of these results would seem to suggest that managers of global IS involving alliances with Korean organizations may need to devote more effort to ensuring that the IS goals are clearly communicated. Thus, managers of global IS involving designers from Korea may find it beneficial to adopt methods to improve communication between users and designers. It would perhaps be beneficial to schedule more frequent meetings like Joint Application Development sessions to allow designers to interact with users and managers and become more familiar with the IS goals. However, managers should refrain from issuing detailed directives outlining IS goals in such meetings, because such behavior is often interpreted by Korean subordinates as indicating that management does not have much confidence in their ability (Chung et al., 1997, p. 158). Instead, to ensure IS goals are properly understood, supervisors should make arrangements for informal discussions with designers during non-business hours where much of the effective organizational communication takes place in Korea firms

Figure 2. Summary of results on rational differences and relationships among variables

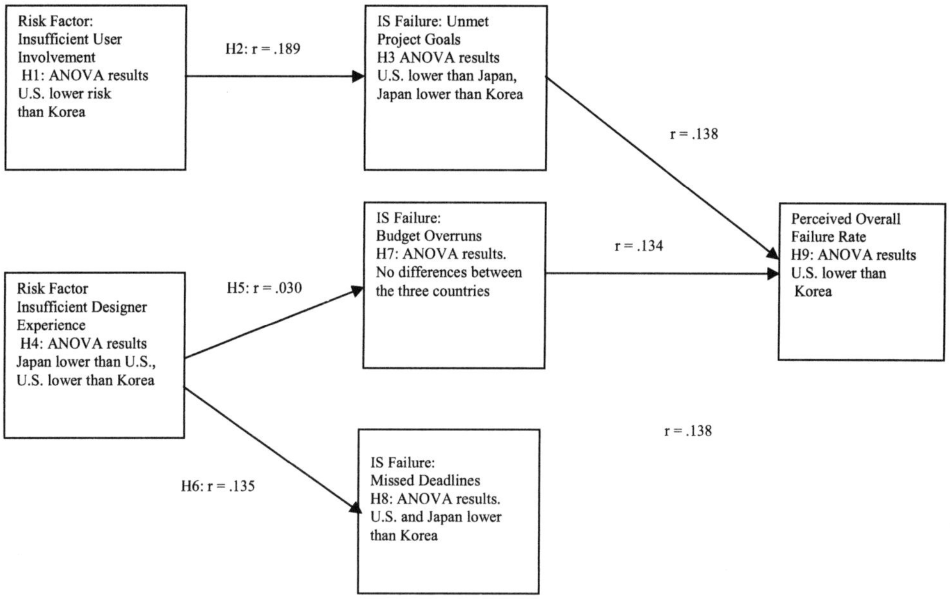

(Chung et al., 1997). In addition to more frequent meetings, managers may wish to consider using some of the proposed techniques that can reduce the risk of unmet goals, such as the use of prototypes or evolutionary and modular approaches (Alter & Ginzberg, 1978). These techniques help designers assess the intended goals and user requirements of an application early in the development process.

The designers from Korea also had less IS experience than the designers from the U.S. This result is consistent with previous studies suggesting that a lack of IS experience represents a greater risk factor in newly industrialized nations than in highly advanced nations (Palvia et al., 2002). The relationship between IS experience and perceptions of budget overruns as a cause of IS failure was not significant. In addition, the difference between the U.S. and Korea in terms of perceptions of IS failure due to budget overruns was not significant. However, the relationship between IS experience and missed deadlines was significant. Inexperienced designers were more likely to report missed deadlines as a cause for unsuccessful projects. In addition, designers from Korea were more likely to view missed project deadlines as a contributor to project failure than designers from the U.S.

These results imply that managers of global IS may find that overly optimistic projections for completion times are common in nations such as Korea where there may be a lack of experienced IS personnel. To overcome problems associated with inexperienced designers and prevent missed project deadlines, IS managers may need to allocate more time and resources to train IS designers in procedures that improve the accuracy of estimated project completion times. Managers may also wish to consider using some of the proposed techniques that might compensate for a lack of IS experience, such as the use of prototypes and the use of evolutionary and modular approaches (Alter & Ginzberg, 1978). To specifically prevent problems associated with overly optimistic projections, global IS managers working with designers from Korea may find it beneficial to devote more effort on monitoring the progress of projects through a formalized review process (i.e., a structured walkthrough) to ensure that the achievement of specific milestones are on schedule. Other recommendations for preventing overly optimistic projections that may be useful include detailed multi-source cost and schedule estimations, incremental development, software re-use, and requirements scrubbing (Boehm, 1991). These methods will not only reduce all types of risks in general, but also help designers establish more accurate budgets and timetables.

Only a few differences were observed between the views of designers from the U.S. and Japan. The designers from Japan perceived unmet project goals as more likely to contribute to project failure than did the designers from the U.S. Although designers from Japan also viewed insufficient user involvement as a greater risk factor than did designers from the U.S., the difference between the two countries was not significant. Thus, it would appear that the differences between the two countries in terms of perceptions on unmet goals may not be attributable entirely to differences in the degree of user involvement. A possible reason unmet project goals may represent a greater problem in Japan than in the U.S. is due to differences in the application knowledge of the IS designers. Although not directly examined in this study, prior literature suggests that IS designers in the U.S. may receive more training in IS applications. Thus, even though user involvement is highly encouraged in the U.S., IS designers in the U.S. are expected to be familiar with business strategies so that they can develop systems that are consistent with the organization's goals (Lee, Trauth, & Farwell, 1995). In the U.S., it is recommended that the training and education of IS designers include knowledge about organizational operations and business functions to provide designers with the ability to align IS with corporate strategies (Nelson, 1991). Perhaps as a consequence of the extensive training in business applications, IS designers in the U.S. often take an active role in ensuring IS goals are achieved and view themselves as service providers offering "technology solutions" to business problems (Bensaou & Earl, 1998).

Conversely, IS designers from Japan generally view themselves as technicians rather than business professionals. They are more likely to have an engineering background and have less training in business management than their American counterparts (Azuma & Mole, 1994). The results of this study tend to confirm these findings as indicated by a comparison of the profile of designers from the U.S and Japan presented in Table 1. In Japan, instead of training IS technicians in business applications, non-IS managers are often responsible for developing an understanding of what information technology can provide for the firm through a job rotation program that requires them to spend two or three years in the information technology department (Bensaou & Earl, 1998). Thus, IS designers from Japan apparently do not have as much responsibility for ensuring that IS goals are achieved as do the designers in the U.S. Rather, IS designers from Japan are more likely to view users and clients as collaborators and work in a cooperative relationship with the users and clients to negotiate and jointly define the goals of the project (Duvall, 1995). However, IS managers in Japan have reported that ISs often do not successfully achieve their goals because designers fail to fully understand the IS applications (Duvall, 1995).

If these assumptions are valid, IS managers of global ISs need to be aware of the possibility that designers in Japan are likely to view themselves as technicians and may have less business application knowledge than designers in the U.S. Thus, strategies to increase user involvement may not prove to be highly beneficial for reducing problems associated with unmet goals. Instead, providing IS designers with additional application training may prove to be more beneficial.

The results of this study revealed that a lack of experienced IS designers did not represent a greater risk factor in Japan than in the U.S. In addition, there were no significant differences between the U.S. and Japan in terms of perceptions of budget overruns and missed deadlines as contributors to unsuccessful IS projects. Thus, it would seem that managers of global IS would be unlikely to encounter any more problems in Japan regarding budget overruns and missed deadlines than is customary in the U.S.

The results for the overall perceived rate of failure on IS projects was higher in Korea than in the U.S. These findings would seem reasonable since the Korean designers reported a higher incidence of missed deadlines and unmet project goals, which may be due to a lack of experienced IS personnel and poor organizational communication. With respect to the comparison of success or failure rates in the U.S. and Japan, some researchers have proposed that Japan may be superior to the U.S. due to the high levels of cooperation between departments, effective organizational communication, high worker motivation, attention to detail, excellence in planning and problem solving, and general perseverance that typifies many Japanese organizations (Belady, 1986; Bensaou & Earl, 1998; Zelkowitz, Yeh, Hamlet, Gannon, & Basili, 1984). However, others have argued that organizations in Japan are at a disadvantage due to excessive government regulation, lack of financing and experienced IS personnel, and an education system that discourages creativity (Friedland, 1994; Rifkin & Savage, 1989). The results of this study do not provide support for either view as the difference in the perceived failure rates between the two countries was not significant. Therefore the results in this study are consistent with empirical investigations demonstrating few significant differences in the quality of software products developed in the U.S. and Japan (Curley, Meyer, & Sorensen, 1996; Cusumano & Kemerer, 1990). Thus, if the perceptions of designers are accurate regarding IS success rates, it may be reasonable for managers of global IS to expect comparable success rates in the U.S and Japan. That is, the difference between reported failure rate in Japan (25%) did not differ significantly from the reported IS failure rate in the U.S. (20%).

Conclusions and Limitations

Although Japan and Korea share many cultural characteristics and both differ dramatically from the U.S. culture, this study found only a few differences between Japan and the U.S. while there were numerous differences between Korea and the U.S. In many ways, the results for the Japanese sample were more similar to the results obtained from designers in the U.S. than the data based on perceptions of designers from Korea. These results would seem to suggest that the vast differences between Western and Eastern cultures do not necessarily imply that there will be large differences in the views of the designers regarding IS development issues. Rather the large observed difference between Korea and both the U.S. and Japan would seem to suggest that most of the differences in IS designers' perceptions may be attributable to differences in the level of economic and technological development among the countries. Thus, managers may find that the need to adapt IS development strategies depends more on a country's classification as technologically advanced or as a newly industrialized nation rather than differences in national culture.

A couple of limitations of this study should be noted. First, in addition to the usual limitations associated with survey research and making inferences based on analyses using correlational techniques, the results of the present study were based on the views of designers from a limited number of organizations in each country. This study relied on previously established contacts with organizations to ensure a nearly 100 percent response rate. This approach avoids the problems associated with a high nonresponse rate that often exceeds 80 percent for studies of a similar nature using mailing lists. However, due to the limited number of organizations in this study, it is possible that the observed differences between the nations could be confounded by the differences in the types of organizations surveyed in each country. As an attempt to minimize this problem, this study included at least eight organizations from each country. In addition, several statistical checks were conducted to examine potential differences between the organizations. For the most part, these tests suggested that perceptions on failure types varied little among the organizations in the study. Still it would be useful to verify the finding in this study through the mailing list approach that would involve designers from numerous organizations. Since both research procedures have their advantages and limitations, the use of both procedures should be complementary in that they could help verify the conclusions found with the other method.

This study also focused strictly on the views of the IS designers. While it is obvious that the views of designers are extremely important, so are the views of users and management. Since IS success is dependent on all three parties and the views of all three groups may differ, similar studies examining the views of users and management might produce data that would provide for an interesting comparison with the results reported in this study.

ACKNOWLEDGMENTS

We would like to thank Dr. Joong Han Kim of Kyonggi University, Dr. Toshiyuki Tamura of Tokyo Metropolitan University in Japan, Yukari Oba, and Sayaka Sakakibara of Southwest Missouri State University for their assistance is gathering the data and translating the survey.

REFERENCES

Alter, S. & Ginzberg, M. (1978). Managing uncertainty in MIS implementation. *Sloan Management Review*, 20, 23-31.

Azuma, M. & Mole, D. (1994). Software management practice and metrics in European community and Japan: Some results of a survey. *Journal of Systems Software*, 26, 5-18.

Baker, B. N., Murphy, D. C., & Fisher, D. (1983). Factors affecting project success. In D. I. Cleland & W. R. King (Eds.), *Project Management Handbook*. New York: Van Norstrand Reinhold.

Belady, L. A. (1986). The Japanese and software: Is it a good match. *IEEE Computer*, (June), 57-61.

Bensaou, M. & Earl, M. (1998). The right mind-set for managing information technology. *Harvard Business Review*, 76, 119-128.

Boehm, B. (1991). Software risk management: Principles and practice. *IEEE Software*, 8, 32-41.

Chen, M. (1995). *Asian management systems: Chinese, Japanese, and Korean styles of business*. New York: Routledge.

Chung, K. H., Lee, H. C., & Jung, K. H. (1997). *Korean management*. New York: Walter de Gruyter.

Clegg, C., Axtell, C., Damadoran, L., Farbey, B., Hull, R., Lloyd-Jones, R., Nicholls, J., Seell, R., & Tomlinson C. (1997). Information technology: A study of performance and the role of human and organizational factors. *Ergonomics Journal*, 40, 851-871.

Cleland, D. I., & King, W. R. (1983). *Systems analysis and project management*. New York: McGraw-Hill.

Curley, K. F., Meyer, M. H., & Sorenson, E. V. (1996). A comparison of U.S., Japanese, and European software development practices and processes. *Journal of Global Information Management*, 4, 18-26.

Cusumano, M. A. & Kemerer, C. F. (1990). A quantitative analysis of U.S. and Japanese practice and performance in software development. *Management Science*, 36, 1384-1406.

DeMente, B. (1991). *How to do business with the Japanese*. Lincolnwood, IL: NTC Books.

Dos Santos, B. L. & Hawk, S. R. (1988). Differences in analyst's attitudes towards information systems development: Evidence and implications. *Information & Management*, 14, 31-41

Duvall, L. M. (1995). A study of software management: The state of the practice in the United States and Japan. *Journal of Systems Software*, 31, 109-124.

Ewusi-Mensah, K. (1997). Critical issues in abandon information systems development. *Communications of the ACM*, 40, 74-80.

Friedland, J. (1994). The information technology gap. *Far Eastern Economic Review*, 157, 46-49.

Gibbs, W. W. (1994). Software's chronic crisis. *Scientific American*, 271, 86-95.

Ginzberg, M. J. (1981). Key recurrent issues in the MIS implementation process. *MIS Quarterly*, 5, 47-59.

Grover, V., Segars, A. H., & Durand, D. (1994). Organizational practice, information resource deployment and systems success: A cross-cultural survey. *Journal of Strategic Information Systems*, 3, 85-106.

Guimaraes, T., Sato, O., & Kitanaka, H. (1999). Comparing U.S. and Japanese companies on competitive intelligence, IS support, and business change. *Journal of Global Information Management*, 7, 41-49.

Hall, E. M. (1998). *Managing risks: Methods for software development*. Reading, MA: Addison-Wesley.

Hassan, S. (1994). Environment constraints in utilizing information technologies in Pakistan. *Journal of Global Information Management*, 4, 30-39.

Hofstede, G. (1991). *Cultures and organizations: Software of the mind.* London: McGraw-Hill.

Humphrey, W. S., Kitson, D. H., & Gale, J. A. (1991). A comparison of U.S. and Japanese software process maturity. In *Proceedings of the International Conference on Software Engineering*, 38-49.

Hunter, M. G. & Palvia, S. C. (1996). Information systems development: A conceptual model and comparison of methods used in Singapore, USA, and Europe. *Journal of Global Information Management*, 4, 5-17.

Jiang, J. J. & Klein, G. (1999). Risks to different aspects of system success. *Information & Management*, 36, 263-272.

Jiang, J. J., Klein, G., & Balloun, J. (1996). Ranking of system implementation success factors. *Project Management Journal*, 27, 49-53.

Jones, C. (1994). *Assessment and control of software risks*. Englewood Cliffs, NJ: Yourdon Press.

Karolak, D. W. (1996). *Software engineering risk management.* Los Alamitos, CA: IEEE Computer Society Press.

Katz, J. P. & Townsend, J. B. (2000). The role of information technology in the "Fit" between culture, business strategy and organizational structure of global firms. *Journal of Global Information Management,* 8, 24-35.

Keil, M., Cule, P. E., Lyytinen, K., & Schmidt, R. C. (1998). A framework for identifying software project risks. *Communications of the ACM*, 41, 76-83.

Keil, M., Mann, J., & Rai, A. (2000). Why software projects escalate: An empirical analysis and test of four theoretical models. *MIS Quarterly*, 24, 631-664.

Kumar, K. & Bjorn-Andersen N. (1990). A cross-cultural comparison of IS designer values. *Communications of the ACM*, 33, 528-538.

Lecht, C. P. (1989). Japanese software no threat. *Computerworld*, (May 8), 21.

Lee, D. M. S., Trauth, E. M., & Farwell, D. (1995). Critical skills and knowledge requirements of IS professionals: A joint academic/industry investigation. *MIS Quarterly*, 19, 313-335.

Liebowitz, J. (1999). Information systems: Success or failure? *The Journal of Computer Information Systems,* 40, 17-20.

Lyytinen, K. (1988). Expectation failure concept and systems analyst's view of information systems failures: Results of an exploratory study. *Information & Management*, 14, 45-56.

Lyytinen, K., & Robey, D. (1999). Learning failure in information systems development. *Information Systems Journal*, 9, 85-101.

McDoniel, P. L., Palko, J., & Cronan, T. P. (1993). Information systems development issues affecting success. *Journal of Computer Information Systems*, 34, 50-62.

Nelson, R. R. (1991). Educational needs as perceived by IS and end-user personnel: A survey of knowledge and skill requirements. *MIS Quarterly*, 15, 503-52.

Odedra, M. (1993). Critical factors affecting success of CBIS: Cases from Africa. *Information Resources Management Journal*, 3, 16-31.

Palvia, P. C., Palvia, S. C. J., & Whitworth, J. E. (2002). Global information technology: A meta analysis of key issues. *Information & Management*, 39, 403-414.

Phan, D., Vogel, D., & Nunamaker, J. (1988). The search for perfect project management. *Computerworld*, (September 26), 95-100.

Pinto, J. F., & Prescott, J. E. (1990). Planning and tactical factors in the project implementation process. *Journal of Management Studies*, 27, 305-327.
Rifkin, G. & Savage, J. A. (1989). Is U.S. ready for Japan software push. *Computerworld*, (May 8), 1.
Schmidt, R., Lyytinen, K., Keil, M., & Cule, P. (2001). Identifying software project risks: An international Delphi study. *Journal of Management Information Systems*, 17, 5-36.
Shore, B. (1998). IT strategy: The challenge of over-regulation, culture, and large-scale collaborations. *Journal of Global Information Technology Management*, 1, 1-4.
Shore, B. & Venkatachalam, A. R. (1996). Role of national culture in the transfer of information technology. *Journal of Strategic Information Systems*, 5, 19-37.
Simon, S. J. & Middleton, K. L. (1998). Asia's pending labor crunch: An analysis of human resource management best practices in IS departments. *Journal of Global Information Technology Management,* 1, 9-25.
van Genuchten, M. (1991). Why is software late? An empirical study of reasons for delay in software development. *IEEE Transactions on Software Engineering*, 17, 582-589.
Zelkowitz, M. V., Yeh, T. T., Hamlet, R. G., Gannon, J. D., & Basili, V. R. (1984). Software engineering practices in the U.S. and Japan. *IEEE Computer,* June, 57-66.

Previously published in the Journal of Global Information Management, 11(3), 19-38, July - September 2003.

APPENDIX

Survey Items on Perceived IS Failure Rate, IS Failure Types, and IS Risks

Answer the following items based only on IS projects on which you have been assigned to in the last five years for your current employer.

- What percent of IS projects would you consider to be a success?

  ____

- What percent of IS projects would you consider to be unsuccessful?

  ____

Considering only the unsuccessful projects:

- What percent of the projects would you consider "failure to achieve the project's goals" as a major cause for the failure of the project?

  ____

- What percent of the projects would you consider "missed project deadlines" as a major cause for the failure of the project?

  ____

- What percent of the projects would you consider "exceeding the project's budget" as a major cause for the failure of the project?

  ____

- To what extent did Lack of User Participation contribute to unsuccessful IS projects?

Not at all					*To a very great extent*	
1	2	3	4	5	6	7

- Please indicate the number of years you have worked in the IS/IT industry?

  ____

Chapter XI

A Taxonomy of Intranet Implementation Strategies: To Make or To Buy?

Jan Karlsbjerg, Aalborg University, Denmark

Jan Damsgaard, Copenhagen Business School, Denmark

Rens Scheepers, University of Melbourne, Australia

ABSTRACT

The mid-1990s marked the widespread adoption of intranets by organizations to facilitate communication between geographically dispersed organizational units. Since then the knowledge barriers to adoption have been lowered by the emergence of advanced development tools and later the availability of ready-made "intranet-in-a-box" packages as well as an elevation of the general awareness and knowledge of Internet/intranet technologies among users. Based on an explorative study of intranet implementations in nine Danish and two South African organizations, this article presents a taxonomy of four archetypes of intranet implementation processes. The dimensions of the framework are sourcing (in-house vs. outsourced implementation) and technology (development tools or packaged intranet products). Using the taxonomy, we classify the strategic choices of the case organizations and make recommendations for organizations using or producing intranet technology products.

INTRODUCTION

Organizations continue to face the communication challenges associated with geographic dispersion. Many have turned towards Internet technologies as a promising avenue to interlink geographically dispersed units with a uniform and rich communication channel.

Organizations with a global presence have been among the first to implement intranets—small versions of the Internet, used purely for communication within the organization itself or even within a subset of its departments (Lyytinen, Rose et al., 1998; Damsgaard and Scheepers, 1999; Newell, Swan et al., 1999). Historically, in-house personnel have developed these intranets using quite basic development tools. Correspondingly, large organizations with plenty of in-house IT and development resources were the first to implement advanced intranets (Jarvenpaa and Ives, 1996; Moeller, 1996; Bhattacherjee, 1997).

Despite the popularity of intranets, the choice of sourcing strategy remains a complex decision. The ubiquity of intranet technology renders implementation decisions (especially those with large-scale implications), a painful and risky area that frequently produces expensive and poor IT systems in organizations worldwide (George, 2000).

During the early 1990s the intranet phenomenon was in its infancy and intranets were developed from scratch as the basic knowledge about the technology had to be "reinvented" (Attewell, 1992) by each organization. Since these humble beginnings, much innovative activity has occurred on the supply side of the technology (Zmud, 1984; Perez and Soete, 1988). First, tools for intranet development and maintenance have increased in availability, diversity, functionality and usability. This has put intranet implementation well within the reach of even small and medium-sized organizations. Second, ready-made "intranet-in-a-box" packages have emerged, enabling—in principle—any organization to implement an intranet without much in-house technical expertise at all. Third, the rise in the use of the World Wide Web has raised awareness and knowledge about Internet/intranet technologies with rank and file employees in most corporations. As such, the question most corporations are confronting is no longer "should we implement an intranet?" but rather "which kind of intranet should we implement?" We would like to suggest that an additional question be asked, namely "How should we implement an intranet?" We shall argue that the latter consideration is especially crucial in the context of globally dispersed organizations.

Similar to corporate Web sites that are routinely re-launched with new designs and functionality, existing intranets are redesigned, multiple efforts are consolidated into a single intranet, or intranets are scrapped in their entirety as organizations roll out new versions of the corporate intranet (Orenstein, 1998; Sliwa, 2000). Our analysis of implementation processes may be of use to both first-time implementers and to organizations with experience from one or several previous versions of intranets in the organization.

In this paper, we propose a framework describing four different intranet implementation strategies based on the choice of implementation process (in-house vs. outsourced) and the intranet architecture (tailor-made using development tools or ready-made using packaged intranet products). We suggest this taxonomy to help implementers contemplate different strategies and we extend recommendations for the implementation of an organizational intranet based on the resources, core competences, and capabilities of the organization.

In the following section, we outline a number of characteristics of intranet technology that are pertinent to conceptualizing the technology's organizational implementation. We then put forth our arguments for the dimensions in our framework—the choice of who implements or customizes the intranet in the organization, and the choice between a tailor-made intranet or ready-made intranet architecture. As such we isolate four different implementation strategies that we explore in detail. We then describe our research methodology, the case organizations we studied, and the data processing and interpretation. Using the taxonomy, we classify the strategic choices of the case organizations, discuss the findings and make recommendations for organizations using or producing intranet technology products. We conclude with perspectives on the application and limitations of the framework.

INTRANET TECHNOLOGY CHARACTERISTICS

In the following section we define and outline a number of characteristics of intranet technology. We do so because intranets are interpreted and integrated in organizations in different ways (Damsgaard and Scheepers 1999; Newell, Swan et al., 1999; Bansler, Damsgaard, et al. 2000; Damsgaard and Scheepers, 2000). We concur with research that argues for specificity about the focal technology in information technology implementation studies (e.g., Kling, 1991; Monteiro and Hanseth, 1995), since the technology's underlying features often significantly impact its implementation.

Intranet Technology

Prior studies have attempted to pinpoint an accurate definition for the term "intranet," but since the technology continues to evolve, such definitions need to be revisited periodically. Here we define an intranet broadly in accordance with our understanding of an evolving technology that is to a large degree socially constructed (Hughes, 1987; Dahlbom and Mathiassen, 1993; Williams and Edge, 1996). At the same time we include a technical definitional component that describes the current protocols, standards and application scope.

Intranets are based on established Internet standards and are often preceded in time by the organization's Internet Web site. The experiences gained in developing the latter means that the technical barriers of intranet implementation are usually low (Attewell, 1992). In addition, intranets have been described as "glueware" or "middleware," since they have the potential to interconnect heterogeneous systems (including legacy applications) through the browser and associated protocols and applications (Lyytinen, Rose et al., 1998). We use the following working definition of an intranet in this paper:

An intranet is an information space that supports exclusive sharing of information among a prescribed community of users (typically members of an organization). The space comprises a number of technical standards and platforms interconnected in a network within well-defined security boundaries. All information exchanges occur via a Web browser using the TCP/IP and HTTP protocols. The intranet integrates text-based information (typically in HTML or XML format), rich multimedia content and dynamic content (e.g., search results, embedded scripts, interactive forums, transactions with other organizational systems). Existing computer-based systems can be integrated with the intranet, with the browser as the primary client interface.

Application Scope

Intranet technology is highly malleable and the application potential of the technology has a very wide scope. Intranets tend to evolve in sophistication over time (Coleman, 1997; Romm and Wong, 1998). Initially the technology tends to be used mainly for publication of static information, but as the organization becomes more familiar with the technology, it may be applied for the more advanced application possibilities. Typical applications include corporate, departmental and staff home pages, group calendars, product and employee directories, knowledge bases and news services (from internal or external sources). Further intranet applications can include the reservation of shared resources such as conference rooms, communication facilities such as embedded e-mail and instant messaging within the organization, and workflow features such as ordering products from suppliers subject to the approval of the employees' supervisors, to name but a few. Personalized to each employee, and with interfaces to the organization's customer databases, product information, etc., the intranet can become an enterprise information portal addressing many employees' information needs (Markus, 2000).

Intranet Implementation

Since the early days of computerization, much has become known about the organizational implementation of computer-based information systems and technologies. This knowledge is rooted in studies of centralized computing systems (e.g., Nolan, 1973; Gibson and Nolan, 1974; Nolan, 1979; Tornatzky and Klein, 1982; Markus, 1983; Kwon and Zmud, 1987; Markus, 1987; Srinivasan and Davis, 1987) and more decentralized information technologies such as office automation, e-mail, groupware (e.g., Hirschheim, 1986; Orlikowski and Gash, 1994; Grudin and Palen, 1995; Orlikowski, Yates et al., 1995; Ciborra, 1996; Dennis, 1996; Karsten, 1996). The advent of the organizational application of Internet-based technologies marks the ubiquitous computing paradigm (Lyytinen, Rose et al., 1998). As an example of this class of technologies, intranet implementation suggests a departure from the traditional IT implementation wisdom. Here we condense the most relevant differences from a number of literature sources.

Intranets may be implemented centrally in the organization as the "corporate" intranet, but units (such as divisions, departments or functional groups) and even individual employees often play an active role in establishing "child intranets" (Bhattacherjee, 1998; Lamb and Davidson, 2000). Thus in terms of scope, various "levels" of intranets can coexist (Ciborra and Hanseth, 1998) and the technology can involve a wide range of organizational actors. Furthermore the different intranet efforts may only be loosely coupled (e.g., by a single hyperlink). In this respect, the notion of an intranet is interpretively flexible (Orlikowski, 1992a) and different users may ascribe very different meanings to what they may perceive to be "the" intranet.

Quite often a "grassroots" intranet implementation effort precedes a formal organizational decision to implement an intranet (Bhattacherjee, 1998; Lamb and Davidson, 2000). In this respect, the starting point of an organizational intranet is often difficult to pinpoint, and the process emerges through a series of implementation initiatives that combine existing initiatives with novel ones. Likewise the implementation seems to be ongoing, since new functionality and possibilities evolve over time as the intranet itself and the supporting technologies mature (Damsgaard and Scheepers, 2000). This is quite different from traditional software implementation processes that Lyytinen et al. (1998)

and many others characterize as a rational lifecycle process that proceeds from determining systems requirements, analysis, design, and technical implementation to systems maintenance. In the following section we describe four different approaches to intranet implementation.

FORMULATING AN IMPLEMENTATION STRATEGY

Implementation of an intranet is an activity that has both technical and organizational aspects. On the technical side, network protocols, Web servers and other server applications must be in place, while on the organizational side users' interest and involvement must be obtained and quality control mechanisms must be established to ensure the value and reliability of the content on the intranet (Markus, 1994; Damsgaard and Scheepers, 1999).

The IS literature abounds with cases of implementation process difficulties caused by misalignment of system goals and the applied means, and organizational power struggles enacted through information systems projects (Markus, 1983, 1994; Orlikowski, 1992b; Hanseth and Braa, 1999; Bansler, Damsgaard et al., 2000). Unlike many other information systems, the use of the organizational intranet is essentially voluntary for the individual user. The quality of the information content often depends solely on the voluntary contributions of individual users and groups of users such as project teams. Along with the technological characteristics of intranets, this makes intranet implementations very sensitive to the circumstances of the implementation process. Thus, extra care should be taken in deciding upon the implementation strategy for an organizational intranet.

We now examine two central facets in the formulation of an implementation strategy. First we examine the issues of whether to outsource the implementation of the intranet or to implement it in-house. Second we examine two intranet architecture choices; one tailor-made from scratch (using development tools) to the organization's specifications; the other a customized version of a standard intranet product (known as "intranet-in-a-box" packages).

These issues lead us to propose a 2-by-2 framework consisting of four implementation strategies. Reflecting on the experiences of the case organizations we studied, we offer advice regarding the strategy to choose given an organization's needs, core competences and available resources.

Implementation Process: In-House or Outsourced

Due to the relative simplicity of intranet technology, many organizations have sufficient in-house competences to address the technical challenges in the implementation of an intranet (see e.g., Jarvenpaa and Ives, 1996; Moeller, 1996; Bhattacherjee, 1997). Exclusive use of internal resources, however, will not readily allow the organization to benefit from the experience, expertise and economies of scale inherent to existing intranet products on the market or vendors specializing in intranet development (Attewell, 1992).

Software make-or-buy decisions need to encompass both the strategic and the tactical level as recommended in the framework suggested by Rands (1993). At the strategic level the organization makes decisions about investment or divestment in capacity. At the tactical level the organization attempts to optimize the allocation of the current capacity by applying internal resources to the projects where internal expertise is most needed, and by purchasing software projects from the outside sources identified in the strategic level.

From a knowledge management perspective, an important issue is whether the organization itself should take on the burden but also reap the potential benefits of the required learning. Scarbrough (1995) outlines such choices as a continuum of tradeoffs between social control over the technical knowledge and economic exchange with other organizations.

Our research shows that the in-house implementation process is the default choice in most cases. Often the option of outsourcing the intranet implementation is not even considered, resulting in a de facto insourcing approach (Lacity, Willcocks et al., 1996). Successful implementation using an insourcing process requires in-house competences in computer networks, Web technology and programming languages used to interlink applications to the intranet. In order to avoid bottleneck situations, staff with the required competencies must be able to dedicate sufficient person-hours, and the organization must take on the project management of the implementation. The in-house implementation process facilitates freedom in the structure and content of the intranet, and the organization reaps the benefit of organizational learning about intranet technology.

Due to the explicitness of the costs, outsourced implementations have a high degree of management ownership. The organization bears the costs of market research in order to find the right contractor, as well as contract negotiation and monitoring before and during the implementation. The technical quality of outsourced intranet implementation projects is often very high as the organization benefits from the experience and economies of scale from the outside contractors. As outsiders, the contractors may offer alternative views of the organization's processes, acting as process consultants. The organization may consider this an advantage, but also an unwelcome interference.

The organization must weigh these factors against each other in the choice of whether to conduct an in-house implementation or to outsource the implementation.

Intranet Architecture: Tailor-Made or Ready-Made

In our research we observed two fundamentally different architectures of intranets, differing in the degree to which they are tailored to the adopting organization. We term an architecture that describes an intranet developed from scratch to suit the organization at hand, the tailor-made intranet. The other architecture describes a commercially available intranet product implemented in—and then customized to—the organization; we call this architecture the ready-made or instant intranet.

Since ready-made intranets support mainly generic and uncomplicated work processes, adapting a ready-made intranet to the organization is much less problematic than the adaptation processes for large standard information systems such as Enterprise Resource Planning systems (Markus, 2000). For many organizations, however, some level of adaptation and alteration will be required, increasing both the initial cost of implementation as well as subsequent maintenance and upgrade costs.

Tailor-made intranets are implemented using a wide range of tools and technologies, and they are usually expensive because the development costs are amortized on a single organization. This intranet architecture requires a high level of technical knowledge and training of the implementers. In most implementations a considerable level of technical knowledge and training is also required of the content providers, as they must use technical writing tools to edit and publish content to the intranet. The tailor-made intranet architecture lends itself well to integration with applications that are already implemented in the organization—the intranet becomes a portal or gateway to existing applications.

The ready-made intranet architecture delivers a collection of well-tested applications with proven functionality. The typical price model is a low system purchase price and additional licenses paid per-user on a bi-annual or annual basis. The licenses then cover incremental upgrades to the standard applications and functionalities. The organization must weigh the benefits of the applications with the fact that it will tie its processes to an inherently proprietary format, and that the organizational learning from the use of the system will not be easily transferable to other information systems. With this architecture it is especially important that the organization has performed an analysis of technical and organizational requirements as well as market research of available products in order to ensure that the intranet package suits the requirements. Ready-made intranets provide an integrated product with a simple form-based user interface for handling the tasks of both intranet administration as well as the content updating, thus reducing the technical skills demands placed on in-house staff.

A Taxonomy of Four Types of Intranets

The two dimensions outlined above can be combined, i.e., for intranets implemented either in-house or outsourced, the organization can choose to implement either a tailor-made or a ready-made intranet. This leads us to suggest the framework of four archetypes of intranet implementation strategies depicted in Table 1.

As nicknames for the four implementation strategies, we chose four analogies to real-world house maintenance tasks such as carpentry or electrical work. The homemade intranet (Type I) resembles the do-it-yourself homebuilder who starts from scratch and builds everything herself. The outside contractor tailoring an intranet to the organization (Type II) resembles the craftsman or the professional builder who draws on professional knowledge, experience and tools to solve the problems *in situ*. The intranet-in-a-box customized by in-house sources (Type III) resembles the engineer who buys a product as an assembly set and carries out the assembling herself. The intranet-in-a-box with outsourced customization (Type IV) resembles the traveling fitter who installs and adapts prefabricated components at customers' locations. In subsequent sections we present the four different strategies for implementing an intranet. The characteristics of the models can be seen as a basis for intranet make-or-buy decisions.

We conducted a field study to learn which of the four strategies was being used. We were also interested in observing any differences between the companies and the strategies they employ. We will first describe our field study and its research design.

RESEARCH METHODOLOGY

A multiple interpretive case study design forms the basis for the findings in this paper (Walsham, 1995). We chose nine organizations in Denmark and two organizations in South Africa. The study was conducted between 1997 and 1999 with informal follow-

Table 1. Four different intranet implementation strategies

	In-house implementation			**Outsourced implementation**
Tailor-made architecture	Homemade intranet *(The do-it-yourself builder)*	**Type I**	**Type II**	Tailor-made intranet implemented by consultants *(The Craftsman)*
Ready-made architecture	Intranet-in-a-box customized by in-house personnel *(The Engineer)*	**Type III**	**Type IV**	Intranet-in-a-box customized by outside contractor *(The Traveling Fitter)*

up studies in some of the case organizations. Because of our intention to study not only the implementation strategy but also the relationship between company characteristics and intranet implementation strategy, the participating organizations in the study varied from medium-sized organizations to large globally dispersed organizations (more than 70,000 employees) with very different core competences. To concentrate our efforts we selected in each organization a single department or geographical location (between 20 and 700 employees) as our locus. We focused on the department/location within the organization that had the most significant influence on the organization's overall intranet strategy. The organizations were selected because they spanned the lines of businesses that were considered to have the potential to be "first-movers" and since they represented diversity in size and geographical scope. The two South African organizations were chosen since they exhibited two of the most advanced intranets in the country at the time of the commencement of this study (1997-1998). Additionally the participating organizations represented both users and providers of technology.

Table 2 outlines the specifics of the interviews that we conducted in the organizations. Depending on the organization (intranet technology user/provider) we interviewed managers and developers in the IT departments as well as end-users where applicable. The inclusion of three intranet provider companies added data regarding development of both consultant-produced intranets and an intranet-in-a-box product. At the time of the interviews, the producers had implementation experiences from approximately 50 customer organizations combined. We conducted the most thorough interview series during the early stages of our research (1997-1998). As we fine-tuned our interview approach, we became more aware of the pertinent research issues (Star and Gerson, 1987), and in the final six organizations we focused our interviews on employees who were the person or persons responsible for the organizational or departmental intranet. By doing so we were able to include a greater number of organizations in the study.

Data Processing and Interpretation

The theoretical reference we drew on for the construction of our data collection instruments was Leavitt's well-known "diamond" model of organizations that presents organizational task, structure, technology and actors as highly interlinked organizational variables that must be addressed in any organizational change process (Leavitt, 1964). If only one or two of these variables are addressed, the results are often unanticipated changes in remaining areas as they compensate for the changes. For example manage-

Table 2. Organizations, dates, and the number of interviewees in each organization

Organization		Line of business (base country)	Country	Dates of interviews	Number of interviewees
1	User	Telecommunications	South Africa	Dec. 1997-Jan. 1998; Nov. 1998	9
2	User	Research and development	South Africa, various international offices	Dec. 1997-Jan. 1998; Nov. 1998	19
3	User	Manufacturing	Globally dispersed, Head office in Denmark	Aug. – Oct. 1998	8
4	User	Software development	Denmark	Dec. 1998	6
5	User	Cellular phone service provider	Denmark	Dec. 1998	5
6	User	Public sector administration	Denmark	Feb. 1999	1
7	User	Software development	Denmark	Feb. 1999	3
8	Provider	Advertising and web	Denmark	Feb. 1999	1
9	Provider	Consulting, web, and intranet	Denmark	Mar. 1999	2
10	User	Software and facility management of large systems	Denmark	Mar. 1999	1
11	Provider	Intranet-in-a-box	Denmark	Apr. 1999	1

ment may introduce new structures in an effort to make production more efficient, but unless the people aspects are addressed, the change may result in human resistance.

For the purpose of this study, we have mapped Leavitt's dimensions as follows. The variable of task is examined according to Porter's organizational processes (Porter, 1985), the variable of structure is examined using Mintzberg's concepts of organizational structure (Mintzberg, 1983), the variable of actors is examined using Schein's work on organizational culture (Schein, 1989), and for the variable of technology we have applied the literature on intranets, specifically intranet characteristics and use modes (Damsgaard and Scheepers, 1999) and the stages of intranet maturity (Damsgaard and Scheepers, 2000). Based on the theoretical concepts from this literature, we designed an interview agenda to capture the intranet implementation process (see Appendix A). All questions were open-ended to allow for a rich and interactive discussion of the topics. The interview agenda was also aligned with other tested interview agendas designed for similar purposes (cf. Scheepers, 1999; Bansler, Damsgaard et al., 2000).

A few pilot interviews were held with practitioners in order to test the interview guide, and as a result minor clarifications were made to a few questions, and the order of two blocks of questions was changed. In the original agenda the questions about the intranet applications came in the chronological order: past, present, future. However, it turned out that the interviewees were usually keener to talk about the current systems first even when asked about the past, so the order of these questions was changed to: present, past, future.

All interviews were tape recorded and transcribed by the authors. Subsequently the transcripts were coded according to a coding scheme developed from the theoretical concepts of intranet and organizations that formed the basis for the interview agenda. The coding process was necessary due to the semi-structured form of the interview (Eisenhardt 1989; Silverman 1993). As the interview progressed, the interviewees often backtracked and clarified issues that had been covered earlier in the interview, or the interviewee's answer to a question prompted the interviewers to ask questions outside

Table 3. Coding scheme based on Leavitt's diamond and intranet literature

1) **Structure:** Statements concerning the organizational structure as per Mintzberg (1983), i.e., concepts such as the way work was coordinated, degree of decentralization, formalization of work, organizational environment, etc.
2) **Processes:** Statements concerning the organizational processes as per Porter (1985), i.e., concepts such as interdepartmental communication, functional division of work, etc.
3) **Culture:** Statements concerning the organizational culture as per Schein (1989), i.e., concepts such as basic values, working environment, etc.
4) **Technology:** Statements concerning the organization's intranet implementation and other information infrastructure. Functionality that is implemented now, and functionality projected or desired for the future:
   - **4a. Intranet use-modes** (Damsgaard and Scheepers, 1999)
   - **4b. Roles in relation to the intranet implementation** (Scheepers, 1999)
5) **Intranet implementation stages:** Statements concerning the organization's level of intranet implementation (Damsgaard and Scheepers, 2000):
   - **5a. Initiation**
   - **5b. Contagion**
   - **5c. Control**
   - **5d. Institutionalization**
6) **Implementation model:** Did the organization employ own personnel and/or outside contractors to implement the intranet?
7) **Methods:** Did the work with the intranet technologies follow a predefined method, or were working methods derived from the experiences with the intranet implementation?
8) **Individual opinions and perceptions:** Did the interviewees express their own personal perceptions of, or attitudes towards the intranet specifically or intranet technology in general?
9) **Experiences and observations from other organizations:** Did the interviewee have experiences from observation of, participation in, or communication with other organizations employing intranet technologies?
10) **Miscellaneous:** Passages in the interview which seem important to the coding researcher, but which do not fit any of the other categories.

of the interview guide, or to encourage the interviewee to elaborate on tangential answers.

During the coding phase a number of measures were taken in order to increase the validity of the findings of the study. Two coders working independently coded each interview transcript. Each coder marked all the sentences or paragraphs in the transcript that contained information relating to one or more of the codes defined in Table 3. Upon completion of this phase, the two coded versions of the transcript were compared, and any differences discussed among the coders until agreement was reached on the final coded version of the transcript. The paragraphs pertaining to the individual codes and sub-codes were then extracted into temporary documents describing in raw form all the findings for each organization according to the final coding scheme. These documents were the foundation of the final case descriptions and were shared with the individual organizations for feedback and validation. Minor corrections were needed in a few cases. The cross-case comparisons and extraction of overall results were based on the case write-ups. The gathering, transcribing, coding, and analysis of the interviews comprised a workload of approximately 24 person-hours per interview.

The technology component (code 4, see Table 3) had two sub-codes to encompass the two different concepts of intranet use-modes (Damsgaard and Scheepers, 1999) and the roles various actors play in the implementation of the intranet (Scheepers, 1999). The sub-codes for code number 5 were created during the coding process, as a considerable amount of information was available on the initiatives taken by the organizations that could be interpreted using concepts from an intranet stage model (Damsgaard and Scheepers, 2000) that became available during the coding process. Code number 9 was originally conceived as a way to capture the work experiences of employees from intranet developing companies, but it turned out to be used in the coding of many of the interviews in general. Many interviewees had prior work experience in other organizations that used intranets, or they participated in professional or social networks with people from other organizations to exchange intranet experiences. When coding for this category, we attempted to filter out hearsay and anecdotal observations based on the context of the statement and the use of language (vague or specific). Code number 10 was crucial in determining whether the codes had captured all vital information, and to flag the odd insightful statement that did not warrant the creation of a whole new category.

Throughout the study there were periods of reflection and interpretation of the evidence (Klein and Myers, 1999). This often meant a number of circular passes of working through the evidence in the light of extant theory, revising the initial interpretations until the final interpretation left no remaining "anomalies" in the case data (Sarker and Lee, 1999). An overview of the field study is depicted in Table 4.

Table 4. Overview of implementation strategies in the 11 case organizations

Organization		Line of business	Implementation strategy
1	User	Telecommunications	Type I, local Type II implementations
2	User	Research and development	Type I, local Type II implementations
3	User	Manufacturing	Experimented with type III, but chose a Type I implementation
4	User	Software development	Type I
5	User	Telecommunication	Type I
6	User	Public administration	Type I
7	User	Software development	Type I
8	Provider	Advertising and web	Type I, vendor of Type II
9	Provider	Consulting, web, and intranet	Type I, vendor of Type II
10	User	Software and facility management of large systems	Type I, local Type II implementations
11	Provider	Intranet in-a-box development	Type III, developer and vendor of Type III and Type IV

THE FOUR INTRANET IMPLEMENTATION STRATEGIES

Type I. Homemade Intranet (Do-It-Yourself Builder)

We found the "homemade" intranet to be the most common of the implementation strategies by far; this was true across the two countries, and across company size and line of business. If the organization chooses to develop a "homemade" intranet, the employees must have certain technical skills and expertise in areas such as web-server technology, web development tools and programming languages. The cost of getting started is often opaque or hidden for this strategy, typically due to a lack of an official budget. The amount of time spent on the project is hard to estimate, and it increases as new ideas and functionality requirements emerge. On the other hand the "homemade" intranet provides a good opportunity to tailor the intranet to specific local requirements, and the organization does not have to accept approximations in terms of adapted standard intranet solutions.

A potential problem we observed with this implementation strategy is that the internal implementers and technically skilled employees focus on their own ideas and perceptions of the organization's intranet needs and tend to lack a holistic view. A countermeasure against this problem is to perform a formalized requirements analysis for the intranet and to assemble an interdisciplinary steering committee as recommended by Damsgaard and Scheepers (2000).

In the larger organizations, we observed a variant of this problem. When the intranet becomes a manifestation of special interest groups in the organization, the result is multiple, often unauthorized intranets used by geographically, functionally or operationally separated groups of employees. The literature also has numerous examples of this phenomenon, for example in pharmaceutical companies (Bansler, Damsgaard et al., 2000; Ciborra, Braa et al., 2000; Damsgaard and Scheepers, 2000).

The homemade intranet strategy seems very inexpensive in the beginning of the implementation process, easing the decision to start the implementation. However the hidden costs from use of internal competences and resources can be high. If the organization holds the required competencies, this strategy provides the highest potential for tailoring the intranet to the organization's requirements. However due to the novelty of the technology, it is unlikely that the internal staff has experience from prior intranet implementation projects. Hence, the resulting intranet functionality does not have the benefit of being tested in other organizations and the implementers are prone to making mistakes that otherwise could have been avoided.

We recommend this implementation strategy for organizations with generic technical competences that wish to develop specific intranet/Internet technology competences, provided that the organization has ample person-hours available for the task. Organizations with non-standard requirements and work-processes may also find this strategy a good choice, as will first-generation implementers who wish to gain first-hand experience with the technology before venturing into a large-scale implementation of intranet technology.

Our findings and recommendations for intranet implementation type I are summarized in Table 5.

Table 5. Intranet implementation strategy type I – homemade intranet

Advantages:	• Inexpensive in the beginning of the implementation process • Potential high degree of tailoring to the organization
Disadvantages:	• High hidden costs of use of internal competences • Experience from similar intranets unlikely • The functionality of the intranet has not been widely tested • Often results in a poor technical design
Organizations that can benefit from choosing type I intranet implementation:	• Organizations with technical competence, that wish to develop their own intranet technology competence • Organizations with non-standard requirements and work-processes • Organizations with high IT-expertise and/or large IT-departments • First-generation implementers who wish to gain first-hand experience with the technology before venturing into a large-scale implementation

Type II. Tailor-Made Intranet Implemented by Consultants (The Craftsman)

We observed this strategy mostly in some of the larger organizations both in Denmark and South Africa. Typically this strategy was followed when the organization needed specialized skills that it did not possess in its current IT department, for example artistic or journalistic skills. This strategy allows the organization to benefit from the consultants' experiences from similar projects aiding in the requirement specification and development of an intranet tailored to the organization's specific needs. The expertise of the craftsman may "rub off" while she works in the organization, whether organized as formal training sessions or through informal interaction, making knowledge transfer a part of the product transfer process (Attewell, 1992). We recommend that organizations that do not have available human resources or in-house skills to develop an in-house solution consider outside consultants. Another reason for contacting expertise from outside contractors may be low availability or high costs of a knowledgeable workforce on the market (King, Gurbaxani et al., 1994).

We believe that a downside to this solution is the potential for lock-in, i.e., becoming too dependent on a single supplier (Shapiro and Varian, 1999b). Most of the organizations we talked to are aware of this risk and seek to maintain control of the project through such measures as ensuring that open standards are used and that the organization has ownership of the source code of the delivered applications, thus making it possible to continue the development either in-house or using different outside consultants. For large implementation projects, the organization may employ a dual sourcing strategy to keep costs down by having at least two consultant companies compete for implementation and maintenance contracts.

We found that the tailor-made intranet strategy is likely to result in a technically well-designed intranet suited for the organization's expressed requirements. Due to the one-off nature of the implementation process, this strategy is best suited for organizations that require specialized intranet applications that are not available in ready-made intranet products.

Our findings and recommendations for intranet implementation type II are summarized in Table 6.

Type III. Intranet in-a-Box Customized by In-House Personnel (The Engineer)

Many of the companies indicated that they were or had been considering buying an intranet-in-a-box and letting in-house personnel customize it (type III strategy). Especially the smaller companies saw this as a promising possibility in the future when replacing their first-generation intranet, though they were concerned with the annual license costs per employee. An intranet-in-a-box solution offers several benefits. The high level of functionality means that the organization's requirements may be met by standard functionality or by simple parameter adjustment of an intranet product, requiring little application development knowledge on the part of the organization. The organization must estimate the extent of adjustments and alterations necessary before the organization's requirements are met, and the result must be weighed against the availability of internal resources.

We observed a danger by committing to a particular intranet-in-a-box solution in that the organization will become dependent on the supplier's proprietary product. This means that the organization can become locked-in and must endure high costs if it later chooses to replace the product (Shapiro and Varian, 1999b). A good example of lock-in for a type III intranet implementation is the training of users and system administrators. Such sedimentation of knowledge (Scarbrough, 1995) about the intranet product reduces the knowledge threshold to intranet adoption and increases the likelihood of success. But brand-specific training also leads to a lock-in to the product's user interface, functions, and features (David, 1985). The investment in building up skills to use the specific intranet represents sunk costs that make a future switch to another intranet product more expensive. Shapiro and Varian (1999a) recommend that this fact be used as bargaining power in the contract negotiations with vendors, since the future switching costs of the customer represents an immediate value to the vendor.

This implementation strategy lends itself to organizations with organizational processes similar to the standard processes supported by the intranet product, provided that the organization has sufficient resources of skilled personnel to conduct the customization and implementation in-house. Most organizations with a nominal complement of IT staff would fall in this category. Because of the users-based pricing model currently associated with most intranet-in-a-box products, this intranet strategy can be quite expensive for larger organizations with many users.

Our findings and recommendations for intranet implementation type III are summarized in Table 7.

Table 6. Intranet implementation strategy type II – tailor-made intranet implemented by consultants

Advantages:	• Little internal expertise is required • Benefits from the consultants' experience and expertise • Likely to be technically well designed • An opportunity to improve the internal competence through education
Disadvantages:	• A risk of becoming too dependent on one supplier • Often an expensive solution
Organizations that can benefit from choosing type II intranet implementation:	• Organizations with insufficient technical competence • Organizations with non-standard requirements • Organizations with no time to develop internal expertise

Table 7. Intranet implementation strategy type III – intranet-in-a-box customized by in-house personnel

Advantages:	• Standard, thoroughly tested functionality • A chance of sedimentation of knowledge (Scarbrough 1995) about the product and its implementation
Disadvantages:	• Not tailored to specific requirements • A risk of becoming too dependent on the vendor product • Can be a relatively expensive solution for large organizations due to licenses • Some internal expertise is required for the implementation process
Organizations that can benefit from implement choosing type III intranet implementation:	• Organizations with well-defined requirements • Organizations with sufficient internal expertise to install and the intranet product • Organizations with standard requirements and work processes that are modeled in the intranet products

Type IV. Intranet In-a-Box Customized by Outside Contractor (The Traveling Fitter)

We observed an intranet technology vendor tailoring its intranet product and expertise to address the needs of customer organizations that purchase both a standard intranet product and outside expertise to perform the installation and implementation. Not surprisingly, for the customer organization this implementation strategy combines many of the advantages and disadvantages from the "tailor-made intranet" (type II) and "intranet-in-a-box" (type III). Due to the architecture choice, the intranet product will be based on thoroughly tested advanced applications, and due to the choice of implementation sourcing, the implementation effort will benefit from the experience of the consultants. This implementation strategy is thus a quick route to advanced, thoroughly tested intranet functionality customized for the organization's requirements.

The consultants can deliver services ranging from minor alterations to the standard intranet product to fairly complex development of new functionality for the intranet. By using this implementation strategy, the organization can adjust its intranet to more demanding requirements without bearing the burden of the required learning. The ultimate manifestation of a type IV implementation strategy is to outsource the maintenance and facility management of the resulting intranet as well, thus further reducing the demands on in-house intranet technology competences. The intranet server can physically be placed at the organization's premises or at a vendor's location, i.e., an application service provider (ASP) setup.

The level of functionality and the speed of implementation make this implementation strategy seem very useful and accessible to many organizations, especially organizations with low levels of technical competence. The strategy can however be very expensive, especially for large organizations with many users. In addition, it holds the double danger of lock-in both to the product and to the consultant company (which may be the same or maintain close ties for mutual benefit) (Shapiro and Varian, 1999a). The intranet-in-a-box provider in our study reported that customers entered into this commitment with open eyes, since for the sake of continuity they would rather be dependent on commercial companies than on the training and retention of in-house technical staff.

Our findings and recommendations for intranet implementation type IV are summarized in Table 8.

Table 8. Intranet implementation strategy type IV – intranet-in-a-box customized by outside contractor

Advantages:	• Advanced standard functionality can be achieved quickly • Thoroughly tested functionality • A chance to either increase the internal competence (professionalism of workers) through education (Scarbrough 1995) or alternatively to outsource all the technical competence • Benefits from the consultants' experience with similar projects • Likely to be technically well designed and implemented
Disadvantages:	• A risk of becoming doubly locked-in to the product and to the supplier (Shapiro and Varian, 1999b) • Can be an expensive solution due to initial purchase costs, product licenses, and consultant fees
Organizations that can benefit from choosing type IV intranet implementation:	• Organizations with demanding requirements regarding quality and functionality • Organizations with low levels of technical competence

DISCUSSION AND RECOMMENDATIONS

For the majority of the case organizations we studied, the "homemade intranet" was espoused as the de facto implementation strategy (see Table 4). We believe this to be a result of several factors. First, the high-tech line of business of most of our case organizations lends itself well to homemade intranets. Second, standard-intranets have only recently become available while the user organizations in our investigation were indeed chosen because they have had intranets for several years. As intranets lose their novelty and the intranet users' functionality expectations increase, we expect that the choice of intranet implementation strategy will become more explicit and deliberate, leading to greater diversity in implementation processes actually applied. Anecdotal evidence suggests that this has not happened yet (Chu, 2002).

The ubiquitous nature (and loose coupling between child intranets) associated with intranet technology means that it is quite feasible for different strategies to be combined within the same organization. Compared with much of the extant literature on information technology implementation, this is an interesting observation. We found such a combination of strategies in four of the case organizations. We attribute this finding to the fact that in global (and even decentralized pockets of the same regional organization), different social contexts apply and hence different intranet strategies may be (and possibly should be) pursued concurrently (cognizant of the advantages/disadvantages as outlined). Although there may be a central espoused approach along the lines of one of the four strategies, due to the ubiquity of intranet technology and the flexibility in meaning attributed to "the" intranet by the various decentralized intranet role players, different strategies can co-exist at different unit levels. One base cause for a multitude of intranets in a single organization is mergers between existing entities, each of which bring in their existing intranet. Other causes are "drift" from the official, espoused strategy as the technology is used in the organization (Cordella & Simon, 1998; Ciborra, Braa et al., 2000), or the simple fact that the enterprise information resources themselves become fragmented (McMahon, 2000). These developments are especially likely in larger, geographically dispersed organizations.

We thus position the four implementation strategies as archetypes. Organizations could use these as a basis to formulate either a dominant or a combined strategy, dependent on contextual considerations. We argue that larger and geographically dispersed organizations would need to factor in a variety of contextual considerations such as available internal technical expertise, core competence, the maturity of the technology and cultural climate in their choice of strategy/strategies.

Some of the intranet implementation projects we encountered were built using the existing information infrastructures. In other cases, however, the intranet projects were implemented in part for reasons other than the intranet itself. In several cases the intranet implementation served to force through infrastructure changes such as unification of network protocols from a heterogeneous environment to a consolidated TCP/IP environment or elimination of legacy applications, or the intranet served as a unifying project to roll out technology training and awareness to all employees. Both of these findings have been reaffirmed in other intranet implementation projects unrelated to our study (see e.g., Zmud & Sambamurthy, 1997; Bhattacherjee, 1998; Cope, 2001).

We did not notice much difference between the large South African companies and the large Danish companies. They all followed the homemade strategy with some assistance of external consultants in some critical areas. One observation worth noticing in the South African companies was the boycott of IT technology during apartheid. We therefore expected that the South African companies had to put a lot of effort in catching up to, for example, Danish companies. To our surprise we found the opposite to be true. The Danish companies had to abandon Lotus Notes in favor of an intranet, whereas the South African companies did not have a large installed base of existing systems and support structures that blocked the way of the intranet. The lower knowledge barrier of the intranet compared to older IT technologies also favors a speedy adoption of intranet technology. This leads us to propose that for many organizations it is more demanding to manage an existing information infrastructure than it is to build a new one from scratch.

Large and small companies initially all followed the homemade strategy, but for different reasons. The small companies follow the homemade strategy because it poses the smallest initial demand on local resources, whereas the large companies follow the homemade strategy because they believe they have the necessary skills and resources to successfully implement a homemade intranet. We expect that the paths of small and large companies will separate later due to the following rationale. All evidence shows that the size and complexity of the intranets grow over time, increasing the resources required for maintenance efforts. In order to curb the complexity and preserve uniformity despite local initiatives, most organizations initiate major upgrade or replacement projects for their intranet. Scheepers et al. (2002) support the observation that periodic intranet replacement/re-launch projects are quite common. The small companies hope for the intranet technology to become a commoditized part of an office suite as the technology matures and the market for intranets grows. In the large companies the continued maintenance of the intranet, including development of applications, will likely require a staff of developers, and thus the costs of maintaining the development in-house will be visible. At some point the intranet will lose its status as a novel technology that is perceived to be easily manageable. Instead the intranet will be viewed as an information system just like any other in the organization's information systems portfolio. The organization will then reconsider whether maintaining intranet development capabilities in-house serves the company best, or if an intranet product should be purchased from an outside vendor. The wide diffusion of intranet technologies has decreased or eliminated the competitive advantage of having intranet implementation knowledge in-house. We suspect that the larger organizations will therefore gravitate towards type III intranet implementations, as they will prefer ready-made intranet suites and utilize in-house expertise to adapt the intranet to the local requirements, and to handle the continued maintenance tasks of the intranet.

An observation that is worth noticing is that the providers we interviewed all explained that their intranet platforms were built upon open standards to satisfy their customers. However the open standards only guarantee that data *in theory* can be extracted from the intranet in its simplest text form, whereby the relationship between data elements will be lost. In reality it may take considerable (down)time and excessive efforts to move the contents from one supposedly "open platform" to another. In further support of this observation, none of the providers we interviewed reported about any customers that had successfully switched from their platform to a competing platform. We also did not encounter any of the user companies expressing the need for an ASP setup. The user companies unanimously stated that the data should be kept on the organizations' own servers. The strong link between the application and the data suggests that companies may think that they control the data, while in actual fact they do not. We therefore caution that a relationship with an intranet-in-a-box provider is "more like a marriage and less like a date."

In terms of cultural differences between the companies, our case studies show that some of the organizations that had the best prerequisites in terms of skills and expertise developed intranets with a poor technical design. This was quite puzzling and at first contradictory to what we expected. A plausible explanation is that the main factor was a relatively low knowledge threshold, which led the implementers to believe they were able to develop an intranet solution without careful analysis and planning. As a consequence, the actual implementation of the intranet often occurred in spare time between scheduled tasks, which led to a poor overall technical design and project management.

Technologies offer limited windows of opportunity where different role players can influence the technology and profit from it (Perez & Soete, 1988). For intranets the time of innovation has long passed, but the technology has not yet become a commodity, and the need for mediating companies has not yet disappeared (Attewell, 1992). Indeed presently, intranet technology appears deceptively simple. However the required expertise (in technology and usability design), implementation experience and the actual person-hours required for a successful intranet implementation are substantial. We therefore recommend that user organizations, which currently face a choice of implementation processes, avoid homemade intranet implementations unless the organization has already built up extensive, successful experience through earlier homemade intranet implementations.

We believe that the commodification of intranets will occur as the evolution of intranet technology continues, resulting in the advent of large, commercial intranet products analogous to the commercial office software suites and Enterprise Resource Planning systems. If these products are complex (similar to ERP systems), their customization to user organizations will be a thriving enterprise for implementation consultants and specialists, whereas the knowledge threshold for user organizations will be all but eliminated if the products get commodified to something similar to office application suites.

Software houses that produce intranet in-a-box solutions have until recently enjoyed a market with only very few participants on the supplier side. This situation has changed, and competition has increased. We believe that conventional market strategies can readily be extended to apply to the intranet market, and accordingly the suppliers should diversify their products with regard to product features and market segments. This way they can seek to avoid commodification and instead achieve customer lock-in (Shapiro & Varian, 1999b).

CONCLUSION

In this paper, we have developed a framework that distinguishes four different intranet implementation strategies. The taxonomy encompasses the architecture of the intranet (standard product or custom built) and the personnel implementing the intranet into the organization (in-house or outsourced). The framework consequently depicts four types of intranet implementation strategies: 1) homemade intranet, 2) tailor-made intranet implemented by consultants, 3) intranet-in-a-box customized by in-house personnel, and 4) intranet-in-a-box customized by outside contractor. In a multiple explorative case study of 11 organizations supported by a literature review, we have found the homemade intranet to be by far the most common.

On a practical level the four intranet implementation strategies have different qualities that make them suited and attractive for different types of organizations. By using our framework, organizations can readily consider which implementation process or strategy is better suited to their situation. Organizations may benefit from alternating between the implementation strategies, and even by pursuing different strategies concurrently in accordance with the available internal technical expertise, core competence, maturity of the technology and contextual considerations. However, we warn organizations of the risk of lock-in inherent in all strategies except the homemade intranet.

We hope to extend the results of this investigation to other areas of emergent Internet technologies. Most modern organizations will be faced with a number of similar choices in the near future as more new communication technologies become available and popular. For example organizations will be faced with sourcing decisions for implementation of the organization's extranet, WAP and/or i-mode services, as well as the question of timing, i.e., if these applications should be bought on the market only when the technology is sufficiently mature or if they should be implemented early in order to achieve a (temporary) competitive advantage over non-adopters.

ACKNOWLEDGMENTS

Thanks to the anonymous referees for many useful comments on earlier versions of the paper. We thank the participating organizations and the interviewees for their time and interest in our research. Thanks also to Jakob Sørensen for participating in some of the initial interviews and to Carsten Olsen who was a co-author of an earlier version of this article. This research was carried out as part of the PITNIT project. PITNIT is supported by the Danish Research Agency, grant number 9900102.

REFERENCES

Attewell, P. (1992). Technology diffusion and organizational learning: The case of business computing. *Organization Science, 3*(February 1), 1-19.

Bansler, J. P., J. Damsgaard, et al. (2000). Corporate intranet implementation: managing emergent technologies and organizational practices. *Journal of the Association for Information Systems (JAIS),* 1, Article 10.

Bhattacherjee, A. (1997). US West Global Village. In T. Goles & R. Hirschheim (Eds.), *Intranets: The next IS solution* (pp. 196-217). Houston, TX: University of Houston.

Bhattacherjee, A. (1998). Management of emerging technologies: Experiences and lessons learned at US West. *Information and Management, 33*(5), 263-272.

Chu, S. (2002). Intranets become 'intramess.' *The Globe and Mail,* B20.

Ciborra, C. U. (1996). Introduction: What does groupware mean for organizations hosting it? In C.U. Ciborra (Ed.), *Groupware & teamwork* (pp. 1-19). John Wiley & Sons.

Ciborra, C. U., Braa, K., et al. (2000). *From control to drift - The dynamics of corporate information infrastructures*. New York: Oxford University Press.

Ciborra, C. U., & Hanseth, O. (1998). From tool to gestell: Agendas for managing the information infrastructure. *Information, Technology & People, 11*(4), 305-327.

Coleman, D. (1997). Collaboration on the Internet and intranets. *Proceedings of the 30th Hawaiian International Conference on Systems Sciences.*

Cope, J. (2001). *New Navy-Marine Intranet Aims to Cut Legacy Apps.* Available at: *http://www.computerworld.com/storyba/0,4125,NAV47_ STO 61252,00.html*

Cordella, A., & Simon, K.A. (1998). Infrastructure deployment: Global intent vs. local adoption. *Proceedings of IFIP 8.7*, Helsinki, Finland.

Dahlbom, B., & Mathiassen, L. (1993). *Computers in context.* Cambridge, MA: Blackwell Publishers.

Damsgaard, J., & Scheepers, R. (1999). Power, influence and intranet implementation: A safari of South African organizations. *Information, Technology & People, 12*(4), 333-358.

Damsgaard, J., & Scheepers, R. (2000). Managing the crises in intranet implementation: A stage model. *Information Systems Journal, 10*(2), 131-149.

David, P. A. (1985). Clio and the economics of QWERTY. *The American Economic Review, 75*(2), 332-337.

Dennis, A. R. (1996). Information exchange and use in group decision making: You can lead a group to information, but you can't make it think. *MIS Quarterly,* 433-455.

Eisenhardt, K. M. (1989). Building theories from case study research. *Academy of Management Review, 14*(4), 532-550.

George, J. F. (2000). The origins of software: Acquiring systems at the end of the century. In R. Zmud (Ed.), *Framing the domains of IT management: Projecting the future through the past* (pp. 263-284). Cincinnati, OH: Pinnaflex Educational Resources.

Gibson, C. F., & Nolan, R.L. (1974). Managing the four stages of EDP growth. *Harvard Business Review,* 76-88.

Grudin, J., & Palen, L. (1995). Why groupware succeeds: Discretion or mandate? *Proceedings of IRIS18, Gothenburg Studies in Informatics.*

Hanseth, O., & Braa, K. (1999). Hunting for the treasure at the end of the rainbow: Standardizing corporate IT infrastructure. *IFIP TC8 WG 8.2 - New Information Technologies in Organizational Processes: Field Studies and Theoretical Reflections on the Future of Work.* St. Louis, MO: Kluwer Academic Publishers.

Hirschheim, R. A. (1986). The effect of a priori views on the social implications of computing: The case of office automation. *Computing Surveys, 18*(2), 165-195.

Hughes, T. (1987). The evolution of large technological systems. In W. Bijker, T. Hughes, & T. Pinch (Ed.), *The social construction of technological systems* (pp. 51-82). Cambridge, MA: The MIT Press.

Jarvenpaa, S. L., & Ives, B. (1996). Introducing transformational information technologies: The case of the World Wide Web technology. *International Journal of Electronic Commerce, 1*(1), 95-126.

Karsten, H. (1996). *Interactions with collaborative technology: Lotus Notes in a networked organization.* University of Jyväskylä.

King, J., Gurbaxani, V., et al. (1994). Institutional factors in information technology innovation. *Information Systems Research, 5*(2), 139-169.

Klein, H. K., & Myers, M.D. (1999). A set of principles for conducting and evaluating interpretive field studies in information systems. *MIS Quarterly, 23*(1), 67-92.

Kling, R. (1991). Computerization and social transformations. *Science, Technology & Human Values, 6*(3), 342-367.

Kwon, T. H., & Zmud, R. W. (1987). Unifying the fragmented models of information systems implementation. In R.J. Boland & R.A. Hirschheim (Ed.), *Critical issues in information systems research* (pp. 227-251). New York: John Wiley & Sons.

Lacity, M. C., Willcocks, L.P., et al. (1996). The value of selective IT sourcing. *Sloan Management Review, 37*(3), 13-25.

Lamb, R., & Davidson, E. (2000). The new computing archipelago: Intranet islands of practice. *Proceedings of the IFIPWG8.2 Working Conference on Information Technology and Changes in Organizational Work.*

Leavitt, H. J. (1964). Applied organizational change in industry: Structural, technical and human approaches. In W. W. Cooper, H. J. Leavitt, & M. W. I. Shelly (Ed.), *New perspectives in organization research* (pp. 55-71). New York: John Wiley & Sons.

Lyytinen, K., Rose, G., et al. (1998). The brave new world of development in the internetwork computing architecture (InterNCA): Or how distributed computing platforms will change systems development. *Information Systems Journal,* 8, 241-253.

Markus, M. L. (1983). Power, politics, and MIS implementation. *Communications of the ACM, 26*(6), 430-444.

Markus, M. L. (1987). Toward a "critical mass" theory of interactive media, universal access, interdependence and diffusion. *Communication Research, 14*(5), 491-511.

Markus, M. L. (1994). If we build it, they will come: Designing information systems that people want to use. *Sloan Management Review, 35*(4), 11-25.

Markus, M. L. (2000). Paradigm shifts - E-business and business/systems integration. *Communication of the Association for Information Systems,* 4.

McMahon, T. (2000). Outsourcing the portal: Another branch in the decision tree. *Information Outlook, 4*(1), 23-26.

Mintzberg, H. (1983). *Structure in fives - Designing effective organizations.* Prentice-Hall.

Moeller, M. (1996). *Boeing network takes flight with pioneering intranet project.* Available online: *http://zdnet.com/pcweek/news/0219/tboing. html*

Monteiro, E., & Hanseth, O. (1995). Social shaping of information infrastructure: On being specific about the technology. *Proceedings of the IFIPWG8.2 Working Conference on Information Technology and Changes in Organizational Work,* Chapman & Hall.

Newell, S., Swan, J., et al. (1999). The intranet as a knowledge management tool? Creating new electronic fences. *Information Resources Management Association International Conference, Managing Information Technology Resources in Organizations in the Next Millennium,* Hershey PA.

Nolan, R. L. (1973). Managing the computer resource: A stage hypothesis. *Communications of the ACM, 16*(7), 399-405.

Nolan, R. L. (1979). Managing the crisis in data processing. *Harvard Business Review,* 115-126.

Orenstein, D. (1998). *Fighting Intranet Flak.* Available online at: *http://www.computerworld.com/cwi/story/0,1199,NAV47_STO34515,00.htm*

Orlikowski, W. J. (1992a). The duality of technology: Rethinking the concept of technology in organizations. *Organization Science, 3*(3), 398-427.

Orlikowski, W. J. (1992b). *Learning from notes: Organizational issues in groupware implementation.* Conference Proceedings on Computer-Supported Cooperative Work, Toronto, Canada.

Orlikowski, W. J., & Gash, D.C. (1994). Technological frames: Making sense of information technology in organizations. *ACM Transactions on Information Systems, 12*(2), 174-207.

Orlikowski, W. J., Yates, J., et al. (1995). Shaping electronic communication: the metastructuring of technology in the context of use. *Organization Science, 6*(4), 423-444.

Perez, C., & Soete, L. (1988). Catching up in technology: Entry barriers and windows of opportunity. In G. Dosi, C. Freeman, R. Nelson, G. Silverberg, & L. Soete (Eds.), *Technical change and economic theory* (pp. 458-479) London: Pinter Publisher.

Porter, M. E. (1985). *Competitive advantage: Creating and sustaining superior performance.* New York: Free Press.

Rands, T. (1993). A framework for managing software make or buy. *European Journal of Information Systems, 2*(4), 273-282.

Romm, C. T., & Wong, J. (1998). The dynamics of establishing organizational web sites: Some puzzling findings. *Australian Journal of Information Systems, 5*(2), 60-68.

Sarker, S., & Lee, A. S. (1999). A hermeneutic interpretation of the effect of computerized BPR tools on redesign effectiveness in two organizations. *Proceedings of the IFIP TC8 WG 8.2, New Information Technologies in Organizational Processes: Field Studies and Theoretical Reflections on the Future of Work.* St. Louis, MO: Kluwer Academic.

Scarbrough, H. (1995). Blackboxes, hostages and prisoners. *Organization Studies,* 991-1019.

Scheepers, R. (1999). *Intranet implementation: Influences, challenges and role players.* Department of Computer Science. Aalborg, Denmark: Aalborg University.

Scheepers, R., Gruhn, V., et al. (2002). Intranet development processes in large organizations: characteristics and management implications. *Proceedings of the Information Resources Management Association International Conference,* Seattle, WA.

Schein, E. H. (1989). *Organizational culture and leadership.* San Francisco, CA: Jossey-Bass Publishers.

Shapiro, C., & Varian, H. R. (1999a). The art of standards wars. *California Management Review, 41*(2), 8-32.

Shapiro, C., & Varian, H. R. (1999b). *Information rules: A strategic guide to the network economy.* Boston: Harvard Business School Press.

Silverman, D. (1993). *Interpreting qualitative data - Methods for analysing talk, text and interaction.* London: SAGE.

Sliwa, C. (2000). Intranet users tackle chaos. Available online at: *http://www.computerworld.com/cwi/story/0,1199,NAV47_STO47442,00.html*

Srinivasan, A., & Davis, J.G. (1987). A reassessment of implementation process models. *Interfaces, 17*(3), 64-71.

Star, S. L., & Gerson, E.M. (1987). The management and dynamics of anomalies in scientific work. *Sociological Quarterly,* 28, 147-169.

Tornatzky, L. G., & Klein, K. J. (1982). Innovation characteristics and innovation adoption-implementation: a meta-analysis of findings. *IEEE Transactions on Engineering Management, 29*(1), 28-45.

Walsham, G. (1995). Interpretive case studies in IS research: Nature and method. *European Journal of Information Systems.*

Williams, R., & Edge, D. (1996). The social shaping of technology. *Research Policy,* 25, 865-899.

Zmud, R. W. (1984). An examination of "Push-Pull" theory applied to process innovation in knowledge work. *Management Science, 30*(6), 727-738.

Zmud, R. W., & Sambamurthy, V. (1997). Justifying visionary uses of information technology. *Journal of Strategic Performance Measurement, 1*(6), 30-42.

APPENDIX A

Interview Agenda

Below is the abridged guide we used for the interviews. The interviewees were given an abbreviated version of the interview agenda well in advance of the interview. The full version of the interview guide is available upon request from the authors.

(a) Information about the interviewee: Name, age, title, organization, department, and contact information (phone and e-mail)
(b) Information about the interview: Date, place, duration, and interviewer
(c) The intranet in the organization: Purpose: To collect information about the current organizational intranet and the interviewee's use of the intranet
(d) The intranet in the past: Purpose: To collect information about the inception of the organizational intranet and the interviewee's first use of the intranet
(e) The intranet in the future: Purpose: To collect information about the perception of the organizational intranet in the future and the interviewee's vision of intranet use
(f) Organizational structure: Purpose: To identify a number of characteristics of the organizational structure and to determine the function and role of the interviewee in the organization
(g) Organizational processes: Purpose: To gather the interviewee's views and opinions on the processes which exist in the organization, i.e., the workflows which lead to the organization's products
(h) Organizational culture: Purpose: To gather information about the organizational culture, values, and basic assumptions on which this culture is based

Previously published in the Journal of Global Information Management, 11(3), 39-62, July - September 2003.

Chapter XII

Cultural Adaptation on the Web: A Study of American Companies' Domestic and Chinese Websites

Nitish Singh, California State University, Chico, USA

Hongxin Zhao, Saint Louis University, USA

Xiaorui Hu, Saint Louis University, USA

ABSTRACT

In the academic literature and the business press, there seems to be a lack of guidance and lack of cross-cultural models to support companies localization strategies on the Web. To address this deficit in literature and to provide marketers and Web designers with insights into website localization, this paper conducted a comparative analysis of the U.S. based international companies' domestic websites and their Chinese websites. A framework to measure cultural adaptation on the Web is presented. Forty U.S.-based Fortune 500 companies are surveyed to investigate the cultural adaptation of their Chinese websites. Content analysis of the 80 U.S. domestic and Chinese websites reveals that the web is not a culturally neutral medium, but it is full of cultural markers that give country-specific websites a look and feel unique to the local culture.

INTRODUCTION

The year 2001 was significant in the history of e-commerce. Despite the economic slow down and the present vulnerability of dot com ventures, online sales revenues worldwide are expected to have reached $550 billion, a 92% increase from 2000 (emarketer, 2001). It is expected that by the year 2004, global e-commerce will generate $3.2 trillion in revenues (e-marketer, 2001). This rapid growth in global e-commerce indicates significant business opportunities abroad. Any delay in launching international websites might prove to be a fatal mistake (Alvarez et al., 1998).

Simon (1999, 2001) emphasizes the need for international websites, as e-commerce expands into Asia, South America, and eventually into Africa. According to Lynch and Beck (2001), U.S. companies lacking an international Web presence are only reaching 5% of the world's population, and less than 25% of the world's purchasing power. However, in order to reach global online consumers, companies have to address a plethora of international issues, including Web page translation, logistics, legal hurdles, and most importantly the cultural and linguistic issue.

However, Web marketers often lack expertise and guidance to adapt their websites to diverse consumer segments. Forrester Research found that 67% of the Fortune 100 companies' websites are standardized and mainly in English. A survey of the top 900 companies' websites listed by Forbes revealed that only 225 companies have localized websites (Singh and Boughton, 2002). It is apparent that there is a need for Web marketers and Web designers to devote their attention to local adaptation in order to better serve local customers. The objective of this study is to examine cultural adaptation on the Web and provide insights to Web marketers on developing culturally adapted websites. Specifically, we applied a modified framework of cultural values based on Hofstede's (1980) four cultural dimensions and Hall's (1976) context dimension to evaluate the degree of cultural adaptation of the selected company websites. The selected company websites are the American companies' domestic websites and their Chinese websites.

The chapter is organized as follows. The next section explains the reason why this study focuses on Chinese cultural adaptation. The following section proposes an analytical framework to measure cultural values on the Web and highlights four hypotheses. The methodology used in this study is described and the results are reported, which is followed by conclusion and managerial implications.

WHY STUDY THE CHINESE CULTURAL ADAPTATION

There are several reasons for choosing China. First, China's online population is 12 million, and it is expected to reach 33 million by the end of 2003, accounting for one-third of the Internet users in Asia (Zhao, 2002). In addition, studies by international market researchers reveal that China generated in excess of $11 billion in online trading in 2001 (Cyberatlas.com, 2001).

Second, the U.S. and China represent two distinct cultures. The U.S. has commonly been used as a 'reference frame' of western culture (Cheng and Schweitzer, 1996; Mueller, 1987) and China represents the epitome of eastern culture (Ralston et al., 1992; Hofstede, 1993, 1998). Third, the U.S. and Chinese cultural values have been shown to be empirically different (Hofstede, 1993). These facts provide us with a good basis to analyze differences in website design between these two distinct cultures.

Finally, previous cross-cultural studies between the U.S. and China have shown significant differences in cultural values in contexts like organizations (Ralston et al.,

1992), print ads (Tse et al., 1989), and T.V. commercials (Cheng and Schweitzer, 1996). Our paper will enrich the existing study of cultural values by extending it to a new marketing medium, the World Wide Web.

ANALYTICAL FRAMEWORK FOR EVALUATING CULTURAL VALUES ON THE WEB

Recent years have witnessed an increasing number of studies in information systems (IS) that have focused on examining the role of national culture in the adoption and successful deployment of information technology in global setting (Myers and Tan, 2002). In a comprehensive review of the related literature, Morrden (1999) classifies the cultural models into three types in terms of dimensionality: single, multiple and historical-social. Studies that fall into the single-dimension model category often contain dyadic characterization of culture. National cultures were classified as high versus low context (Hall, 1960, 1976), monochromic versus polychronic (Lewis, 1992), high versus low trust (Fukuyama, 1995), ideocentric versus allocentri (Triandis, 1995) and monochromic versus polymorphic (Bottger, Hallein, and Yetton, 1995). The multiple dimension models, on the other hand, tend to abstract cultural elements into more cultural facets. The typical two-dimension categorizations include pragmatism-idealism and rationalism-humanism (Lessem and Neubauer, 1994) and free will/determinism and accumulation of wealth/"just enough" (Kluckhohn and Strodbeck, 1961). The cultural models that possess more than two dimensions include Newman, Summer, & Warren's (1977) five-dimension, Hampden-Turner & Trompenaars' (1994) seven-dimension and Hofstede's (1980) four-dimension models. Historical-social models classify culture at the regional level rooted in the shared social heritage and similarity of historical evolution.

Given the large number of cultural models presented in the existing literature, we choose Hofstede's (1980, 1991) four-dimension and Hall's (1960, 1970) two-dimension models as our basic cultural value framework to categorize websites. There are two reasons why we adopt these two models. First, while we recognize the shortcomings of Hofstede's model (Myers & Tan, 2002), this model is the one that has been most extensively applied and validated in a variety of cultural context and management research in general. We believe that this model that is repeatedly used and tested shows a higher degree of validity and legitimacy as compared with other alternative models. Second, the parsimony of both models also presents themselves as a more viable framework since both offer analytical flexibility which is desired in measuring culture (Pollay, 1983).

Despite criticism of Hofstede's cultural typology regarding methodology and context (Fernandez et al., 1997; Huang, 1995; Myers and Tan, 2002), it can be argued that Hofstede's typology is appropriate for several reasons. First, Hofstede's cultural typology has been extensively replicated and supported as an important part of cultural theory. Second, according to Clark (1990), there seems to be an overlap among different typologies of culture, and their dimensions correspond to Hofstede's typology. Third, Hofstede's framework has been found to be a valid basis for the analysis of regional differences, and as a means through which Web marketers could adapt their websites to local cultures (Simon, 1999). Finally, Hofstede's dimensions have been empirically developed and verified, while cultural typologies by Trompennars (1994), Hall (1976), and Kluckhohn and Strodtbeck (1961) have yet to be empirically validated.

To extend Hofstede's framework (1980) and Hall's dimension (1976) to study the specifics of Web communication, we took two steps to generate Web-specific cultural elements. The first step is to identify the existing cultural typologies that have been used in cultural studies. For this purpose, an extensive review of major cultural typologies in the business literature was conducted.

- Cultural values proposed by Hofstede (1980).
- A set of 42 cultural value appeals in advertising developed by Pollay (1983).
- Trompenaars (1994) has proposed cultural values like- Universalistic vs. Particularistic Orientation, Affective vs. Neutral Orientation, Specific vs. Diffuse Cultures, Achievement vs. Ascription Orientation, Time Orientations, and Relationship with Nature.
- Other cultural values include High Context and Low Context cultures (Hall, 1976), and the six cultural value dimensions proposed by Kluckhohn and Strodtbeck (1961).

In addition, the literature pertaining to the use of these cultural typologies was reviewed to see how cultural value dimensions have been operationalized and empirically tested (Alibers-Miller and Gelb, 1996; Cheng and Schewitzer, 1996; Borden, 1991; Fock, 2000; Gregory and Munch, 1997; Gundykunst, 1998; Hang and Shavitt, 1994; Ju-Pak, 1999; McCarty and Hattwick, 1992; Mueller, 1987; Tansey, Hyman and Zinkhan, 1990; Triandis, 1982; Tse, Belk, and Zhou, 1989; Zandpour et al., 1994). Finally, all the major cultural typologies were closely evaluated in the context of Web communication to identify Web-related cultural traits. A random sample of 50 Fortune 100 companies selected from www.fortune.com was utilized to investigate how well the cultural typologies apply to the websites. Each of the cultural dimensions from various different typologies was individually tested to see how well it can explain the web features and how different Web features can be developed or used to accentuate that cultural dimension in Web communication. For example, Achievement vs. Ascription Orientation was carefully analyzed in the context of Web communication, but it was found to overlap with the power-distance dimension of Hofstede; use of titles, status-orientation, and emphasis on Hierarchy were common features in both ascription-oriented and high power distance country websites (Japan and India). Thus, based on exclusivity of cultural dimensions, four dimensions of Hofstede and the context dimension of Hall were included in the study.

The second step was to develop the website cultural-coding categories for six cultural dimensions. To operationalize Hofstede's and Hall's cultural dimensions on the Web, we first developed a list of major interactive or multimedia features commonly presented on the websites (clubs, newsletters, FAQs, security policy, privacy policy, free stuff, downloads, graphics, hyperlinks, and others), and evaluated which feature would be preferred more in which culture. To help this conceptualization, the work of Albers-Miller and Gelb (1996) was consulted, as they have empirically tested which of the 42 cultural values appeals by Pollay (1983) are reflective of each of Hofstede's four

Table 1. Cultural categories

DIMENSION: COLLECTIVISM *CATEGORIES:* • Community relations • Clubs or chat rooms • Newsletter • Family theme • Country specific news • Symbols and pictures of national identity • Loyalty programs • Links to local websites **DIMENSION: UNCERTAINTY AVD** *CATEGORIES:* • Customer service • Secure payment • Guided navigation • Tradition theme • Local stores • Local terminology • Free trials or downloads • Customer testimonials • Toll-free numbers **DIMENSION: POWER DISTANCE** *CATEGORIES:* • Company hierarchy information • Pictures of CEOs	• Quality information and awards • Vision statement • Pride of ownership appeal • Proper titles **DIMENSION: MASCULINITY** *CATEGORIES:* • Quizzes and games • Realism theme • Product effectiveness • Clear gender roles **DIMENSION: HIGH CONTEXT** *CATEGORIES:* • Politeness • Soft sell approach • Aesthetic **DIMENSION: LOW CONTEXT** *CATEGORIES:* • Hard sell approach • Ranks or prestige of the company • Explicit comparisons • Use of superlatives • Terms and conditions of use

dimensions. Meanwhile, the works of Borden (1991) and Gudykunst (1998) were consulted to understand what societal norms are exhibited by the cultures studied by Hofstede (1980). Finally, other studies that have used and operationalized Hofstede's typology and Hall's context dimension were also evaluated (Cheng and Schweitzer, 1996; Cho et al., 1999; McCarty and Hattwick, 1992; Mueller, 1987; Zandpour et al. 1994). Based on this extensive review, Web-specific cultural traits were categorized into Hofstede' four main cultural dimensions and Hall's context dimension. The categories are presented in Table 1 and justification for including certain Web elements under each cultural dimension is elaborated in Table 2.

To test the reliability of the cultural categories, four doctoral students in a U.S. business school were asked to assign a random list of category items under the cultural dimension they best represented. A total inter-judge reliability of 85% was achieved. This procedure was repeated once more after a year and the inter-judge reliability was 88%.

Table 2. Explanation for cultural categories

Collectivism

- *Community Relations, Symbols of group identity and Newsletters:* This category was included because collectivist societies place emphasis on community-based social order (Hofstede, 1991). Furthermore, collectivist societies pay more emphasis on group well being (Cho et al., 1997) and preserving the welfare of others (Gudykunst, 1998).
- *Family theme and Clubs or Chat rooms:* In collectivist societies, there is emotional dependence of individuals on organizations and the society (Hofstede, 1980), thus people need forums, places or clubs where they can share their concerns, views, and emotions. Pollay (1983) emphasizes the nurturance theme, wherein members of a group provide each other with support and sympathy.
- *Loyalty programs and links to local websites:* Building lasting relationships and engendering a sense of loyalty are important to people in collectivist cultures (Trompenaars, 1994). Having links to local websites shows that the company is well connected in the country, and has legitimate relationships with local companies.

Uncertainty Avoidance

- *Customer Service and Navigation:* Uncertainty avoidance cultures show higher levels of anxiety (Gudykunst, 1998) and avoid ambiguous situations (Hofstede, 1980). Thus, people in such cultures value advice and help from experts in the field.
- *Tradition theme:* High uncertainty avoidance societies are "tight societies" that value conservatism and traditional beliefs (Hofstede, 1980).
- *Local Stores and local terminolgy:* Depicting local store locations on websites helps to reduce anxiety by letting people know the store locations nearest to their city. Since high uncertainty avoidance cultures are rooted in traditions and value ritual behavior (Hofstede, 1980), local metaphors, puns and idioms are widely used.
- *Free Trials and Downloads:* People in high uncertainty avoidance cultures are less risk-taking. For them, "what is different is dangerous" (Hofstede, 1991, p. 119). Thus, free trials help reduce purchase uncertainty and facilitate purchase.

Power Distance

- *Hierarchy Information and Pictures of Important People with titles:* High power distance societies emphasize hierarchical structures and stress coercive and referent power (Gudykunst, 1998; Hofstede, 1980).
- *Quality Assurance and Awards and vision statements:* As people in high power distance societies value referent power, they stress a popularity and recognition theme (Pollay, 1983). Thus, certifications and awards are viewed as symbols of universal recognition.
- *Pride of Ownership appeal:* It emphasizes the status appeal, which is valued in high power distance societies (Hofstede, 1991; Mueller, 1987). Mueller (1987) found that Japanese (Japan is high on power distance) advertisements emphasized status appeal.

Masculinity

- *Quizzes and Games and Realsim theme:* Ads in masculine cultures emphasize the basic value of enjoyment (Pollay, 1983), mastery over nature, and reflect adventure theme, thrill and gaming (Cheng and Schweitzer, 1996).
- *Product effectiveness:* In masculine cultures, achievement, performance and growth are emphasized (Hofstede, 1991; Gudykunst, 1998).

Further, a sample of 30 Global Fortune 500 company websites from www.fortune.com was selected for a pilot study. The pilot study helped us to remove three coding categories, namely, celebrity endorsement, product return policy, and viral marketing option, as these categories were either not on the websites or have shown some overlap with other categories. Finally, a list of 35 operational categories was retained for content analysis of websites.

HYPOTHESES

As it is mentioned in the previous section, the aim of this research is to extend the research on cross-cultural analysis of cultural values to a new global communication medium, the World Wide Web. No previous study to our attention has used cultural values to study the information and advertising content on the Web. The extent that advertising and information content prominently reflects local cultural values is shown in the previous research on television and print media (Albers-Miller and Gelb, 1996; Cheng and Schweitzer, 1996; Mueller, 1987; Tse, Belk, and Zhou, 1989). We anticipate that cultural adaptation is likely to exist if significant differences in web-specific cultural dimensions between a firm's domestic websites and its foreign websites are manifested. Specifically, we conjecture a high degree of culture adaptation on the websites by the

Table 2. Explanation for cultural categories (cont.)

- *Clear Gender role:* In masculine societies, gender roles are clearly differentiated and men dominate in most settings (Hofstede, 1991).

High Context Culture

- *Politeness and soft sell approach:* In high context cultures, the messages are implicit and indirect. Direct comparisons are not viewed favorably (Mueller, 1987). Ads in such cultures try to be polite and make friends with the customer rather than directly selling them products.
- *Aesthetics:* Communications in high context culture emphasize harmony, beauty, and oneness with nature (Cho et al., 1999). Use of art, designs, beautiful scenery, and nature appeals are emphasized (Mueller, 1987, 1992; Gudykunst, 1998).

Low Context Culture

- *Hard sell approach and Explicit Comparisons:* Communications in low context cultures are direct, explicit, and even confrontational (Hall, 1976). Thus, use of direct and confrontational appeals in the form of discounts, sales promotions and aggressive selling is common in such cultures (Cutler and Javalgi, 1992).
- *Rank and Prestige and use of superlatives:* Advertisements in low context cultures also use implicit comparisons to highlight the benefits and the prestige of the products and the company brand (Mueller, 1987). Statements like "the number one company" and "the leader of the industry" are common.
- *Terms and conditions:* Low context cultures emphasize clarity, background information, and directness. The information is vested in the explicit code (Hall, 1976). Thus advertisements and communications in such cultures tend to be informative (Cho et al., 1999).

American companies if their U.S. domestic websites prominently depict western cultural values, and their Chinese websites are designed to retain and portray rich eastern cultural values. To investigate this overarching proposition about the website cultural adaptation, we propose five hypotheses that correspond to the six cultural dimensions proposed in the framework.

Individualism-Collectivism: The bi-polar individualism-collectivism dimension has been widely studied in cross-cultural research (Gudykunst, 1998; Hofstede, 1980, 1991, 1993; Triandis, 1982). This cultural dimension focuses on an individual's relationship with society or other individuals. In individualist societies, ties between individuals are loose, personal freedom is valued and individual decision-making is encouraged. On the contrary, in collectivist societies, in-group ties are strong, societal norms are valued, and group decision-making is encouraged. Australia, Canada, the UK, and the U.S. score high on individualism, while China, Columbia, Japan, and Mexico score high on collectivism.

On the individualism-collectivism dimension of Hofstede (1993), the U.S. scores the highest (91), while China's estimated score is much lower (20) as Chinese society is more collectivist in orientation. Thus it is hypothesized that on the Web:

H1) Fortune 500 companies Chinese websites will show a higher frequency of collectivist features than their U.S. domestic websites.

Uncertainty Avoidance: According to Hofstede (1980), the degree to which societies can tolerate uncertainty and ambiguity differs among cultures. Some cultures are high on the avoidance of uncertainty, and value security and low risk situations, while others are low on the avoidance of uncertainty, and have greater tolerance for ambiguity and risk. Countries like Greece, Japan, and Mexico rank high on uncertainty avoidance, and their citizens are generally more risk averse, need security, and look for direction. On the other hand, countries like Canada, Australia, U.K. and the U.S. have been found to have greater tolerance for uncertainty.

Hofstede's (1993) study found that the American culture is a comparatively low-risk avoidance culture (uncertainty avoidance score = 43) as compared to the high uncertainty avoidance culture of China (60). The Internet, being a new communication and transaction medium, has a certain degree of risk and uncertainty associated with it. A study by Lynch and Beck (2001) found that Asian consumers are less secure when shopping online. This may be due to the high uncertainty avoidance of three major Asian countries: China, Japan, and Taiwan. Thus, we hypothesize that on the Web:

H2) Fortune 500 companies' Chinese websites will show a higher frequency of uncertainty avoidance features compared to U.S. domestic websites.

Power Distance: According to Hofstede (1980) the bi-polar power distance dimension explains how different societies treat inequalities in social structure. Societies like Malaysia, Mexico, China, and India, that are high on power distance, accept power and hierarchy in the society, and are low on egalitarianism. The emphasis in high power distance societies is on status, referent power, authority, and legitimacy, while countries like Canada, Denmark, the U.S., and Sweden that are low on power distance dimension are more egalitarian, respect equal rights, and discourage status symbols. The U.S. score is far below (40) the estimated score of China (80) on power distance (Hofstede, 1991, 1993). Thus, we hypothesize that on the Web:

H3) Fortune 500 companies' Chinese websites will show a higher frequency of features relating to power distance compared to their U.S. domestic counterparts.

Masculinity-Femininity: The Masculinity-Femininity dimension proposed by Hosfstede (1980) explains how gender roles are allocated in different cultures. "Masculine" cultures value assertiveness, ambition, success, and performance. To such cultures, big and fast is beautiful, masochism is acceptable, and clear gender roles are the norm. On the contrary, "Feminine" cultures value beauty, nature and nurturance, and the ambiguity of gender roles. Countries like Japan, Austria, and Mexico are examples of masculine cultures, while most of the Nordic countries score high on femininity. The U.S. is relatively a more masculine society (62) than China (50). Thus, it is hypothesized that on the Web:

H4) The U.S. domestic websites of the Fortune 500 companies will show a relatively higher masculine orientation compared to their Chinese websites.

High and Low Context Cultures: Context is the information that surrounds an event; it is inextricably bound up with the meaning of that event (Hall and Hall, 1990, p. 6). In high context cultures the information is already embedded in the context of the communication, and very little is explicitly coded. On the contrary, in low context cultures mass of information is vested in the explicit code, and is directly shared in the act of communication. Low context cultures emphasize directness and clarity in communication. According to Hall (1976) and Hall (1990), eastern cultures like China are high-context in orientation, while the U.S. is low-context in orientation. Thus, it is hypothesized that on the Web:

H5) The U.S. domestic websites of the Fortune 500 companies will show a relatively higher low-context orientation, while their Chinese websites will show a high-context orientation.

METHODOLOGY

Content Analysis of the Websites

Content analysis is considered a reputable and widely utilized tool for conducting objective, systematic, and quantitative analysis of communication contents (Berelson, 1952; Kassarjian, 1977). However, the use of content analysis to study information content on the Web is still at an infancy stage; only few studies have adopted this technique to analyze Web content (Rafaeli and Sudweek, 1997; Ju-Pak, 1999). Since content analysis is regarded as an appropriate technique for analyzing values, norms of behavior, and other elements of a culture (Cheng and Schweitzer, 1996; Huang, 1995; Ju-Pak, 1999; Mueller, 1987; Tse et al., 1989) from the verbal material produced by people

(Berelson, 1952), we use it to systematically analyze the cultural values depicted on the websites of the U.S. and China.

Sample

The sample for this study was generated from the list of top Fortune 500 global companies featured at www.fortune.com. From a total of 500 listed company websites, only 50 companies had country-specific websites for both the U.S. and China. Most companies of the Fortune 500 have a website in a Chinese WWW domain; however, the website itself is written in English. We define the country-specific website as a website that is written in Chinese. To control for industry-specific effects in website design, we selected only the companies in the computers and electronics industry. This selection process yielded a total sample of 40 U.S.-based international companies. The unit of analysis was the domestic and Chinese websites of these 40 U.S.-based companies.

Reliability

Through a series of intensive sessions, two coders, fluent in English and Chinese, were trained in the coding scheme. A random sample of 25% from each of the U.S. and Chinese websites was used for reliability testing. Both coders analyzed the full websites to count the occurrence or non-occurrence of items under each of the 35 categories proposed in the framework. When an item of disagreement occurred, a third coder reviewed and evaluated the website, and a majority rule was used to determine the final coding. Inter-judge reliability was then calculated by using the per item agreement method suggested in the literature (Kassarjian, 1977). Overall reliability for the U.S. domestic websites was 89%, and for the Chinese websites was 83%. The intra-judge reliability is also tested. The authors coded a random sample of 25% of the U.S. and Chinese websites again after a month. The intra-judge reliability score of 90% was achieved in the case of the U.S. domestic websites, and 89% in the Chinese websites. Another intra-judge reliability test after a year resulted in a reliability score of 88%.

RESULTS

To test for the differences in the depiction of cultural values between a U.S. companies' domestic and Chinese websites, a chi-square test was performed. All the 35 cultural value categories were compared via cross-tabulations to test for significant differences in the way they were depicted.

In order to test the proposed hypotheses, the overall score for each of the six cultural dimensions (Collectivism, Uncertainty Avoidance, Power Distance, Masculinity, and High & Low Context) is needed. An aggregate variable for each dimension is generated in the following way. We calculate the average presence of each of the sub-categories within each dimension, and recorded it as an aggregated score for that cultural dimension. For example, the aggregated score for the cultural dimension collectivism is calculated by counting the presence of all eight sub-categories of collectivism on a website: Community Relations, Clubs or Chat Rooms, Newsletter, Family Theme, Country-Specific News, Symbols and Pictures of National Identity, Loyalty Programs, and Links to Local websites, and then divided the number by eight

A simple one-way analysis of variance test was then implemented. The results of cross-tabulations and chi-square analysis are shown in Table 4, while the ANOVA results for the aggregated variables are shown in Table 3. The aggregated data for each dimension is generated to represent each cultural dimension.

Collectivism: For the collectivism dimension, the percentage of occurrence of collectivist features on Chinese websites (53%) was comparatively more than on the U.S. domestic websites (40.6%). An F-test reveals that the collectivism cultural dimension has been adapted in those companies' Chinese websites ($F=8.00$, $p<.006$) (Table 3), which makes it distinguishable from the domestic websites. Thus Hypothesis 1 that collectivist features in Chinese websites will be more prominently depicted is supported.

In addition, the chi-square analysis (Table 4) also revealed that under some specific sub-categories, cultural adaptation is taking place. Categories such as clubs and chat rooms (Chinese: 35% vs. U.S.: 7.5%, Chi-square=9.03, $p<.003$), family themes (Chinese: 57.5% vs. English: 10%, chi-square=20.18, $p<.000$), newsletters (Chinese: 80% vs. English: 47.5%, chi-square=9.14, $p<.002$), and symbols specific to Chinese society (Chinese: 62.5% vs. English: 32.5%, chi-square=7.21, $p<.007$) were significantly depicted in these U.S. companies' Chinese websites compared to their domestic websites.

Uncertainty Avoidance: It was hypothesized that Chinese websites would show more uncertainty avoidance features compared to the U.S. domestic websites due to the high uncertainty avoidance nature among Chinese people.

The results show that the overall frequency of the occurrence of uncertainty avoidance was relatively higher in Chinese websites (65.8%) than in the U.S. domestic websites (58.6%) ($F=3.90$, $p<.052$). Thus Hypothesis 2 is supported.

Meanwhile, Table 4 shows that under some categories, such as traditional theme (Chinese: 55% vs. English: 32.5%, chi-square=4.11, $p<.03$), local store locations (Chinese: 85% vs. English: 60%, chi-square=6.27, $p<.01$), and local terminology (Chinese: 57.5% vs. English: 27.5%, chi-square=7.36, $p<.006$), there existed a significant distinction between the U.S. company's domestic websites and its Chinese websites.

Power Distance: Hypothesis 3 proposed that China, being a high power-distance society, would show more status and power distance-oriented features on its websites compared to the U.S. counterparts. The analysis of variance revealed that Chinese websites were high on power distance (53.6%) compared to the U.S. domestic websites (33%) ($F=12.76$, $p<.001$) (Table2). Thus Hypothesis 3 is supported. Table 4 shows that the hierarchy information (Chinese: 55% vs. English: 12.5%, chi-square=16.5, $p<.000$) and the pride of ownership appeal (Chinese: 62.5% vs. English: 17.5%, chi-square=16.87, $p<.000$) were much more prominently depicted in Chinese websites.

Masculinity: It was hypothesized that Chinese websites would score lower on Masculinity compared to U.S. domestic websites. The results of the analysis of variance reveal that Chinese websites depicted much lower masculine values (36.8%) than the U.S. domestic websites (63.1%) ($F=50.66$, $p<.000$). The depiction of realism theme (Chinese: 30% vs. English: 62.5%, chi-square=8.49, $p<.003$), product effectiveness (Chinese: 55% vs. English: 95%, chi-square=17.06%, $p<.000$), and clear gender roles (Chinese: 40% vs. English: 60%, chi-square=3.20, $p<.059$) were found to differ significantly. Thus, Hypothesis 4 is supported.

High and Low Context Cultures: It was hypothesized that China being a high context society will show high context-oriented features more prominently, while U.S being a low context society will depict more of low context-oriented features. The Chinese

Table 3. ANOVA results for four cultural dimensions

Dimension	Chinese Website	English Website	F-Value	Significance
	Mean values (%)			
Collectivism	53%	40.6%	8.00	.006
Uncertainty Avoidance	65.8%	58.6%	3.90	.052
Power Distance	53.6%	33%	12.76	.001
Masculinity	36.8%	63.1%	50.66	.000
Low Context	27.5%	58%	74.00	.000
High Context	42.5%	28.1%	8.21	.005

ns: non-significant at p< 0.1

websites were much more colorful, fanciful, and depicted a softer sell approach compared to the U.S. domestic websites. The U.S. domestic websites, in contrast, were high on explicit comparisons (Chinese: 0% vs. English: 22.5%, chi-square=10.14, p<.001), use of superlatives (Chinese: 15% vs. English: 92.5%, chi-square=48.32, p<.000), and terms and conditions of use (Chinese: 47.5% vs. English: 92.5%, chi-square=19.28, p<.000). Thus Hypothesis 5 is supported.

DISCUSSION

The results show a clear evidence of cultural adaptation being practiced by the sample firms on their Chinese (international) websites. All five hypotheses are supported. However, the degree of cultural adaptation depicted on those Chinese websites is at an early stage. For example, only 53% of the Chinese websites exhibited collectivist features, even though China is a very collectivist society. Low scores were similarly reported for the other four cultural dimensions (Table 3).

Nonetheless, in comparison to the domestic websites, Chinese websites had a very different structure and appearance. The Chinese websites exhibited bold colors and animation, representing high contextuality and more of oneness with nature and traditions theme. The Chinese websites were also "pregnant" with Chinese cultural symbols such as the Great Wall of China, Chinese festivals, the Chinese Flag, and other Chinese Landmarks. In contrast, the U.S. domestic websites depicted more of a realism theme, where direct, to the point information is presented with less fantasy and imagery. In addition, frequently used superlatives such as " we are the leader," "world's largest," and a hard sell approach, such as promotions and coupons, are common features on the U.S. domestic websites.

One striking theme in most U.S. domestic websites was the depiction of patriotic phrases in response to the September 11th attacks. For example, the website of Bell South Corporation had a streaming video telling, "Our one true triumph over this tragedy is our unity as a nation."

Table 4. Frequency of items under cultural value categories: A comparison of Chinese and English websites

Frequency (%) Items		Chinese	English	Chi-sq	(p<)
Collectivism					
1.	Community Relations	70.0	82.5	ns	
2.	Clubs or Chat Rooms	35.0	7.50	9.03	(.003)
3.	Newsletter	80.0	47.5	9.14	(.002)
4.	Family Theme	57.5	10.0	20.18	(.000)
5.	Country Specific News	80.0	90.0	ns	
6.	Symbols & Pictures of National Identity	62.5	32.5	7.21	(.007)
7.	Loyalty Program	10.0	10.0	ns	
8.	Links to Local Websites	30.0	45.0	ns	
Uncertainty Avoidance					
1.	Customer Service	92.5	90.0	ns	
2.	Secure Payment	67.5	87.5	4.58	(.030)
3.	Guided Navigation	97.5	85.0	ns	
4.	Tradition Theme	55.0	32.5	4.11	(.035)
5.	Local Stores	85.0	60.0	6.27	(.011)
6.	Local Terminology	57.5	27.5	7.36	(.006)
7.	Free Trails and Downloads	50.0	40.0	ns	
8.	Customer Testimonials	12.5	25.0	ns	
9.	Toll Free Numbers	75.0	80.0	ns	
Power Distance					
1.	Company Hierarchy Information	55.0	12.5	16.15	(.000)
2.	Pictures of CEO's	60.0	47.5	ns	
3.	Quality Information and Awards	47.5	55.0	ns	
4.	Vision Statement	52.5	40.0	ns	
5.	Pride of Ownership Appeal	62.5	17.5	16.87	(.000)
6.	Proper Titles	42.5	30.0	ns	
Masculinity					
1.	Quizzes and Games	22.5	35.0	ns	
2.	Realism Theme	30.0	62.5	8.49	(.003)
3.	Product Effectiveness	55.0	95.0	17.06	(.000)
4.	Clear Gender Roles	40.0	60.0	3.20	(.059)
Low Context					
1.	Rank or Prestige of the Company	35.0	32.5	ns	
2.	Hard Sell Approach	40.0	50.0	ns	
3.	Explicit Comparisons	0.00	22.5	10.14	(.001)
4.	Use of Superlatives	15.0	92.5	48.32	(.000)
5.	Terms and Condition	47.5	92.5	19.28	(.000)
High Context					
1.	Politeness	57.5	47.5	ns	
2.	Soft Sell	45.0	10.0	12.28	(.000)
3.	Aesthetics	67.5	55.0	ns	

ns: non-significant at $p < .05$

The findings from this study confirm that the Web is not a culturally neutral medium. Instead there are significant differences in depiction of local cultural values on the Web. By providing evidence of cultural adaptation on the Web, this research enhances the validity of previous studies that have confirmed the use of a localized-specialized approach to advertising in print and broadcast media but not on the Web (Albers-Miller and Gelb, 1996; Cutler and Javalgi, 1992; Tse, Belk and Zhou, 1989). Furthermore, this research also serves as a guide for future studies attempting to measure cultural values on the Web.

To summarize, our study supports the notion that the Web is not a culturally neutral medium. Instead the Web is full of cultural markers (Barber and Badre, 1998) that give country-specific websites a look and a feel often unique to that local culture. However, the extent of cultural adaptation exhibited on the country-specific websites is still at an early stage and Web designers need to use both etic and emic approaches to understanding local cultures and local websites.

CONCLUSION AND IMPLICATIONS

The objective of this article was to apply research on cultural values to Web content. Based on the results of this study, three main managerial implications have been identified. First, the framework proposed in this study would be useful for marketers and website designers to develop country-specific websites. Second, if applied this framework cannot only make the websites culturally adapted, but it can also enhance website usability and increase the global competitiveness of companies using global web communications. Studies by Barber and Badre (1998), Fock (2000), and Simon (1999) suggest that country-specific and culturally sensitive Web content enhances usability, reach, and website interactivity leading to more Web traffic and business activity on the Web. Culturally congruent Web style reduces the anxiety associated with a new medium like the Internet and makes navigation easier. Moreover, it is easier for local consumers to relate to the websites that have a local feel (Barber and Badre, 1998). Finally, studying cultural content on websites can also provide insights into the cultural and societal characteristics of a particular national culture and help marketers to avoid cultural *faux pas* when marketing globally.

As the proposed framework only measures the attitudinal or value aspect of cultural reality, future studies are needed to incorporate a much more exhaustive list of cultural dimensions that measure perceptual, attitudinal, behavioral, and symbolic aspects of cultural reality. For example, research in the area of cross-cultural perceptual categorization and information processing (Detweiler, 1978; Hoosain, 1987), has identified cultural variables like:

- *Language translatability:* because language structures the way we think about or conceptualize something (Steinfatt, 1989).
- *Language structures:* Kaplan (1966) identified different logical language structures among different cultures.
- *Color perceptions and color categories* (Segal et al., 1966).

- *Ecological perceptual styles:* perception of geometric shapes, open spaces, depth. An Example is the Muller-Lyer illusion (Segal et al., 1966).
- *Field independence:* the ability to see objects separately from the context, which leads to cross-cultural spatial illusion (Segal et al., 1966).

The above-mentioned variables can be used as a basis for developing websites that locally adapt the spatial orientation of the Web content, layout of the Web pages, text length, navigational modes, translation equivalence, language dialects, and color categories.

In terms of the symbolic aspects of cultural reality, some variables can be studied:

- *Codes of society:* codes are systems into which signs of the society are organized (Fiske, 1990). For example, aesthetic codes, codes of conduct, and non-verbal codes.
- *Metonyms:* linguistic structures that make a part stand for the whole.
- *Myths:* myths stimulate us to construct the rest of the chain of concepts that constitute the myth (Fiske and Hartley, 1978).
- *Color:* Cultures ascribe symbolic meaning to different colors (Leach, 1976).

Thus, the cultural sensitivity of websites at this level of symbolic analysis can be analyzed in terms of country-specific symbols, icons, and color symbolism.

In addition to incorporating the perceptual and symbolic variables to study the web content, the framework also needs to be tested and validated in other countries. It would be interesting to assign weights to each cultural dimension depending on the degree of importance of it in a particular culture, and additional reliability testing may enhance the reliability of the framework.

Due to the length limitation of the paper, this research only compares U.S. domestic websites and their Chinese websites in studying the cultural content of websites. Future studies could extend such a comparison to websites created in other countries' languages. Furthermore, we analyzed the websites of the U.S. Fortune 500 companies; future research could extend such an analysis by comparing the Chinese websites of top global firms with the websites of Chinese local firms. It would be interesting to study the level of cultural adaptation practiced by Chinese local firms in their websites, and compare it with that of the U.S. companies' Chinese websites.

In conclusion, we believe this study will help marketers and Web designers in their website localization efforts. This research can also serve as a crucial starting point for future studies in the area of the cultural analysis of website design. Finally, by showing that country-specific websites differ in the depiction of cultural values, our study proves that the Web is not a culturally neutral medium.

REFERENCES

Albers-Miller, N.D. & Gelb, B.D. (1996). Business advertising appeals as mirror of cultural dimensions: A study of eleven countries. *Journal of Advertising, 25*(Winter), 57-70.

Alvarez, M.G., Kasday, L.R. & Todd, S. (1998). How we made the web site international and accessible: A case study. Available online at: *http://www.research.att.com/conf/hfweb/*

Andren, G. (1980). The rhetoric of advertising. *Journal of Communication, 30*(August), 74-80.

Barber, W. & Badre, A. (1998). Culturability: The merging of culture and ysability. Available online at: *http://www.research.att.com/conf/hfweb/*

Berelson, B. (1952). *Content analysis in communication research.* New York: Free Press.

Berthon, P., Leyland, P., & Watson, G. (1996). Marketing communications and the World Wide Web. *Business Horizons, 29*(September/October), 24-32.

Bloom, H., Calori, R., & de Woot, P. (1994). *Euromanagement.* London: Kogan Page.

Borden, G. A. (1991). *Cultural orientation: An approach to understanding intercultural communication,* Englewood Cliffs, NJ: Prentice Hall.

Bottger, P., Hallein, I., & Yetton, P. (1985). A cross-national study of leadership: Participation as a function of problem structure and leader power. *Journal of Management Studies, 22*(4), 358-368.

Chen, M. (1995). *Asian management systems.* London: Routledge.

Cheng, H. & Schweitzer, J.C. (1996). Cultural values reflected in Chinese and U.S. television commercials. *Journal of Advertising Research*, (May/June), 27-45.

Clark, T. (1990). International marketing and national character: A review and proposal for an integrative theory. *Journal of Marketing, 54*(October), 66-79.

Cragg, C. (1995). *The new Taipans.* London: Century Business.

Cutler, B. D., & Raj, S.G.J. (1992). A cross-cultural analysis of visual components of print advertising: The United States and European community. *Journal of Advertising Research, 32*(Jan/Feb), 71-80.

Detweiler, R.A (1978) Culture, category width, and attributions. *Journal of Cross-Cultural Psychology, 9*(3), 259-284

Emarketer. (2001). The eGlobal report. Available online at: *http://www.emarketer.com*

Ferranti, M. (1999). From global to local. *Infoworld, 21*(41), 36-37.

Fernandez, D., Carlson, D.S., Stepina, L.S. & Nicholson, J.D. (1997). Hofstede's country classification 25 years later. *Journal of Social Psychology, 137*(1), 43-54.

Fiske, J. (1990). *Introduction to communication studies.* New York: Routledge.

Fiske, J. & Hartley, J. (1978). *Reading television.* London: Methuen.

Fock, H. (2000). Cultural influences on marketing communication on the World Wide Web. A Paper Presented at the *Multicultural Marketing Conference*, Hong Kong, September.

Fukuyama, F. (1995). *Trust: The social virtues and the creation of prosperity.* London: Hamish Hamilton.

Gilly, M. (1988). Sex roles in advertising: A comparison of television advertisements in Australia, Mexico, and United States. *Journal of Marketing,* 52, 75-85.

Gudykunst, W. B. (1998). *Bridging differences: Effective intergroup communication* (3rd ed.) Sage Publications.

Hall, E.T. (1960). The silent language in overseas business. *Harvard Business Review,* (May-June).

Hall, E.T. (1976). *Beyond culture.* New York: Anchor Press.

Hall, E. T. & Hall, M.R. (1990). *Understanding cultural differences.* MI: Intercultural Press.

Hampden-Turner, C., & Trompenaars, F. (1994). *The seven cultures of capitalism.* London: Piatkus.

Han, S.-P. & Shavitt, S. (1994). Persuasion and culture: Advertising appeals in individualistic and collectivistic societies. *Journal of Experimental Social Psychology, 30*(July), 8-18.

Harris, R., & Davison, R. (1999). Anxiety and involvement: Cultural dimensions of attitudes toward computers in developing countries. *Journal of Global Information Management, 7*(1), 26-38.

Hofstede, G. (1980). *Culture's consequences: International differences in work-related values.* Beverly Hills, CA: Sage.

Hofstede, G. (1991). *Culture and organizations: Software of the mind.* London: McGraw-Hill.

Hofstede, G. (1993). Cultural constraint in management theories. *Academy of Management Executive, 7*(1), 81-94.

Hofstede, G. & Bond, M.H. (1998). The Confucius connection: From cultural roots to economic growth. *Organizational Dynamics, 16*(4), 5-21.

Holsti, O.R. (1969). *Content analysis for the social sciences and humanities.* Reading, MA: Addison-Wesley.

Hornik, J. (1980). Comparative evaluation of international versus national advertising strategies. *Columbia Journal of World Business, 15*(Spring), 36-45.

Hoosain, R. (1987). Perceptual Process of the Chinese. In M. H. Bond (Ed.), *The psychology of the Chinese people.* Hong Kong: Oxford University Press.

Huang, J.-H. (1995). Cultural values as manifested in U.S. and Taiwan television advertising. In B. B. Stern, & G. M. Zinkhan (Eds.), *Proceedings, AMA Educators Conference* (pp. 155-160). Washington, DC.

Internet.com. (2001). Big picture geographics. Available online at: *http://cyberatlas.internet.com/big_picture/geographics/article/*

Jacob, L., Keown C., Worthley, R. & Ghymn, K. (1991). Cross-cultural colour comparisons: Global marketers beware. *International Marketing Review, 8*(3), 21-30.

Ju-Pak, K.H. (1999). Content dimensions of Web advertising: A cross-national comparison. *International Journal of Advertising, 18*(2), 207-231.

Kale, S.H. (1991). Culture-specific marketing communication: An analytical approach. *International Marketing Review, 8*(2), 18-30.

Kassarjian, H. H. (1977). Content analysis in consumer research. *Journal of Consumer Research, 4*(June), 8-18.

Kluckhohn, F.R. & Strodtbeck, F.Z. (1961). *Variations in value orientations.* WestPort, CT: Greenwood Press.

Leach, E. (1976). *Culture and communication: The logic by which symbols are connected.* New York: Cambridge University.

Lessem, R., & Neubauer, F. (1994). *European management systems.* London: McGraw-Hill.

Lewis, R. D. (1992). *Finland: Cultural lone wolf – Consequences in international business.* Helsinki: Richard Lewis Communications.

Lynch, P.D., & Beck J.C. (2001). Profiles of Internet buyers in 20 countries: Evidence for region-specific strategies. *Journal of International Business studies, 32*(4), 725-748.

Morden, T. (1999). Models of national culture – A management review. *Cross Cultural Management – An International Journal, 6*(1), 19-44.
Mueller, B. (1987). Reflections of culture: An analysis of Japanese and American advertising appeals. *Journal of Advertising Research*, (June/July), 51-59.
Myers, M D. & Tan, F.B. (2002). Beyond models of national culture in informational system research. *Journal of Global Information Management*, (January/March), 24-32.
Newman, W. H., Summer, C. E., & Warren, E. K. (1977). *The process of management.* Englewood Cliffs, NJ: Prentice Hall.
Pollay, R.W. (1983). Measuring the cultural values manifest in advertising. In Leigh, J.H. & Martin, C.R. (Eds.), *Current issues and research in advertising* (pp. 72-92). Ann Arbor, MI: University of Michigan Press.
Ralston, D.A., Gustafson D.J., Elsass P.M., & Cheung, F. (1992). Eastern values: A comparison of managers in the United States, Hong Kong, and the People's Republic of China. *Journal of Applied Psychology, 77*(5), 664-671.
Segal, M. H., Campbell, D.T., & Herskovits, M.J. (1966). *Influence of culture on visual perception.* Indianapolis, IN: Bobbs-Merrill.
Simon, S.J. (1999). A cross-cultural analysis of web site design: An empirical study of global web users. *Paper Presented in Seventh Cross-Cultural Consumer Business Studies research Conference.* Cancun, Mexico.
Simon, S.J. (2001). The impact of culture and gender on websites: An Empirical Study. *Database for Advances in Information Systems, 32*(1), 18-37.
Steinfatt, T.M. (1989). Linguistic relativity: Toward broader view. In S. Ting-Toomey, & F. Korzenny (Eds.), *Language communications and culture.* Sage.
Triandis, H.C. (1972). *The analysis of subjective culture.* New York: John Wiley & Sons.
Triandis, H.C. (1982). Review of culture's consequences: International differences in work-related values. *Human Organization,* 86-90.
Triandis, H.C. (1994). *Culture and social behavior.* New York: McGraw-Hill.
Triandis, H. C. (1995). *Individualism & collectivism.* Boulder, CO: Westview Press.
Triandis, H.C. (1999). Cross-cultural psychology. *Asian Journal of Social Psychology, 2*(1), 127-143.
Trompennars, F. (1994). *Riding the waves of culture: Understanding diversity in global business.* New York: Professional Publishing.
Tse, D.K., Belk, R.W. & Zhou, N. (1989). Becoming a consumer society: A longitudinal and cross-cultural content analysis of print ads from Hong Kong, The People's republic of China, and Taiwan. *Journal of Consumer Research, 15*(March), 457-472.
Violino, B. (2001). E-business lurches abroad. *Internet Week*, (March 19). Available online at: *http://www.internetweek.com*
Zandpour, F. & (a team of 10 researchers). (1994). Global reach and local touch: achieving cultural fitness in TV advertising. *Journal of Advertising Research, 34*(5), 35-63.
Zhang, Y., & Gelb, B.D. (1996). Matching advertising appeal to cultures: The influence of product's use conditions. *Journal of Advertising, 25*(Fall), 29-46.
Zhao, H. (2002). Rapid Internet development in China: A discussion of opportunities and constraints on future growth. *Thunderbird International Business Review, 44*(1), 119-138.

Previously published in the Journal of Global Information Management, 11(3), 63-80, July - September 2003.

Chapter XIII

Doing Business in Paradise: Infrastructure and Information Issues for Small Business in a Developing Island Economy[1]

Michelle L. Kaarst-Brown, Syracuse University, USA

Chingning Wang, Syracuse University, USA

ABSTRACT

This chapter presents findings from a case study of the Turks and Caicos Islands (TCI) in the British West Indies area of the Caribbean. TCI is a tax haven that has worked to attract offshore financial firms such as trust, insurance, and financial management companies. All of these firms, which qualify as "information intensive," are small in size (average 11 employees), engage in business on a global basis, and yet must compete while dealing with local infrastructure challenges. TCI is presented as the developmental context in which small businesses (largely owned or managed by foreigners from other cultures) must interpret and cope with national infrastructure challenges in this very small, young, rapidly growing island nation. Not surprisingly, we found that these firms share similar challenges with those in other developing countries; however, the perceptions of these challenges and how these small firms cope provide insights into the importance of small firms, small-scale foreign investment, and cross-national transfer of technology expectations.

INTRODUCTION

Some organizations, such as financial services firms, traditionally have had high requirements for information processing while also having high information content in their products and services. More recently, however, the pervasiveness of information and Internet technologies has resulted in many organizations becoming increasingly information-intensive and information-technology-reliant (Glazer, 1991; Proudlock, Phelps, & Gamble, 1998). The technical challenges for these information-intensive firms are many, even in technologically advanced countries. Although there is recent evidence that the speed of IT diffusion among Latin American and Caribbean countries is "quite high and comparable to other countries," this diffusion is dependent on the country's IT infrastructure and institutional factors, such as regulatory policies (Bagchi, Hart, & Cerveny, 2002). These challenges increase exponentially when operating as a smaller enterprise in developing countries with uncertain infrastructures (Bingi, Leff, & Shipchandler, 2000; Davy & Allgood, 2002; Mansell & Wehn, 1998).

This chapter presents findings on how such small, information-intensive firms perceive and attempt to cope with these challenges through a case study of organizations in TCI, one of the lesser-known international financial centers. The next section provides our research design and data collection methods, followed by background on TCI and our findings. The chapter closes with a discussion of implications for research and management practice.

RESEARCH DESIGN AND INITIAL MODEL

The research for this case study began in 1999. During a visit to TCI, the first author was trying to check e-mail and could not connect to TCI Cable and Wireless, the monopoly firm providing telecommunications and Internet services to TCI. Upon inquiring how long the network would likely be down and anticipating a response of "fifteen minutes to an hour," the first author was told, "the server is probably down, and so it could be a few hours or a few days before it comes back up." When clarification was sought, it was explained that the technician who dealt with the server was on the mainland (USA) for holiday celebrations with family. His return would depend on "how good a time he is having and whether he makes the flight back tomorrow." This initiated the formal case study investigation into how local firms coped with this infrastructure uncertainty, especially since many of the firms relied heavily on information and communications technologies to provide products and services or to process transactions.

Data Collection

As suits case study research, a variety of methods were used to collect and verify data over several years of TCI's history (Lee, 1989; Lincoln & Guba, 1985; Miles & Huberman, 1994; Yin, 1994). This research relied on a combination of (1) intensive, semi-structured interviews with 12 participants in fall 2001, spring 2002, and fall 2002; (2) informal discussions with locals, residents, tourists, and business owners from fall 1999 to fall 2002; (3) periodic direct and participant observations over that same three-year period; and (4) analysis of public documents and research literature.

Three methodological issues were critical to this study: access, discretion, and confidentiality. It should be noted that of the 16 firms approached, three chose not to participate, expressing that they did not want to speak publicly on issues of any challenges, necessity for coping, or anything else that could come back to haunt them. One firm did not return our calls or e-mails.

TCI is both old and new. It has a history that goes back hundreds of years, and yet it really only began its major economic development in the last two decades. Further, its history includes more than a few skeletons that it would like forgotten. TCI is also sensitive to the divide between native "belongers" and their growing population of immigrants (temporary and permanent residents, both Caucasian and black) (Bolderson, 2003). The belonger status is very important, as only belongers can vote and have an unrestricted right to work as well as reside in TCI. Given the politics of a developing country, it was very important that we guarantee our discretion, not only in how we used information, but also in what we asked. We stressed that we were not seeking to criticize the current government or policies, or uncover anything outside of exactly what we stated as the research questions provided.

Interview Guide and Initial Model

Figure 1 provides the basic interview model, based on early discussions and the literature. Questions were grouped into five categories: demographic, information intensity, technology needs, challenges and coping strategies, and perceived outcomes.

First, basic demographic and historical information was gathered about the history of the firm and the interviewee. This included questions about why the firm located in TCI, individuals' past experiences in developing countries, education, exposure to IT, and length of time in TCI. There is increasing evidence that firm level and individual level factors will influence computer adoption, IT investment, and satisfaction with systems in a developing country (Abdul-Gader & Kozer, 1995; Ballantine, Levy, & Powell, 1998; DeLone, 1988; Montazemi, 1988; Petrazzini & Kibati, 1999; Raymond, 1985; Zinatelli, Cragg, & Cavaye, 1996).

Second, information was gathered to determine the exact nature of information activities and information intensity of the firm. For example, questions focused on the number of transactions, how information was gathered and disseminated, types of information used by the businesses, and the criticality of information for the business. This type of questioning is supported by the view that information intensity addresses

Figure 1. Initial model

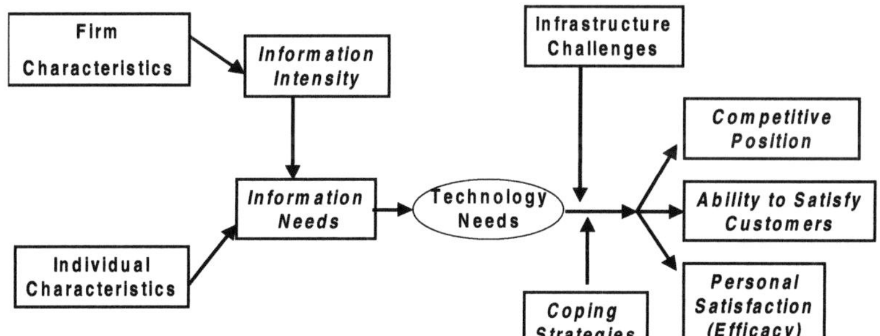

the whole "information value chain" (Glazer, 1993) and is reflected in the products/services, processes, or decision-making of the organization (Glazer, 1991, 1993; Proudlock et al., 1998).

Despite Glazer's assertion that truly information intensive firms are less concerned with technical issues, it was assumed that there was some relationship between the types of technologies the firm would require, the firm's information intensity in terms of transactions and information behaviors, and their specific information needs (Ballantine et al., 1998; Zinatelli et al., 1996). The third area of questioning, therefore, focused on the exact types of technologies that the firm used to gather, store, analyze, or disseminate information either internally or to external clients and agencies.

The fourth category of questions related to general infrastructure challenges encountered in TCI and those that were specific to technology. The rationale for this separation was that some people were anticipated to focus on technical issues without considering that skilled labor or weather issues might be equally important to technology infrastructure. There are many challenges faced by developing countries (Agha &

Table 1. Summary of selected literature on challenges for firms and/or countries

	Bingi et al. (2000) **Challenges Experienced by Firms Operating in Developing Countries**	**Davey & Allgood (2002)** **Challenges for Developing Countries**	**Agha & Akhtar (1992)** **General Challenges For Countries**
Economic & Cost Challenges	Cost justification of IT investment difficult given uncertain systems capabilities	IT development may not be seem as a national priority	Limited national budgets
	Replacement of labor by capital	Stereotype image of third world countries may limit investment desirability	
	During recessionary times, IT development becomes lower priority	Cost efficiencies not available if small population	
	Lack of effective infrastructure increases costs		
	Concerns for sovereignty and trade restrictions limit business capabilities and/or availability of hardware/software from foreign countries		
Cultural Challenges		Cultural differences create conflict	
Labor Related Challenges	Lack of IT personnel	Lack of skilled pool of English-speaking developers	Understaffed or undertrained IT personnel
	IT departments too small, so lack upward mobility		
	Unsuitable educational systems leads to lack of computer literacy among workers		
	Lack of the computer sophistication among the users		
Information Related Challenges	Problems with quality of input data		IS systems not well utilized by traditional services
	Government-owned information businesses		Unavailable data and information

Akhtar, 1992; Bingi et al., 2000; Davy & Allgood, 2002). Small island economies may be especially challenged. These issues included cost-justification of IT investment; lack of skilled personnel; economic recession; and the lack of an effective, nationwide infrastructure (i.e., power sources, hard currency, telecommunications services). These challenges continued as we moved into more advanced Internet-based electronic commerce (Kaarst-Brown & Evaristo, 2002; Petrazzini & Kibati, 1999; Travica, Johanson, & Leon-Pereira, 2000; Van Beveran & Thomson, 2002). Table 1 summarizes some of the key technical and non-technical challenges identified for firms operating in developing countries or for developing countries in general. After identifying challenges, respondents were asked how the firm coped in response to the challenges raised.

The final area of questioning asked respondents about their relative satisfaction with TCI's infrastructure in terms of enabling their firm to be competitive, their perceived ability to satisfy customers' expectations, and their personal efficacy (Zinatelli et al., 1996). This outcome category was divided in this way to allow for expression of *comparative* outcomes. As an example, perhaps the firm's customers were satisfied, but the individual felt the firm was hampered from future growth.

Intensive Sample: 12 Firms

In addition to informal discussions over a three-year period, 12 small firms were identified to pursue in greater depth. All of the firms interviewed were located on Providenciales, the most populated and well developed island. Our sample included a senior manager at TCI Cable & Wireless. For additional validation and perspective (Lincoln & Guba, 1985), we also interviewed the past-president of the Chamber of Commerce, a belonger with a long family history on the islands spanning more than 200 years. Two technical support firms were chosen because of their comprehensive knowledge of the technical history of the islands and their contact with most of the small businesses outside of our sample. The international art gallery was included because it was reported to rely heavily on IT to market to global customers. The owner was also the current president of the Chamber of Commerce, which is very

Table 2. Initial interview sample of TCI firms

Int.#	Firms Types	Interviewee Status
1.	Trust & Asset Management company ("Boutique" firm)	TRP
2.	Commercial Loan Officer at subsidiary of International Bank	TRP
3.	Trust & Asset Management company ("Boutique" firm)	TRP
4.	Trust & Asset Management company ("Boutique" firm)	PRC
5.	Trust & Asset Management company ("Boutique" firm)	TRP
6.	General Manager of TCI Cable & Wireless (Subsidiary)	TRP
7.	International Art Gallery Owner/Operator and Current President TCI Chamber of Commerce (Awarded citizenship and "belonger" status 1999)	B
8.	Owner/operator Technical Services Firm (Hardware)	TRP
9.	Credit Card Issuer	TRP
10.	Venture Capital Risk Management Consultant (new resident to island)	TRP
11.	Owner/Operator Technical Services Firm (Software)	TRP
12.	Past President TCI Chamber of Commerce and Owner of Contracting firm ("Belonger," born in TCI, with family history going back over 200 years)	B
Legend: Belonger Status (B), Permanent Resident Certificate (PRC), Temporary Resident Permit (TRP)		

active in lobbying on behalf of small business. The remaining firms were in financial services. Table 2 provides details on the firms interviewed and their owners' statuses as belonger, permanent resident with annual work permits, or temporary resident with annual work permits. Other than Interview #12, all of these firms were viewed to be "information intensive" (Glazer, 1993; Proudlock et al., 1998).

Information intensity was initially implied based on the perceived criticality of information for business success, and the expected volume of information seeking or dissemination activities. Nearly all had between one and 15 employees. It should be noted that some of the financial firms had assets under management in excess of $100 million USD (source concealed). In the trustee industry, this is considered a small "boutique" firm. It highlights the strategic importance of information management for firms of this type.

Country Setting: The Turks & Caicos Islands (TCI)

It is important to put the technical challenges of these smaller businesses into the context of a young developing country. TCI has had only 20 years to evolve from an infant island with virtually no industry to an economy capable of supporting tourism and offshore finance.

TCI lies in the British West Indies area of the Caribbean. While the capital (Cockburn Town) is located on Grand Turk, approximately 80% of the population and the international airport are on Providenciales, which is a one and a half hour flight from Miami and a direct flight from London, New York, and Toronto. With a population growth from 800 people to approximately 20,000 people[2] in less than two decades, this is an extremely small but rapidly developing economic environment that seeks to become an elite tax haven environment for reputable offshore financial services firms, while developing into an upscale tourist paradise (L'agefi, 2000) (Table 3). The island has struggled to balance the roles of government, monopoly service provider, and small local businesses in supporting infrastructure development.

Table 3. Overview of Turks and Caicos Islands (TCI) (Sources: CIA Worldbook 2002, 2004; EIU Country Reports)

Population	19,956 (est. July 2004); Median age 27.2 (2004 est.)
Main Islands	8 Inhabited Islands in a chain of 40 islands and small cays
Land Mass	430 Square Miles; (approximately 2.5 times the size of Washington, D.C.)
Language	English
Currency	US Dollar
Primary Industry	Tourism and Offshore Financial Services
Annual Budget	Revenues: $ 47 Million USD; Expenditures: $33.6 Million USD
Economic Aid Recipient	$7.1 Million USD (2001)
Real GDP grown rate	3.4% (2002)
Exports	$8.7 Million USD; mostly to US & UK (2002)
Imports	$177.5 Million USD; mostly from US & UK (2002)
Dependency Status	Overseas Territory of United Kingdom; Constitution ratified 1972; withdrawn in 1986 after senior government officials were charged with drug dealing and money laundering; reinstated 1988)
Labor Force	4,848
Labor Force by Occupation:	33% Government; 20% Agriculture & Fishing; Significant numbers in tourism, financial and other services

TCI still has the title of British Overseas Territory (Sadler, 1997). Despite the political affiliation and oversight provided by the United Kingdom, TCI's development is relatively independent of the technological and other resources of the UK (EIU, 2004; Sadler, 1997). Only belongers or UK appointed officials hold government office.

The island chain has almost no natural resources and uses the American dollar as currency. Cost of living is high, given that nearly all food and capital goods are imported. The benefits of the protected tax status and the natural beauty of the environment draw many small financial services firms, insurance companies, and trust companies. IT adoption among government agencies has been uneven. "Treasury is excellent, but across the street at the Department of Immigration, things are still sorted by hand. They are working to become fully computerized" (respondent identity concealed). Most of the firms reported in our research compete globally and yet must deal with local technological, labor, and natural limitations.

TCI is a zero tax jurisdiction. In addition, TCI enacted legislation designed to maintain financial privacy by an emphasis on nondisclosure for individuals seeking confidentiality in their financial affairs. The privacy-friendly policy and tax regulation impact businesses' information seeking and disseminating behaviors, as will be noted later (Dodsworth, 2000).

Despite being a so-called tax haven, there are still numerous license and stamp taxes such as import duties, immigration fees, vehicle licenses, and business licenses. In August 2001, a sudden and unexpected increase in business and residency fees and changes in residency policy created upheaval among local businesses, belongers, and residents. One government official called for the resignation of the Chief Minister because of the unexpected fee hikes. Another headline read, "Native business community plans to petition Govt. on tax increases: Drop taxes or we'll shut the country down" (Anonymous, 2001b).

As an example, the application fee for Permanent Residence Certificates (PRC) for the self-employed person increased from $10,000 to $50,000 USD (Anonymous, 2001b), adjusting back to $10,000 in 2002 (TCInvest, 2003a, 2003b). Application does not mean that PRC status will be granted, and refusal can be idiosyncratic, even for long-term residents with significant investment in the island (Bolderson, 2003). Overnight, work permits for professionals, managers, or directors also jumped from $2,500 to $7,000 USD annually per person. Skilled laborers' work permits increased from $1,000 to $5,000 USD annually per *person* (Anonymous, 2001a).

FINDINGS

We will first review findings for each element of the model and then follow with discussion on implications for research and practice.

Firm Characteristics: Corporate ROI vs. Entrepreneurship

As shown in Table 2, two of the firms were subsidiaries of larger firms, while the remainder were owner-operated businesses—entrepreneurial firms. Table 4 summarizes firm characteristics that were made available to us. Consistent with Table 1, which

Table 4. Firm characteristics

Business	Yr Est.	Firm Age 2003	Business Vol. (in US $)	# Empl.	# Temp. Resident Empl.	Locations	Operations
Trust (01)	1994	9	>$100 Mill	7	3	Provo	Global & Local
International Bank (02)	1988	15	>$45 Mill	42	6	*Subsidiary:* (52 Countries)	Local
Trust (03)	1995	3	--	9	5	Provo	Local, Canada, US, UK,
Trust (04)	1978	25	>50 Families	6	0	2 Offices TCI (Main office Provo)	US, UK, Canada
Trust (05)	1995	8	--	15	8	Provo	US, UK, Canada, Local
Service Provider (06)	--	100+	--	45	3**	*Subsidiary:* (International; Caribbean)	Local
Intern'l Art Gallery (07)	1990	13	--	2	0	Provo	Global & Local
Tech. Services (08)	1989	14	--	3	2	Provo	Local
Credit Card (09)	1999	4	--	10	3	Provo	Canada, US
Consultant (10)	2001	2	--	1	1	Provo	US, UK, Asia
Tech. Services (11)	1999	3	--	1	1	Provo	Local
Contracting (12)		~30	--	25	0	Provo	TCI

**Interview 12 has been retained for comparison purposes.*
*** As of 2004, TCI C&W indicates 100% Belonger Staff*

summarized infrastructure challenges for firms in developing countries, the subsidiary firms reported themselves as having varying influence on the overall IT strategies of their parent firms. Investment strategy of the parent firm constrained subsidiary IT investment, including simple hardware and software upgrades, because the focus was on return-on-investment (ROI) provided by the subsidiary. The bank in particular reported that hardware and software were older than that available in larger North American branches. As noted by Bingi et al (2000), companies operating in developing countries share the difficulty of translating intangible benefits of proposed systems into dollar values. They recommended that the emphasis should be more on value added over the long term, rather than immediate contribution to the bottom line, but this is further complicated in TCI because general managers or other management may be temporary residents on relatively short-term rotations. As such, they are willing to simply cope with lesser technology and will not question technology decisions overseen by the foreign Head Office.

The bulk of the small firms in this study were what we termed the entrepreneurial firms. These owner-operated firms were similar in orientation to parent firms in that they invested for the long-term; however, they reflected a different learning orientation that required them to stay on top of IT development both locally and globally (Fuller & Jenkins, 1994). With a commitment to success, they needed to understand everything that impacted their business and were free to make the best technology choices for their needs.

Individual Characteristics: Commitment—A Two-Way Street

Most of the individuals, whether owner-operators or employees, were attracted to TCI because it is a young, developing country with a desirable tax structure and climate. Bermuda and the Caymans are viewed as being over developed for offshore financial services. Several came to TCI on temporary rotations because job opportunities were offered. It was not unusual for our interviewees to report prior experience in developing countries that made it easier for them to adapt to living in TCI. One interviewee stated that he selected TCI in which to locate his business because he grew up in the Caribbean and had a desire to move back to the region.

While two interviewees had engineering degrees, the others acquired IT knowledge primarily through previous work experience. *"I was working in London when use of technology in business was slowly developing. I was forced to use my own initiative to learn computers."* (Trust)

It is perhaps not surprising to find differences in commitment to island infrastructure development. Due in part to the restrictions TCI placed on residency permits, many of those who live and work in TCI struggle to reconcile their somewhat precarious status with a long-term view of island or technology infrastructure development. Two distinct views emerged. The first was a clear commitment to TCI, regardless of personal long-term residency plans. The other view was very colonialist and almost entitled—happy so long as things were going their way but quick to criticize when things became challenging. This is a small sample; however, there was consistency between lack of prior experience in developing countries and the colonialist view. Table 5 summarizes basic individual characteristics.

For those we classified as committed, their strategy for coping with residency uncertainties was to become more involved in island development by hiring locally, training and promoting from within, and taking a long-term view of the islands' development and their own personal contributions while on the island. For those with what we refer to as a colonialist view, their coping strategies seemed to revolve around hiring outside labor, training only low level positions, and leaving the business field when the going gets really tough. Table 6 presents a sample of these perspectives with quotes from the interview data.

As noted earlier, the owner of the local (but international) art gallery, also the current president of the Chamber of Commerce, is among a very small group of individuals not born in TCI who has been awarded belonger status for his contributions to the island country. He keeps this certificate proudly displayed on the wall of his shop. He explained that technology adoption fit with his goals and his commitment to the people of TCI.

"In 1989, the original retail operation, the art gallery, was the first of its type in the Turks and Caicos. I decided very quickly that what it would need on the artistic side was to look at the very young generation.... We went out and gave lectures at the local schools. One thing lead to another and, as a result, we have a fairly good number of talented local artists who we not only show, but our concept—and this is where I think technology comes in—was also to provide them with the tools to launch their careers in a domestic market and also internationally." (Interview #7)

Table 5. Summary of individual respondent characteristics

Interviewee & Status	Age	Years Lived in TCI	Prior History with IT	Why TCI?	Prior Experience Developing Countries	Status	Orientation
Trust (01)	30-39	9	Self-trained	BDO; Climate; Tax Status	Asia, E. Europe	TRP	Committed
Int'l Bank (02)	30-39	1	Accounting.	JO	Middle East; S. America	TRP	Committed
Trust (03)	50-59	3	Self-trained	BDO; Tax Status	No	TRP	Colonialist
Trust (04)	40-49	25	Legal background Self-trained	JO	No	PR	Committed
Trust (05)	50-59	4	Mainframe; self-trained	BDO	No	TRP	Colonialist
Service Provider (06)	30-39	1	Engineering Degree	JO	Caribbean	TRP	Committed
Int'l Art (7)	30-39	13	Learned basics **	BDO; Returning	Caribbean	B	Belonger
Tech. Services (08)	40-49	14	Electronic Engineering	BDO; US Proximity	E. Europe; Caribbean	TRP	Committed
Credit Card (09)	60-65	4	Self-trained	JO; Start-up Opportunity	Caribbean; E. Europe	TRP	Committed
Consultant (10)	40-49	1	Self-trained	Friends; US Access	No	TRP	Colonialist
Tech. Services (11)	40-49	3	IT Certificates	BDO	No	TRP	Colonialist
Contractor (12)	--	Born in TCI	Self-trained	Returned to TCI	Caribbean	B	Belonger

** Hired Web consultant to develop their Web site

Legend: BDO=Business Development Opportunity; JO=Job Offer

Table 6. Sample quotes illustrating commitment to infrastructure development

Committed to Island Development	"Colonialist View" of Personal or Firm Benefits
Hire Locally *"We've hired the class valedictorian three years in a row. Our youngest staff is 17 and she is sharp..."* *"We hire the best students and so we get good basic skills."*	Hire top people from outside *"People here have basic education – math skills may be weak."* *"Government policy on immigration is no longer business friendly. We were limited to a 2 year permit for a recent hire and then we have to replace him with a local."*
Train and Educate Locals; Promote from within *"We train them. We encourage them to go to university, even if we can't get them to come back."*	Train for low level jobs *"You cannot create a trust manager or an accountant overnight. Many don't want to leave the island for training, and if they do, they don't come back."*
In difficult times – "Hunker down for the long-haul" *"I'm not worried. This is all part of the cycle. It is normal for developing countries to have hiccups. This is one of them."*	In difficult times – "Seek greener pastures" *"I've let my business license expire. We are looking at development opportunities in other places."*

The respondents who expressed least frustration and most satisfaction with the infrastructure on TCI were those who felt they were working to resolve this dilemma by investing in the training of locals and holding a long-term perspective on building that candidate pool. Others simply relied on a revolving door of short-term expertise and on mainland services where needed.

Information Intensity and Information Use: Influence on Perceptions of Technology Challenges and Coping Strategies

"We were always quite technology oriented. We realized that it would make life much easier in terms of our ability to service customers, if we could automate." (Trust #4 – 25 years on island)

Our interview questions focused on the number of transactions, information seeking and information dissemination behavior, and how this related to perceived challenges. This is consistent with the broader view of information-intensive firms; that is, those who seek to maximize their competitive use of information, rather than those who simply seek better bandwidth (Glazer, 1991, 1993; Proudlock et al, 1998). Table 7 summarizes our assessment of the information intensity of the firms. We used both objective and subjective measures. The objective measures included the estimated monthly volume of e-mail, phone, Internet search, and fax transactions performed, but excluded hits to the firm's Web site. Even the smallest firm engaged in over 600 information transactions each month (international art gallery). The highest volume of transactions exceeded 100,000 a month (credit card issuer).

Table 7. Information intensity

Company	Average Monthly Transactions	Information Seeking	Information Dissemination	Product/ Service Information Intensity	Process Information Intensity
Trust (01)	4000	H	H	4	4
Bank (02)	>25,000	H	L	4	4
Trust (03)	4000	H	L	4	4
Trust (04)	2000	H	M	4	4
Trust (05)	4000	H	L	4	4
C & W (06)	Not provided	H	H	Carrier	4
Int'l Art Gallery (07)	600	L	M	4	4
Technical Support (08)	2500	M	M	--	4
Credit Card (09)	100,000	L	H	4	4
New Venture Consultant (10)	1000	M	M	4	4
Technical Support (11)	4000	H	H	4	4

H=High; M=Medium; L=Low
The contracting firm (#12) has been excluded from this comparison due to lack of relevant data.

Figure 2. Information use

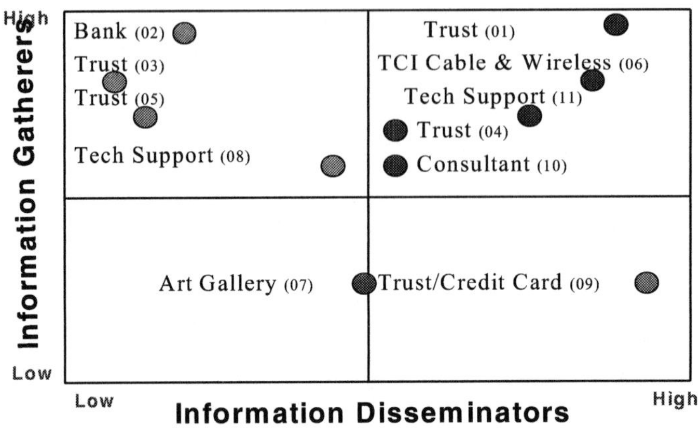

Table 7 also includes our subjective ranking of the firms as high, medium, or low information seekers or disseminators, and whether their information intensity is related to product/service or process. All of the firms rely on information in the *process* of transacting their business. In most cases, information is also a critical part of each firm's *product*, in that their product is expertise disseminated to various global clients using various information or communication technologies. As a global seller of Haitian and Caribbean art, the art gallery is reliant on e-mail and Web-based technologies to market and provide service to customers around the world. Other firms were emphatic in their emphasis that they were heavier information *seekers* as part of investigating and verifying the legitimacy of potential clients. This assessment of information intensity and information behavior is further illustrated in Figure 2.

Technology Use, Challenges, and Coping Strategies

The owner of one of the trust firms casually mentioned that his day had been hectic, since his hard drive had crashed that afternoon and he had raced out to buy a new computer. When asked why he had not simply replaced the hard drive, he replied that it was too much hassle and *"much easier to simply buy an entire new CPU than try to find a hard drive on the island for the existing machine"*. Technical support could be a problem, as there were only two technical support firms providing services in 2000; the younger firm closed in 2002. Facing limited technical support in the islands, businesses are inclined to simply replace their machines with a new one rather than attempt repairs. The smaller size of most firms was actually viewed by the technical experts as an advantage for coping with this problem, as there were fewer machines to be replaced.

We found that these firms relied on a variety of "information" technologies, including the television and some physical transportation technologies. *"We have a television on all the time.... Cable will get you news from the US, Canada, and BBC. That is how I heard about 9/11.... Someone was watching and called me over"* (Trust #3). Whereas some firms had retired the television from their offices with the advent of the Internet, others reported that there was always a television on so that it could be glanced at. Several of the older firms talked about how they evolved from telegraph and early facsimile technologies in the 1980s to e-mail and PC-fax in the 1990s. The Internet was widely touted to be a critical technology for their businesses, despite the costs.

The tropical weather of TCI is an invisible challenge to the nationwide infrastructure, to TCI Cable & Wireless, and for local business. The humidity and salty air greatly decrease the life span of computers and telecommunications technologies. Cable & Wireless addresses hurricanes and gales with towers able to withstand winds in excess of 200 miles per hour. They continually seek ways to reduce maintenance and repairs of corrosion to copper cable, cabinets, and joints. In order to avoid paper loss from hurricanes and to streamline the information management process, some firms expected to adopt document management software. Other firms simply replace equipment and monitor weather during hurricane season.

With the emphasis on information-intensive firms that provided services rather than physical products, the concept of delivery technologies again implies information and communication technologies. We tend to associate transportation technologies and supply chain issues with manufacturing and retail industries rather than with information intensive firms. This means technology requires frequent updating and repairs. Air access for tourists is also a critical lifeline for the financial sector, but it has not always been so easy. It should be noted that TCI's access to regular and frequent air flights only took off in the past decade, and transportation by sea can take several days from the closest port of Miami. Prior to the mid-1990s, shipping by sea was a major delay and, for information-intensive firms, was not always fast enough or convenient. We were told that even now, deliveries by boat are only once or twice a week.

Despite Cable & Wireless' best abilities, the cost of making even local calls is significantly higher than that found in the mainland U.S., the peer reference used by most islanders. The significance of the higher telephone costs in TCI varied from those firms who made relatively few calls daily to those who felt that the historically high costs shaped their culture and communication relationship with customers. To avoid the high cost of a phone call, many firms turned to facsimile or the Internet (primarily e-mail) as a means of communication with clients. Recent DSL capabilities encouraged some firms to turn to Internet telephony, but Cable & Wireless tried to block access to this competitive service.

TCI's tax and information privacy policies also had an impact on small businesses. Security of information was mentioned as a significant issue and one that was pursued to the full degree of available technology, including frequent updating of machinery and software. Nearly every financial firm had formal backup procedures, firewalls, and hacker warning systems.

"We are in an information sensitive business; confidentiality is critical. We need contingencies for loss of office, loss of systems, so we make daily backups and store them in a separate location on island.... Probably the catastrophic situation would be a hurricane that would destroy the office ... or a fire. We face the same problems as anywhere else really. And just as they back up data elsewhere, we backup data here, but perhaps in a less sophisticated manner." (Trust #1)

The level of technological sophistication surprised us. Most of the smaller firms had well under 10 employees, and yet they were all organized around LAN or client/server technologies, and understood the key technology issues on the island and time tables for infrastructure plans. Mostly self-trained, they kept themselves up-to-date through seminars, periodicals, the Internet, and close communication with Cable & Wireless.

The increase in fees for business licenses and work permits and the changes in immigration policy were also highlighted as new and major challenges to business in TCI, as it reduced the pool of skilled labor. While the trust and other financial services companies were quiet "in print," they had their own concerns about restrictions on professional residencies and were lobbying more discretely. These small financial businesses face a paradox in that they require only a handful of employees yet require very specialized skills not easily found in a country with a limited labor pool and tight immigration policy. This human challenge was addressed by spreading the strategic risk of lower skill levels or untrained (and in some cases, untrainable) workers.

Table 8 draws upon selected quotes to summarize and illustrate the main elements of technology infrastructure challenges and coping strategies. It is a long table, but helps match challenges and coping in the voice of the respondents.

Table 8. Quotes illustrating technology challenges and coping strategies

Selected Quotes to Illustrate Challenges	Technology Challenge	Coping Strategy	Selected Quotes to Illustrate Coping Strategies
"Technical support used to be all telephone based. We used to pay two or three thousand dollar monthly phone bills back in 1992,'93,'94. I can show you almost $100,000 US in telephone bills in those years. You are hurting under those conditions. Any technical support call used to cost $50, $60, or $100." (Computer Services Firm). *"Our telephone costs are still $3,000 to $6,000 a month, and that is only for our trust division. The other units have about the same." (Trust)*	High Telephone & Internet Costs	Technology Use & Communication Culture:	*"Telephone rates have come down steeply since last year. We had a culture that said, 'We don't call clients – clients call us.' We still tend to do that. You telex or fax clients to call you. We communicate in writing instead of voice.... Now it's so ingrained, it's not really going to go away until some of us retire." (Trust)*
"We couldn't get things here quickly enough before." (Trust)	Transportation & Import Problems	Pioneer Self-Sufficiency	*"That is initially how and why I got into flying. We bought a plane and shipped our own hardware or parts in, as well as building materials, office supplies.... We established our own import/export business in Miami." (Trust)*
"Power surges and outages are normal."	Power Outages	Culture of Technical Urgency	*"When the alarms go off it sounds like we are under an armed attack. We tell the staff they have 12 minutes to back out of everything and shut everything down. They really have 15 minutes. They do it in seven minutes. The last time was six weeks ago. The staff is trained, cross-trained, and tested on this." (Credit Card Issuer describing what happens during a power outage.)*
"Security is the biggest concern we have.... The risk of viruses... or if anybody was ever to hack into our system. There is a lot of confidential information." (Credit Card Issuer) *"We get pinged and I get notified when a hacker tries to break in. Last time I checked it is about 25 attempts a day, but not one gets in."*	Security of Data; Customer Privacy	Risk Assessment Standards & Protocols Self-Reliance	*"We are constantly assessing the risks. We probably upgrade our firewalls every three months." (Credit Card Issuer)* *"We back up daily and keep disks off-site." (Trust)* *"We change the on-island location where we store our backups every day. Only two of us know where they are."* *"In the event of a hurricane, we usually have at least 24- hour notice. That is enough time to lift the hard drives and fly to Miami." (Trust)*

Table 8. Quotes illustrating technology challenges and coping strategies (cont.)

Selected Quotes to Illustrate Challenges	Technology Challenge	Coping Strategy	Selected Quotes to Illustrate Coping Strategies
"Even with a low amount of rain, the humidity is very high, very salty. Here, if you don't keep the computers in the right air conditioned environment, if it is near the window and the humid air blows directly on top... well, the computer craps out in one year.... Corrosion of the board level component." (Technical Services Firm)	Oceanic Tropical Climate ("Salt Air") and *Firm* Computers	Acceptance of Early Technical Obsolescence	*"Because the businesses are small, they don't have as many concerns with changing or replacing the computer..."(Technical Services Firm)*
"Cable & Wireless has to deal with weather type issues, you know, the salt, the storms, all the rest of it. This is an island nation." (Technical Services)	Oceanic Tropical Weather and Country-*Level* Technology Infrastructure	Invest in Technical Innovations	*"Fiber optics, heavy duty towers, bite the bullet and maintain, maintain, maintain."* *"Remotes will help because we'll put fiber between the remotes and the exchange and the fiber is impervious to water. Fiber optics is laser light as opposed to copper electricity. As I mentioned earlier, it reduces the propensity for error and faults." (Cable & Wireless)*
"One of the biggest problems facing everybody on this island is that the government's recent policy with respect to immigration has changed very adversely to business. As an example, in the past, if you were a business owner you were pretty much assured a work permit for five years. At the end of five years, they would give you another one for another five years. Now, at the end of five years, your permit is gone and you have to leave unless you do not need to work. It used to be that at the end of 10 years you could convert to what is known as a permanent resident certificate (PRC) with a right to work here. So the PRC was a big feature." (Trust)	Increased Business Fees and Limited Skills Retention	Hunker down for the long haul OR Seek greener pastures	*"I'm not worried. This is all part of the cycle. It is normal for developing countries to have hiccups. This is one of them." (Trust)* *"I've let my business license expire. We are looking at development opportunities in other places." (concealed)*
"You've got a very small pool of labor here." (Technical Services)	Limited Labor Pool; Lack of Technical Skills	Train & Retain Contacts and Connections	*"What labor we have, we try to keep hold of because we educate them and we pay them quite well." (Cable & Wireless)* *"We have a local guy who supports us. He's here within five minutes and on call 24 hours a day." (Credit Card)*
"We have to employ local people. So we are in a position where we don't necessarily have fully qualified staff for certain positions. We have to allow for that in the way we manage the company." (Trust) *"I think that the only thing that we go against here is assuming too much. We are very happy if someone has good word processing, and are delighted if they have spreadsheet skills." (Trust)*	Lack of Technical Skills	Spreading Strategic Risk Realign Expectations	*"You just expect lower standards. Your information needs do not change because of available skills. You have your trust officers, who know what they are doing. They handle all your client stuff so that you don't mess up your clients. Then you have other tasks looked after by the less competent people because that is your risk and you can take that risk." (Trust)* *"Probably most are looking for people who are just general computer savvy. Somebody who is familiar with Windows and who knows their way around. We can find that." (Trust)*

What emerges from this analysis is a "pioneer self-sufficiency" typical of many entrepreneurial firms and small owner-operated businesses, but taken to a higher level. Unlike larger firms where there might be an IT department to phone when there are problems, these small businesses had to figure out their own ways to preempt disaster and quickly guarantee assistance when they needed it. For an information-intensive firm in a developing island nation, time is of the essence.

Competitive Position, Customer Satisfaction, and Self-Efficacy

Generally speaking, these companies are content with their ability to compete on a global basis. Information, communication, and Internet technologies have enabled a new freedom that was previously unavailable when similar tax havens were established, such as the Cayman Islands or Bermuda.

"I think in this day and age the differential has narrowed very substantially. When I first arrived here, there was no Internet; everything was done by fax. So there were problems associated with communication and delivery of information. Now you can scan a document and e-mail it in five minutes. I think the differential between us and anyone else has narrowed. We are on par with New York." (Trust)

As noted in the literature, even firms not traditionally thought of as information-intensive have increasingly become creative with and then dependent upon their use of IT. In a developing island nation, the isolation may, in fact, be working contrary to other research findings (Abdul-Gader & Kozer, 1995), as geographical isolation and transportation difficulties encourage faster adoption by small businesses of alternative technologies for customer relationship management.

"Collectors are serious about their collections. Years ago, prior to '96 or '97, we would have to do things the old fashioned way with photos mailed out. It would take many weeks. Now of course, with the help of the Internet, the jpeg attachment goes off to the Ivory Coast and a collector knows within minutes what we have. I think that is a good example how a small island company can reach out far beyond its borders and find an audience out there. As I said earlier, we started 12 years ago and retail became e-commerce—or e-commerce was added to retail—simply because the technology was suddenly there. It wasn't because we were terribly clever." (International Art Gallery)

Customers were also reported to be satisfied with the firms' services, but brought an outsider's level of expectation for reliable technology.

"Fax line quality has been atrocious here.... It led to a lot of client frustration and a lot of frustration in the office here trying to send and receive faxes that are not going through.... We had a lot of difficulty getting and maintaining dial-up connections for quite a period of time." (Trust)

The major complaints over the quality of phone service reflect the impact of the telecommunication systems on customer service. How to find relatively inexpensive ways to communicate with the customers has been an important issue in customer relationship management in TCI. Many of the firms' owners still travel regularly to visit clients.

In terms of personal self-efficacy, literature would suggest that perceived lack of technical support on the island would affect personal satisfaction, in that firms need to rely on off-island support for purchasing and maintenance, which, in turn, causes difficulty in managing time and spending. However, not everyone perceived the technological infrastructure in this developing island country as a constraint. Many had simply evolved and developed self-sufficiency or tolerance for the idiosyncrasies of a Caribbean infrastructure. Some held optimistic views and reported that they have everything they need to do business, and so the infrastructure of the island does not affect their satisfaction or their clients'.

"TCI has telecommunications. It has Internet access. It has e-mail access. It has magazines.... Cable TV will get you news from the U.S., Canada, and the BBC. I am not sure what else one would need." (New Venture Consultant)

The president of the Financial Services Commission in TCI summed up the entrepreneur's necessity to remain competitive through services and technology.

"Legislation can bring in business, but the essential thing is having the right product: it is the quality of the financial services industry which ultimately brings people to TCI." (Borsuk, 2001).

CONCLUSION

This case study of coping strategies for small information-intensive firms in the TCI adds to the growing body of case studies and empirical research on developing countries. It also provides data for comparisons to other Caribbean or small developing countries. The value of this study is somewhat limited in the small number of firms involved; however, it should be noted that about 15% of the current members of the Chamber of Commerce were included. In addition, the longitudinal nature and multiple data collection methods helped ensure a balanced and comprehensive view of the challenges faced by these businesses in TCI. Only page constraints limit richer description and stories. The first author has lived in the Caribbean; the second author grew up in another developing country; and so the data was also interpreted through both an academic lens and through the eyes of personal experiences. The generalizability of the study applies well to other countries facing similar challenges and offers insights to those wishing to understand the plight and contributions of smaller information-reliant firms. Perhaps the most significant contribution, however, is in capturing the historical infrastructure development of this small island country that has grown from infancy to adolescence in less than two decades.

Implications for Research

Despite the reference to "paradise" in the title, for smaller organizations there is considerable creativity required to cope with the uncertainty of a developing island economy and the growth pains associated with infrastructure development.

This study provides several lessons for researchers interested in how developing countries adopt or diffuse information technologies. These lessons include: appreciating the challenges of smallness; considering the role of smaller business entities in infrastructure development; being sensitive to dynamism; and exploring how information and communication technologies are becoming critical drivers and saviors for many firms not normally included as information intensive.

As much as was said during interviews, there were many instances where people held back. In developing countries, especially small ones, researchers must be sensitive to the subtle network of influence and politics within which businesses must function. There are close relationships that exist between businesses, and there are often even closer familial ties with government officials. Gossip travels quickly and can have negative ramifications. In these settings, personal face-to-face contact may be much more effective than large-scale surveys. This has implications for costs and the breadth of data collected; however, it will result in a richer understanding of the country and firm dynamics. Outside sources of evidence and repeated exposure may also be needed to supplement what is said, or not said, even in face-to-face interviews.

As researchers of information technology and infrastructure issues, we too often look at the large software development projects, the ERP implementations, and the massive technological changes faced by large organizations and governments. We forget, however, that the largest portion of the Gross Domestic Product (GDP) in nearly every developed country in the world comes from the efforts of smaller, entrepreneurial business enterprises. With the pervasiveness of information, communication, and Internet technologies in developed countries, to a large extent we seem to have given up our study of end-user sophistication and individuals' roles in infrastructure development (Abdul-Gader & Kozer, 1995; Igbaria, Zinatelli, Cragg, & Cavaye, 1997; Mansell & Wehn, 1998; Zinatelli et al., 1996). In developing countries, the pioneering self-sufficiency and entrepreneurial orientation of individual business owners may be just as important to IT diffusion as existing IT infrastructure and governance policy. It behooves us to expand our study of the roles that these smaller businesses and educational facilities play, especially in smaller developing countries that may not sustain large manufacturing firms.

We refer to developing countries because they are, in fact, evolving—often at a rapid and unstable pace. The unexpected increase in business licenses and residency permits in TCI caused a sudden and unanticipated instability in an already dynamic economic environment. As a result, over half of the original interviewees left the island 18 months later. It was necessary to continually maintain contacts through electronic means and in person. This is a time-consuming and expensive, but necessary, process for researchers.

The impact of information intensity, and specifically information seeking versus information dissemination behavior, also may be an important distinguishing element in future research on infrastructure development in developing organizations. It has been noted that information and communication technologies can help developing countries leapfrog developmental stages (Steinmueller, 2001). Those with the highest concerns and the highest level of coping strategies were those who were higher information seekers, as well as those who were high information disseminators.

Implications for Practice

Despite technological advances and an upscale clientele, an old-fashioned pioneering spirit is evident in TCI among these small businesses. Even those who question how long they will stay in TCI are clear on one thing: you have to be self-sufficient. As with most small, entrepreneurial firms, a few people have to be able to do a lot of different things. In the case of doing business in a developing island nation, however, self-sufficiency extends to understanding information security, supply chain management, and IT evolution on a broader country level.

In defense of those categorized as colonialists, the uncertainty felt by those seeking to build a life in TCI has only increased and needs to be considered. For example, after losing a large contract due to the economic downturn of her major client (prior to September 11, 2001), one entrepreneur, who could potentially have helped train and develop island workers, instead let her business license lapse. Newspaper articles abounded with pleas from local business owners for recognition of the difficulties they faced. It was too late for many of the smaller businesses by the time some in TCI were lowered some months later. Many permanent resident certificates are still being denied, even for those who have committed significantly to the island for 10 years or more. The contributions and value provided by smaller firms need to be assessed by the government.

Due to a dependence on imports, including inventory, equipment, and maintenance, the geographical proximity to the mainland (e.g., Miami) was viewed as an advantage. Like the other island nations (or city-states) such as Ireland, Singapore, and Hong Kong, TCI's proximity to larger neighbors and effective sharing of Cable & Wireless' adjoining Caribbean and mainland resources have leveraged TCI's technical capabilities (Martinsons, 1997; O'Higgins, 2002). The dependency on imports, however, also results in higher costs and vulnerability to outside forces and events. As a recent example, TCI was completely cut off during the September 11 crisis.

An issue associated with self-sufficiency was the importance of developing technology support connections to ensure firm survival. All of the firms explained that they had a good relationship with a reliable off-island or on-island technical supplier and/or alternative technical support that could be called upon 24 hours a day. The importance of being able to keep their systems up 24/7 again highlights the true information intensity of these firms. The proposed deregulation of the long-distance market may lead to greater challenges as TCI's Cable & Wireless may lose its incentive to invest so generously in supporting such a small population (Miller, 2003). The government has relied heavily on Cable & Wireless to help them leapfrog into the 21st century. The urging of smaller businesses may be needed to maintain pressure for a high level of ongoing infrastructure support.

Surprising similarities exist when looking at the challenges faced by large organizations and small countries. There may be lessons to be learned from the smaller, information-intensive businesses trying to innovate and grow in a developing country. There are significant governance challenges faced by large organizations in building the parts while coordinating the whole. These governance challenges are replicated when looking at monopoly IT infrastructure firms such as TCI's Cable & Wireless and the smaller businesses that have to develop their own IT strategies and firm infrastructure. There also may be lessons found for these smaller countries in research on large organizational IT governance; and in turn, there may be lessons for the larger firms in the challenges and successes of smaller countries.

It has been said, "It takes a village to raise a child." We think, in the case of TCI, it takes a community to raise the nation. Rather than seeing government, large international business, small business, indigenous populations, and newcomers as being more or less responsible for the well-being of the child, or more or less committed, it makes sense to look at each stakeholder as providing a rich and varied background to nurture this adolescent island nation. Having a fondness for the island and for the people there, we see a variety of individuals and groups who admittedly have differing levels of commitment to TCI's development; however—and it is a large "however"—we do not believe that the interests of these groups are diametrically opposed. On the contrary, even personal self-interest can be turned to the benefit of the island through supportive policy.

EPILOGUE

The first author continued to gather data in TCI into 2004, including additional interviews with three businesses in the tourism and restaurant industries. These three firms had been on the island for 12 to 17 years and employed 35 to 55 workers. Their experiences were similar to those reported by the other firms, with most of their skilled workers doing short-term rotations on the island, and locals (including belongers) filling non-technical positions. The impact of work permit license fees has continued, as has the idiosyncratic and highly politicized nature of work permit approvals (Bolderson, 2003). By mid-2004, only three of the firms we initially studied remained in TCI.

Cable & Wireless is no longer an official monopoly provider of telecommunications and Internet services (Miller, 2003); however, smaller providers still face challenges from competition with this incumbent provider that has over 100 years of history in TCI and the Caribbean. Increased competition has allowed for broader cell phone and Internet services, but this has not solved the labor problems or lack of skilled workers to support automation for information-intensive small businesses or government offices. Labor pressures continue as the government promotes the island as having a skilled labor base, yet policy restricts technology workers and perpetuates an unstable and unfriendly employment climate.

Hotels, resorts, and restaurants are recovering from the downturn caused by global economic changes and the aftermath of the September 11, 2001, terrorist attacks in the U.S. Unfortunately, many firms import their skilled workers on six-month rotations or do their information processing and reservation services through U.S. mainland offices. Many of the resort operations are all-inclusive facilities that do not encourage their guests to patronize local small business. Better access to telecommunication and Internet services has only made it easier for firms to use mainland services.

Ironically, these larger companies have significantly larger investments and can influence government and win concessions due to the higher employment they provide and the greater volume of work permit revenues they bring to island coffers. Regrettably, these firms often do not contribute in direct or lasting ways to technology skill development or to IT infrastructure development in TCI. The small businesses that could be part of changing the culture of TCI are increasingly struggling to remain in business.

ACKNOWLEDGMENT

Our sincere thanks to the participants for their time and sharing of their experiences, to the School of Information Studies at Syracuse University for their financial support, and to the anonymous reviewers who have shared their insights and suggestions.

REFERENCES

Abdul-Gader, A., & Kozer, K. A. (1995). The impact of computer alienation on information technology decisions: An exploratory cross-national analysis. *MIS Quarterly, 19*, 535-559.

Agha, S. S., & Akhtar, S. (1992). The responsibility and the response: Sustaining information systems in developing countries. *Journal of Information Science, 18*(4), 283-292.

Anonymous (2001a, August 9-23). Hot taxes on the hot plate. *Turks & Caicos Free Press*, Front Page.

Anonymous (2001b, August 11-17). Native business community plans to petition gov't on tax increases. *TC Weekly News*, p. 3.

Bagchi, K., Hart, P., & Cerveny, R. (2002). Information technology diffusion in Latin America: Some empirical results. *The 8th Americas Conference on Information Systems*, pp. 1519-1523.

Ballantine, J., Levy, M., & Powell, P. (1998). Evaluating information systems in small and medium sized enterprises: Issues and evidence. *European Journal of Information Systems, 7*, 241-251.

Bingi, P., Leff, L.G., & Shipchandler, Z. E. (2000). Critical IT implementation issues in developed and developing countries. *Information Strategy, 16*(2), 24-34.

Bolderson, C. (2003). Paradise lost: Anthony Graham moved to idyllic Turks and Caicos for a dream new start. Ten years on he's bitter and back home, a victim of the islands' rising xenophobia. *The Voice*, pp. 14.

Borsuk, K. (2001). Quality assurance. *TCI News.*

Davy, H., & Allgood, B. (2002). Offshore development, building relationships across international boundaries: A case study. *Information Strategy, 18*(3), 13.

DeLone, W.H. (1988). Determinants of success for computer usage in small business. *MIS Quarterly, 12*(1), 51-61.

Dodsworth, P.J. (2000). Turks & Caicos Islands: Offshore insurance bastion. *The International Magazines of Turks & Caicos Islands*, Vol. 2002.

EIU (2004, February 10). *Turks and Caicos Islands: Country outlook,* pp. n/a.

Fuller, T., & Jenkins, A. (1994). *The learning needs associated with adoption of IT by small businesses* [Paper]. Internationalising Enterprise Education & Training, Durham University Business School.

Glazer, R. (1991). Marketing in an information-intensive environment: Strategic implications of knowledge as an asset. *Journal of Marketing, 55*(4), 1-19.
Glazer, R. (1993). Measuring the value of information: The information intensive organization. *IBM Systems Journal, 32*(1), 99-110.
Igbaria, M., Zinatelli, N., Cragg, P.B., & Cavaye, A.L. (1997). Personal computing acceptance in small firms: A structural equation model. *MIS Quarterly, 21*, 279-305.
Kaarst-Brown, M.L., & Evaristo, J.R.E. (2002). International cultures and insights into global electronic commerce. In P. Palvia, S. Palvia, & E. Roche (Eds.), *Global information technology and electronic commerce*. Marietta, GA: Ivy League Publishing.
L'agefi (2000, July). Special issue on Turks and Caicos Islands.
Lee, A.S. (1989). Case studies as natural experiments. *Human Relations, 42*(2), 117-137.
Lincoln, Y.S., & Guba, E.G. (1985). *Naturalistic inquiry*. Thousand Oaks, CA: Sage Publishing.
Mansell, R., & Wehn, U. (1998). *Knowledge societies: Information technology for sustainable development*. New York, NY: Oxford University Press.
Martinsons, M.G. (1997). IT policies & information infrastructures: Comparing Hong Kong to the Singapore model. In J. Burn, & M.G. Martinsons (Eds.), *Information technology and the challenge for Hong Kong* (pp. 27-60). Hong Kong: Hong Kong University Press.
Miles, M.B., & Huberman, A.M. (1994). *Qualitative data analysis* (2nd ed.). Thousand Oaks, CA: Sage Publications.
Miller, D.J. (2003). Economy -Carib: Open market telecom era brings new challenges. *Global Information Network, 1*.
Montazemi, A.R. (1988). Factors affecting information satisfaction in the context of the small business environment. *MIS Quarterly, 12*(2), 239-256.
O'Higgins, E.R.E. (2002). Government and the creation of the celtic tiger: Can management maintain the momentum? *Academy of Management Executive, 16*(3), 104-120.
Petrazzini, B., & Kibati, M. (1999). The Internet in developing countries. *Communication of the ACM, 42*(6), 31-36.
Proudlock, M.J., Phelps, B., & Gamble, P.R. (1998). IS decision-making: A study in information-intensive firms. *Journal of Information Technology, 13*, 55-66.
Raymond, L. (1985). Organizational characteristics and MIS success in the context of small business. *MIS Quarterly*, 9(1), 37-52.
Sadler, H. E. (1997). *Turks island landfall: A history of the Turks & Caicos Islands*. Kingston, Jamaica: United Cooperative Printers, Inc.
TCInvest (2003a). TC investment getting started: Work permit. Vol. 2003.
TCInvest (2003b). TC investment opportunities: The information age has arrived. Vol. 2003.
Travica, B., Johanson, E., & Leon-Pereira, M. (2000). Caribbean electronic commerce: Exploration of Costa Rica and Venezuela. *AMCIS Conference*, Long Beach, CA.
Van Beveran, J., & Thomson, H. (2002). The use of electronic commerce by SME's in Victoria, Australia. *Journal of Small Business Management, 40*(3), 250-253.
Yin, R.K. (1994). *Case study research: Design and methods*. Thousand Oaks, CA: Sage Publications.

Zinatelli, N., Cragg, P.B., & Cavaye, A.L.M. (1996). End user computing sophistication and success in small firms. *European Journal of Information Systems, 5*(3), 172-181.

ENDNOTES

1 An earlier version of this chapter was published in Palvia, Palvia, & Roche (2002) "Global Information Technology and Electronic Commerce". Ivey Publishing.

2 We discovered population figures ranging from 20,000 to 24,000. What was often added during discussions is the fact that only a small portion of these people really belong here, referring to the "belonger" status.

Chapter XIV

National Environment and Internet Banking Adoption: Comparing Singapore and South Africa

Irwin Brown, University of Cape Town, South Africa

Adrie Stander, University of Cape Town, South Africa

Rudi Hoppe, University of Cape Town, South Africa

Pauline Mugera, University of Cape Town, South Africa

Paul Newman, University of Cape Town, South Africa

ABSTRACT

The widespread diffusion of the Internet globally has prompted most retail banks to offer Internet banking services. A recent study in Singapore identified attitudinal and perceived behavioral control factors as having an influence on consumer adoption of Internet banking. It is expected that the national environment will also impact this process. The aim of this study, therefore, was to replicate the Singapore study in South Africa, compare the results between the two countries, and explain differences in adoption process in terms of the national environment. The results confirm that attitudinal and perceived behavioral control factors influence adoption in South Africa as in Singapore, but with differences in the number of determinants and the degree of influence of certain determinants. These differences were explained in terms of three environmental dimensions: socio-economic conditions, the state of Internet diffusion, and government information and communication technology (ICT) policies, respectively.

INTRODUCTION

The Internet and the World Wide Web (Web) have had a big influence on retail banks, with many now providing products and services via the Internet (Aladwani, 2001). A brief perusal through several bank web sites reveals that consumer services available via Internet banking include the ability to access accounts in order to check balances; view, download, or print statements; transfer funds; and pay bills. Additional services include facilities for loan applications, share trading, and secure online shopping, among others. Many of the services offered are also available via other channels; thus, the primary benefit of this channel may be its convenience (Leonard, 2002).

The success of Internet banking depends to a large extent on consumer acceptance of this technology, and, to this end, Tan and Teo (2000) identified several factors that influence the adoption of Internet banking in Singapore; namely, attitudinal and perceived behavioral control factors. Characteristics of the national environment may also influence the adoption process, giving rise to differences between countries. The objectives of the study are, therefore, to:

1. examine factors influencing the adoption of Internet banking in South Africa;
2. compare these findings with Singapore; and
3. explain differences in the Internet banking adoption process in terms of national environmental characteristics.

It is important to capture the influence of the national environment. Contrary to early popular opinion, the Internet and electronic commerce have not leveled the playing field between developed and developing nations. Instead, those countries with well-developed infrastructures, resources, and skills have gained the initial advantage, giving rise to a digital divide between developed and developing nations (Licker & Motts, 2000; Mbarika, 2002). It is pertinent to note also that "to what a user has access to, and why depends on the specific legal, economic, political, and social conditions that surround that user (Wolcott et al., 2001, p. 3). In terms of social conditions, the influence of national culture has gained widespread attention in studies of technology adoption (Anandarajan et al., 2002; Straub et al., 1997). However, it is clear that national culture is but one element of a complex, multi-faceted phenomenon. The contribution of this study, therefore, is to seek other descriptors of national environment apart from culture and use these to explain differences in the adoption of Internet banking between Singapore and South Africa.

Singapore and South Africa are of interest, as they lead their respective regions in terms of technological advancement, while also having significant geographic and socioeconomic differences. Important lessons may be learned, particularly for South Africa, since Singapore already has a successful, highly developed economy with a sophisticated ICT infrastructure. In contrast, South Africa is a middle-income developing country, striving to bridge the economic, social, and digital divide between different societal groups (Licker & Motts, 2000).

In the next section, the environmental factors likely to impact Internet banking adoption processes in Singapore and South Africa are compared. This is followed by a brief description of the Internet banking framework employed by Tan and Teo (2000) before the research framework and hypotheses for this study are outlined. The results of the South African study are reported and compared with Singapore, leading to a discussion of the major findings. Limitations of the study and ideas for future research are detailed before the chapter is finally concluded.

NATIONAL ENVIRONMENTAL CONTEXT

Internet banking, and banking in general are targeted at those with financial means. Thus, in a national context, socioeconomic conditions, which affect general income level and levels of affluence, will be relevant to this innovation. Similarly, Internet adoption will depend on the necessary technological infrastructure being in place and the ability of consumers to use this technology. Thus, the state of Internet diffusion in a nation becomes relevant. Government ICT policies and plans, if successfully implemented, can promote the adoption of innovations such as Internet banking, as evidenced in Singapore (ITU, 2001c). These three environmental dimensions—socioeconomic conditions, Internet diffusion, and government ICT policy—will be used to describe and compare the national environment in Singapore and South Africa. Given that Internet banking adoption is the central focus of the study, the general state of Internet banking also will be discussed.

Socio-Economic Profile (CIA, 2004)

Socioeconomic indicators for Singapore and South Africa are shown in Table 1. In brief, there are differences in terms of geographic area, population size, and economic status. While Gross Domestic Product (GDP) is about four times higher in South Africa, GDP per capita is about half that of Singapore. There are similarities in that both have multi-ethnic societies in which English is one of the main languages used for official purposes.

Government ICT Policy

The Singapore government is proactive in supporting and promoting ICT usage, positioning itself as the region's high-tech hub (CIA, 2001; ITU, 2001a). The telecommunications market has been gradually liberalized with full competition introduced in 2000 (ITU, 2001c). The recent IT2000 plan, which aimed to transform Singapore into an

Table 1. Socio-economic profile: Singapore and South Africa (CIA, 2004)

Item	Singapore	South Africa
Land Area	682.7 sq km	1,219,912 sq km
Population	4.3 million	42.7 million
Ethnic Groups	Chinese (76.7%); Malay (14%); Indian (7.9%); Other (1.4%)	Black (75.2%); White (13.6%); Colored (8.6%); Indian (2.6%)
Official Languages	Chinese, English, Malay, Tamil	Afrikans, English, Ndebele, Pedi, Sotho, Swazi, Tsonga, Tswana, Venda, Xhosa, Zulu
Literacy Rates	93%	86%
Economy	Highly developed and successful free-market economy.	Middle-income developing. Overcoming disparities created by apartheid.
GDP (PPP)	USD 109.1 billion	USD 456.7 billion
GDP Per Capita (PPP)	USD 23,700	USD 10,700

intelligent island, has been largely fulfilled. Consequently, the national information infrastructure includes a nationwide broadband network, reaching nearly every household and institution (IDA, 2001). South Africa has a sophisticated communications infrastructure with an emphasis on providing universal telecommunications access (ITU, 2001b). The monopoly on fixed line telecommunications provision officially ended in May 2002, but, as yet, with no second national operator in operation (Goldstuck, 2002a). An Electronic Communications and Transactions (ECT) bill was passed in June 2002 with some controversy surrounding the perception that government is aiming to control the Internet domain name administration (*The Star*, 2002). A more recent government announcement on the liberalization of the telecommunications sector has been met, however, with strong support and approval from a variety of sources (*BusinessDay*, 2004).

Internet Diffusion

National Internet diffusion can be described in terms of the Global Diffusion of the Internet (GDI) framework, which consists of six dimensions: pervasiveness (number of Internet users per capita); sophistication of use; geographic dispersion; sectoral absorption (a measure of Internet use in the four sectors of commerce, education, health, and public service); connectivity infrastructure (extent and robustness of the physical Internet structure); and organizational infrastructure (number of ISPs and the competitive environment) (Wolcott et al, 2001). A more detailed comparison between South Africa and Singapore along these dimensions is described in Appendix 3, with a high-level overview presented below.

While there are up to one in two Singaporeans accessing the Internet, in South Africa it is estimated at one in 15 (3 million) (Goldstuck, 2002a). Furthermore, a comparison of online consumer sales figures indicates the use of the Internet for electronic commerce transactions is greater in Singapore (Goldstuck, 2002b; SaVoix, 2002). Achieving wide geographic Internet dispersion has been easier for Singapore, given its smaller size. Sectoral absorption also may be greater in Singapore, with the high profile government ICT strategies having an impact. In South Africa, absorption is gradually increasing across all sectors, but falls behind Singapore, especially in the public sector. Much government attention has been on addressing historical inequalities in society, however (*BusinessDay*, 2004). In terms of Internet connectivity and organizational infrastructure, Singapore once more may be ahead, as evidenced by the greater international bandwidth and widescale broadband access to consumers (ITU, 2001a). Table 2 displays some common ICT statistics.

Table 2. ICT statistics (Singapore and South Africa)

Item	Singapore	South Africa
Telephone lines	1,949 million (ITU, 2002)	4,969 million (ITU, 2002)
Main lines per 100	47.2 (ITU, 2002)	11.35 (ITU, 2002)
Avg. local call cost (USD/3 min)	0.0 (World Bank, 2004)	0.1 (World Bank, 2004)
PCs per 100 people	50.83 (ITU, 2002)	6.85 (ITU, 2002)
Internet Hosts per 10,000	479.2 (ITU, 2002)	54.5 (ITU, 2002)
Internet Users	1.5 million (ITU, 2002)	3.07 million (ITU, 2002)

State of Internet Banking

All major retail banks in South Africa offer Internet banking services, with a total just over 1 million accounts (Goldstuck, 2004). In Singapore, an A.C. Nielson survey put the number of Internet banking users at 391,000 in 2001 (CNETAsia, 2002).

FACTORS INFLUENCING INTERNET BANKING ADOPTION

The Internet banking adoption framework used for this study was based on the framework developed by Tan and Teo (2000), as shown in Figure 1. As with other studies that have examined consumer adoption of online banking (Liao et al., 1999), the framework was derived from existing theories, mainly the theory of planned behavior (Ajzen, 1985) and the diffusion of innovations theory (Rogers, 1983). The sound theoretical underpinning of this framework was a major factor in deciding to use it, as was its comprehensiveness. Using an existing framework furthermore allows for easy comparisons to be made between studies.

Figure 1. Factors influencing Internet banking adoption (amended from Tan and Teo, 2000)

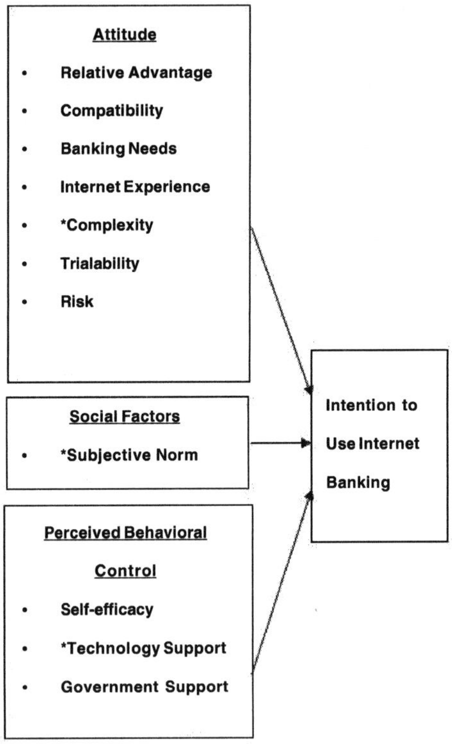

** There was subsequently no empirical support for the influence of these factors on Internet banking adoption.*

Figure 1 illustrates that consumer intentions to use Internet banking are determined by attitudinal, social, and perceived behavioral control factors, with those shown to empirically influence adoption highlighted in bold. A brief discussion of the framework follows.

Attitudinal Factors

Attitudinal factors include the following:

- **Relative Advantage.** Relative advantage is defined as the extent to which a person views an innovation as offering an advantage over previous ways of performing the same task (Rogers, 1983, cited by Agarwal & Prasad, 1997). Agarwal and Prasad (1997) found relative advantage to be a major influence on intentions to use the WWW, which concurs with the findings of Tan and Teo (2000) on Internet banking.
- **Compatibility.** Compatibility has been defined as the degree to which an innovation is viewed as being consistent with the existing values, needs, and experiences of a user (Rogers, 1983, cited by Taylor & Todd, 1995). Tan and Teo (2000) found that the greater the compatibility of Internet banking with user values, the more likely that it would be adopted.
- **Banking Needs.** It is expected that those who use a wide variety of banking products and services would be more likely to adopt Internet banking (Tan and Teo, 2000).
- **Internet Experience.** Prior similar experience of using a technology is likely to influence further adoption of similar types of technology (Tan & Teo, 2000). Tan and Teo (2000) assessed Internet experience in terms of number of years of use, frequency and intensity of use, diversity of use, and Internet skill level, and found that the greater the experience with using the Internet, the more likely that Internet banking would be adopted.
- **Complexity.** Complexity can be defined as the degree to which a person believes that using a particular technology will demand a great deal of effort—the inverse of perceived ease of use as defined by Davis (1989) in the technology acceptance model (TAM). Jiang et al. (2000) state that a rapidly diffusing, user-friendly technology such as the Internet/WWW makes complexity of no great significance to adoption decisions. Tan and Teo (2000) also found no significant relationship between complexity and adoption.
- **Trialability.** Trialability is defined as the extent to which users would like an opportunity to experiment with the innovation prior to committing to its usage (Rogers, 1983, cited by Agarwal & Prasad, 1997). Tan and Teo (2000) demonstrated that the greater the trialability, the more likely that Internet banking would be adopted.
- **Risk.** Risk may be defined as the consumer concern about security and privacy when transacting over the Internet (Tan & Teo, 2000). Aladwani (2001) reports Internet security to be a key challenge from the bank management perspective. Tan and Teo (2000) also showed consumer security concerns to negatively influence Internet banking adoption, justifying bank management attention.

Social Factors

The social factors dimension is represented in the framework in Figure 1 by subjective norm, defined as a person's perception that most people who are important think that he or she should perform a behavior (Tan & Teo, 2000). The influence of subjective norm on adoption of Internet banking was not proven in the Tan and Teo (2000) study.

Perceived Behavioral Control

The perceived behavioral control dimension was assessed by the following constructs:

- **Self-efficacy.** Self-efficacy may be defined as an individual's self-confidence in their ability to perform a behavior (Taylor & Todd, 1995). Tan and Teo (2000) demonstrated that the greater the self-efficacy towards using Internet banking, the more likely that Internet banking would be adopted.
- **Technology Support.** Tan and Teo (2000) hypothesized that the greater the extent of technology support in the environment, the more likely that Internet banking would be adopted, but no empirical support for this relationship was demonstrated.
- **Government Support.** It was demonstrated that the greater the extent of government support for Internet commerce, the more likely that Internet banking would be adopted (Tan & Teo, 2000).

DEVELOPMENT OF THE RESEARCH FRAMEWORK AND HYPOTHESIS

While there is conceptual justification and empirical support for the framework in Figure 1, in differing environmental contexts, the relative influence of factors may vary. Straub et al (1997), for example, demonstrated that national culture influenced the technology adoption process. Therefore, it is expected that national environment would also affect consumer Internet banking adoption. Cultural values are but one element of the overall environment. As discussed in Section 2, other dimensions include socioeconomic conditions, Internet diffusion, and government ICT policies. Theories supporting these dimensions as possible environmental influences on technology adoption processes are discussed next.

General socioeconomic conditions determine the range of banking products and services that would be in demand and whether the Internet would be a feasible channel of delivery. This assertion is consistent with the diffusion of innovations theory in which it is stated that adoption depends, among other things, on the social system of potential adopters (Rogers, 1995, cited by Wolcott et al., 2001).

The feasibility of the Internet as a channel for banking would depend on the state of Internet diffusion in a country. Wolcott et al (2001) state that their "Internet diffusion model does not start with the elements that directly influence the adoption decision, but looks at elements one or more levels back in the chain of determination, in order to find which factors make the most difference in any given country" (p. 35). Thus, this model

Figure 2. Research framework (amended from Tan and Teo, 2000)

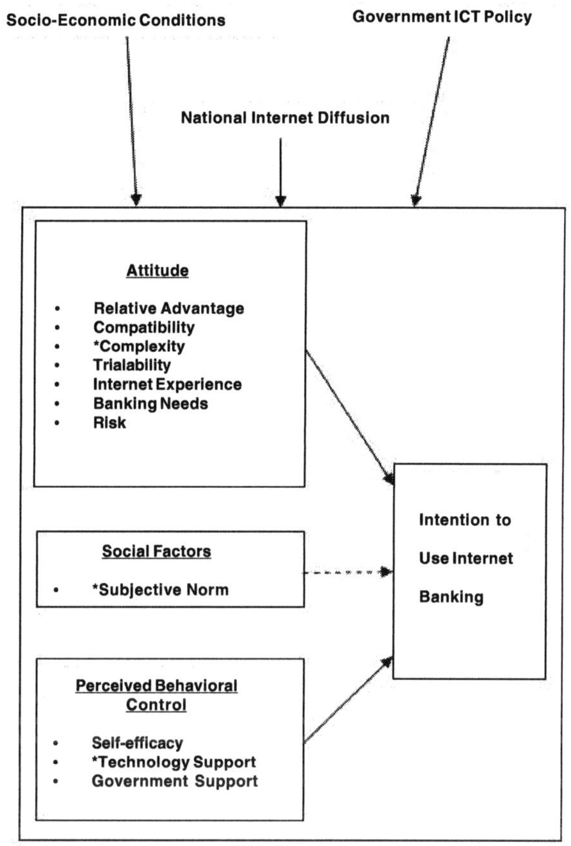

** No influence on Internet Banking adoption is expected*

is ideal for explaining differences in Internet banking adoption processes between countries.

Government policy could aid or hinder Internet diffusion (Mbarika, 2002). This is consistent with the national systems of innovation theory, which posits that government policies may encourage or mandate technology development and adoption (King et al, 1994, cited by Wolcott et al, 2001).

The conceptual framework, illustrating these influences is shown in Figure 2, and their expected impact on Internet banking adoption processes in Singapore and South Africa are then discussed.

Hypotheses

Relative Advantage – The advantage and convenience of using Internet banking from home or office are likely to have broad appeal. However, most consumer Internet banking would take place from home through a dial-up connection to an ISP. Both Singapore and South Africa offer broadband Internet connectivity to the home, but it was only recently made available in South Africa and, thus, has not been widely diffused yet among home users. Local call costs for dial-up subscribers are also cheaper in Singapore (see Table 2), which enhances the perception of convenience. As a result, the relative advantage of Internet banking may be more salient to Internet banking adoption in Singapore, leading to the following hypothesis:

H1A: The influence of relative advantage on Internet banking adoption will be greater in Singapore than in South Africa.

Compatibility – The compatibility of Internet banking with user values will be greater among those who use the Internet a great deal for work, study, or pleasure. White-collar professionals and students are more likely to fit into this category. In South Africa, the Internet population is largely made up of young, affluent, educated professionals, as well as a sizeable number of tertiary-level students (Goldstuck, 2001). In Singapore, the younger age group also predominates (Sa Voix, 2002), but the use of the Internet is spread across a larger proportion of the population, and, as such, there may be greater diversity of Internet users. This, then, leads to the following hypothesis:

H1B: The influence of compatibility on Internet banking adoption will be greater in South Africa than in Singapore.

Internet Experience – Singaporean Internet users are posited to be more sophisticated in their usage than South Africans. Online consumer sales in Singapore, for example, far exceed those in South Africa, and the Singapore government's e-citizen initiative encourages transacting over the Internet (IDA, 2001). Thus the experience of using the Internet for transactional purposes is greater than in South Africa. Given that the nature of most Internet banking is transactional, the prior similar experience may be greater for a user in Singapore, leading to the following hypothesis:

H1C: The influence of Internet experience on Internet banking adoption will be greater in Singapore than in South Africa.

Banking Needs – Those with larger incomes will require and will be eligible to use a more diverse array of banking products and services. The Singapore GDP per capita is about two times higher than that in South Africa (See Table 1). However, South African Internet users are relatively wealthier than the general populace (Goldstuck, 2001). Nevertheless, given that Singapore's GDP per capita is among the highest in the world (CIA, 2001), relative wealth and subsequently banking needs still may be higher. Thus, the following hypothesis:

H1D: The influence of banking needs on Internet banking adoption would be greater in Singapore than in South Africa.

Complexity – In Singapore, complexity was found to have no influence on Internet banking adoption. Given the relative user friendliness of the Internet, associated applications, and the education profile of the South African Internet user, it is expected that also in South Africa, complexity will not be factored into the Internet banking adoption decision. Thus, the hypothesis is:

H1E: Complexity will have no influence on Internet banking adoption in South Africa, as in Singapore.

Trialability – Trialability was found to be the most influential factor in Internet banking adoption in the Singapore study (Tan & Teo, 2000). In South Africa, it is expected also that users would want to experiment and try out Internet banking first before committing to adoption. Thus, no major differences are hypothesized in this regard.

H1F: Trialability will be a major influence on Internet banking adoption in South Africa, as in Singapore.

Risk – The South African Internet user is typically young and well educated, which may translate into a user that is less risk-averse. This is evidenced by the continued strong growth in Internet banking numbers, despite a widely reported security breach in 2003 (Goldstuck, 2004). In Singapore, on the other hand, there may be a broader Internet user profile (See Section 2) encompassing a wider cross-section of people, some of whom may be more concerned about the risks. Thus the hypothesis is:

H1G: The negative influence of risk on Internet banking adoption would be greater in Singapore than South Africa.

Subjective Norm – Given that consumer banking is a relatively private matter, the opinions of friends, peers, and family may not be that significant in channel adoption decisions. Furthermore, information may be easily available through "Frequently Asked Questions" (FAQ) pages on bank Web sites (Tan & Teo, 2000). This would apply equally in South Africa and Singapore. Thus the hypothesis is:

H2: Subjective norms will have no influence on Internet banking adoption in South Africa, as in Singapore.

Self-efficacy – Self-efficacy towards using Internet banking was a major factor influencing Internet banking adoption in Singapore (Tan & Teo, 2000). Having the self-confidence to use Internet banking also is likely to influence adoption in South Africa, with no major differences hypothesized, thus:

H3A: Self-efficacy towards using Internet banking will be an influence on Internet banking adoption in South Africa, as in Singapore.

Technology Support – The lack of influence of technology support in Singapore was attributed to the probability that since the necessary technology for conducting Internet banking is already in place, users may tend to take it for granted; hence, the indifference (Tan &d Teo, 2000). Furthermore, the effect of technology support on adoption may be indirect through other variables. For example, the relative advantage of

Internet banking could be improved by faster Internet access speeds. These arguments could be equally valid in South Africa; thus, it is hypothesized:

H3B: Technology support will have no significant direct influence on Internet banking adoption in South Africa, as in Singapore.

Government Support – The government ICT policies in Singapore have been widely acclaimed, due particularly to the success in converting plans into reality (ITU, 2001c). In South Africa, the government is endeavoring to facilitate universal telecommunications access. However, an ECT bill only recently has been introduced, and controversy surrounding parts of it got widespread media coverage (*The Star*, 2002). Therefore, in South Africa, perceived government support for Internet commerce is not expected to influence Internet banking adoption (Note: More recent efforts to liberalize the telecommunications sector may have changed perceptions, but these were announced only recently, long after this study was conducted):

H3C: Perceived government support for Internet commerce will have no significant influence on Internet banking adoption in South Africa, unlike in Singapore.

RESEARCH METHODOLOGY

Research Procedure

Both primary and secondary data were gathered in this study, with the primary data collected in two phases. In the first phase, an online questionnaire was used to collect empirical data (Hoppe et al., 2001). This was considered to be most appropriate, given that the questions, for the most part, were directed at Internet users, and the hypotheses were best tested using a sample of this group rather than the general South African population. A small pilot study suggested changes to the questionnaire (e.g., item TSUP3 of the Technology Support construct in Appendix 1).

The second phase was intended to bolster the number of responses. Hard copies of the questionnaire were produced, targeted specifically at students in an MBA program. This ensured a higher response rate than would have been the case with the online questionnaire, as the questionnaire was completed during a class session. The rationale was that this group was typical of the South African Internet user, who is between 18 and 35, educated, relatively well-off, and in a position to adopt Internet banking (de Villiers & van der Merwe, 2001; Goldstuck, 2001). In addition, as MBA students, all respondents either were working or had some work experience, unlike typical undergraduate students.

A seven-point multi-item Likert scale was used for the constructs in the questionnaire (except banking needs), ranging from 1 (strongly disagree) to 7 (strongly agree). For banking needs, respondents were asked to indicate which of a range of banking products and services they used. Adding together the number of banking products used by each respondent produced a score. The online questionnaire was hosted at a South African university and advertised through e-mail lists and in print media. Questionnaire items are shown in Appendix 1.

Secondary data were gathered about the general socioeconomic profile, state of Internet diffusion, government ICT policies, and banking environment in both Singapore and South Africa, as reported in Section 2.

Demographic Profile

A total of 102 complete responses were received in the first phase of primary data collection and 40 (out of 80 targeted) from the second phase. The total number of responses (142) were, therefore, less than in the Singapore study (454 respondents). The greater incentives offered to respondents in Singapore may have contributed to the higher number of responses, as well as the possibly wider publicity of that study. The demographic profile of respondents is shown in Table 3 below, together with the profile from the Singapore study. As can be seen, the sample was quite diverse. In regard to gender, males dominated (63%) as compared to females (37%). This domination was, however, not as great as the Singapore study, which was 81% male but compares favorably with the South African Internet user profile, which is estimated at about 56% male (Goldstuck, 2001). In regard to age, the majority was in the 20-29 age bracket (67%), as with the Singapore study, which had 64% in this age bracket. Considering educational levels, most had a bachelors degree or equivalent (68%, including those currently in an MBA program), which is slightly different from the Singapore study, in which the majority had junior college/polytechnic qualifications (47%). The number of student respondents was 47% (37% excluding MBA students), comparable to the 39% in the Singapore study. The overall profile of the respondents approximated somewhat the profile of the South African Internet user in terms of gender (slight majority of males), age (18–35 predominate), education level (tertiary qualified), and income (above average) (de Villiers and van der Merwe, 2001; Goldstuck, 2001;).

Intentions to Use Internet Banking

An examination of banking needs revealed that about 37% of respondents had already used online banking at the time of the study. Intentions to use Internet banking were also high overall, with a sample mean of 5.6 reported (on a scale of 1 to 7). As can be seen from Table 4, the great majority (85% of respondents) had at least some intention of using Internet banking ($x > 4$), with 45% having strong intentions of doing so ($x > 6$). Using intentions as the dependent variable, rather than actual usage, allowed even those who had not yet adopted Internet banking to have an opportunity to respond.

Validity and Reliability Tests

Prior to analysis of the results, the research instrument was tested for validity and reliability to confirm the psychometric properties in a different environmental setting. The validated model suggested by the Tan and Teo (2000) analysis was used, comprising 34 of their original 37 item measures. To test for validity of the constructs, factor analysis with varimax rotation was used. For the items in this study, 10 factors were expected (excluding banking needs), so a 10-factor structure was suggested with an eigenvalue of greater than 1. Items were expected to load at greater than 0.5 on their own factors, and less than 0.4 on all other factors to demonstrate construct validity (Hair et al, 1992, cited by Tan & Teo, 2000). The meaning of the item names can be seen in the questionnaire in Appendix 1, and the results of factor analysis outlined in Appendix 2. The same nine-

Table 3. Demographic profile of respondents

	Singapore (Tan & Teo, 2000)		South Africa	
	Frequency	Percent	Frequency	Percent
Gender				
Male	367	80.8	90	63.4
Female	87	19.2	52	36.6
Age				
Under 20	43	9.5	1	0.7
20-29	291	64.1	95	66.9
30-39	92	20.3	27	19.0
40-49	23	5.1	9	6.3
50-59	3	0.7	9	6.3
Over 59	2	0.4	1	0.7
Highest Education				
Primary School	5	1.1	0	0.0
Secondary School	53	11.7	19	13.4
Junior College/ Polytechnic	215	47.4	8	5.6
Bachelor's Degree	135	29.7	56	39.4
Master's Degree	36	7.9	11	7.7
MBA (Enrolled)	Not Reported	Not Reported	40	28.2
Doctorate Degree	3	0.7	3	2.1
Other	6	1.3	3	2.1

	Singapore (Tan & Teo, 2000)		South Africa	
Current Profession				
Student	176	38.8	* 67	* 47.2
Professional	153	33.7	29	20.4
Academic	9	2.0	20	14.1
Self-employed	13	2.9	3	2.1
Manager	35	7.7	13	9.2
Executive	30	6.6	4	2.8
Technician	17	3.7	1	0.7
Retiree/Housewives	2	0.4	1	0.7
Other	19	4.2	3	2.1
Income (excludes students with little income)			**Approximate equivalent**	
Less than SGD1, 000	12	4.3%	13	16.7%
SGD1, 000 – 1,999	45	16.2%	17	21.8%
SGD2, 000 – 2, 999	97	34.9%	17	21.8%
SGD3, 000 +	124	44.7%	31	39.7%

** 53 (37.3%) excluding MBAs*

factor structure described by Tan and Teo (2000) was found in this study, as relative advantage and compatibility again loaded on the same factor. The variance inflation factors (VIFs) for these and other independent variables were all less than 3.9, however, which is an indication that multicollinearity would not pose a problem (Kleinbaum et al., 1998, cited by Tan & Teo, 2000).

The constructs were tested for reliability using the Cronbach alpha test. The score for each construct should be greater than 0.6 for it to be reliable (Nunally, 1967). The results of this test are shown in Table A2 of Appendix 2. All constructs have coefficients

Table 4. Scores for adoption (South Africa)

Intentions to Use (Between 1 and 7)	Count	Percent
x=1	2	1.4
1<x<=2	2	1.4
2<x<=3	5	3.5
3<x<=4	12	8.5
4<x<=5	28	19.7
5<x<=6	29	20.4
6<x<=7	64	45.1

greater than 0.6, except complexity and technology support. Recommendations for improvements to the instrument measures are outlined in Appendix 2.

Hypotheses Testing

In order to test the hypotheses, multiple linear regression was used, with intention to use Internet banking being the dependent variable. The results of regression are shown in Table 5 below, together with those from the Singapore study for easy comparison. The beta values were, in general, of the same order of magnitude in each study, with factors significant at $p < 0.05$ highlighted.

In summary, results of hypotheses testing are as follows:

H1A: There was only partial support for this hypothesis, as although relative advantage did not significantly influence Internet banking adoption in South Africa (beta = 0.15, $p > 0.05$) as it did in Singapore, the beta coefficient was of the same order of magnitude (beta = 0.14, $p < 0.05$ in Singapore).

Table 5. Results of multiple regression analysis

		Singapore (Tan & Teo, 2000)			South Africa	
Hypotheses	**Supported?**	**Independent Variable**	**Beta**	***p***	**Beta**	***p***
H1A	Partial	Relative Advantage	**0.14**	**0.006**	0.15	0.232
H1B	Yes	Compatibility	**0.15**	**0.008**	**0.37**	**0.004**
H1C	Yes	Internet Experience	**0.10**	**0.004**	0.04	0.528
H1D	Partial	Banking Needs	**0.09**	**0.005**	0.08	0.221
H1E	Yes	Complexity	-0.03	0.397	-0.01	0.900
H1F	Yes	Trialability	**0.32**	**0.000**	**0.23**	**0.001**
H1G	Yes	Risk	**-0.08**	**0.017**	-0.04	0.554
H2	Yes	Subjective Norm	0.03	0.431	-0.04	0.534
H3A	Yes	Self-Efficacy	**0.16**	**0.000**	**0.19**	**0.019**
H3B	Yes	Tech. Support	-0.02	0.559	-0.04	0.559
H3C	Yes	Govt Support	**0.11**	**0.004**	0.01	0.837

H1B: This hypothesis was clearly supported, as the influence of compatibility on Internet banking adoption was stronger in South Africa (beta = 0.36, $p < 0.05$) than Singapore (beta = 0.15, $p < 0.05$).

H1C: Internet experience showed a greater influence on adoption in Singapore (beta = 0.10, $p < 0.05$) than South Africa, with the beta value in the South African study being small and insignificant (beta = 0.04, $p > 0.05$), thus supporting this hypothesis.

H1D: The influence of banking needs was statistically significant in Singapore (beta = 0.09, $p < 0.05$), but not in South Africa (beta = 0.08, $p > 0.05$). The beta coefficients were of the same order of magnitude, however, thus lending only partial support to this hypothesis.

H1E: Complexity was not a factor of influence in South Africa as in Singapore, supporting this hypothesis.

H1F: In both South Africa and Singapore, trialability was a major factor of influence, as hypothesized.

H1G: Risk was of no significant influence in South Africa, while having a small but significant negative impact in Singapore, thus supporting the hypothesis.

H2: Subjective norm showed no influence on adoption in both South Africa and Singapore, as hypothesized.

H3A: Self-efficacy was a major influence on adoption in South Africa as in Singapore, supporting the hypothesis.

H3B: In both Singapore and South Africa, technology support had no influence on adoption, supporting the hypothesis.

H3C: Government support was a factor of influence in Singapore, but not in South Africa, as hypothesized.

DISCUSSION AND IMPLICATIONS

The findings show that in both the Singapore and South African studies, Internet banking adoption was predicted by attitudinal and perceived behavioral control factors, but not by subjective norm. In Singapore, the factors of influence included relative advantage, compatibility, banking needs, Internet experience, trialability, risk, self-efficacy, and government support. In South Africa, only compatibility, trialability, and self-efficacy were statistically significant. Generalizability and interpretation of findings, however, is limited by the small sample size in this study, the over-representation of students, and the differences in timing between studies.

The strong influence of trialability in both Singapore and South Africa may reflect the relatively recent introduction of this innovation in both countries. Self-efficacy is also a major influence in both countries, reflecting its importance to technology adoption across cultures. The particularly strong influence of compatibility in South Africa could be a reflection of the narrow white-collar professional profile and large student population among Internet users, both groups finding Internet banking compatible with their values and lifestyle. These groups also may have been overrepresented in the sample, further amplifying the influence of this factor.

The advantage and convenience of Internet banking in South Africa may be enhanced through cheaper ISP and call charges for dial-up subscribers as well as affordable broadband access to the home, as is the case in Singapore. To a large extent, these imperatives depend on the introduction of greater privatization and competition in the fixed line telecommunications market (Mbarika, 2002). Nevertheless, an innovative way of promoting Internet banking has been demonstrated by Absa bank in South Africa. Their provision of a free ISP service in 2001 unsettled the ISP market, and although Absa will now start charging for Internet access, they have managed to gain the largest market share (36%) in Internet banking as a result (News24, 2002). They also may have contributed to a drop in ISP charges to consumers.

In Singapore, the government vision of creating an intelligent island and its successful implementation have encouraged the use of technological innovations such as Internet banking (ITU, 2001c). South Africa lags behind Singapore in this respect, in that an ECT bill only recently has been signed into law, and not without some controversy (*The Star*, 2002). Thus, the influence of government support on Internet banking adoption may be perceived as weak at the time of the study. It is necessary to point out, however, that the South African government faces many other pressing issues (i.e., high unemployment, poverty reduction,), with the need for provision of basic infrastructure sometimes taking precedence.

The attention given to creating an intelligent, wired island in Singapore furthermore translates into a greater awareness and acceptance among residents of services such as banking being delivered via the Internet. As such, where there are needs (i.e., banking needs), an Internet-based solution is seen as feasible. Notwithstanding, it is also pertinent that Singapore is more affluent than South Africa, which may translate into a greater number of banking products and services being used by Singaporeans. This, in turn, has an influence on Internet banking adoption.

The one factor discouraging Internet banking adoption in Singapore is the perceived risk of conducting transactions online. This factor is surprisingly of no effect among the South African sample. This anomaly may be explained in terms of the generally youthful profile of the South African Internet user, who is less risk-averse, and the education profile, which implies a user more knowledgeable or aware of such risk. Recent evidence of strong growth in Internet banking numbers, despite widely-publicized media reports of a few recent security breaches, confirms this assertion (Goldstuck, 2004).

LIMITATIONS AND FUTURE RESEARCH

With the rapid rate of change within the IT environment and the rapid diffusion of the Internet globally, timing of studies becomes critically important. The study by Tan and Teo (2000) was published in 2000, with data having been gathered earlier. The study in South Africa was conducted in late 2001 and early 2002, and thus, Internet banking would have been available a few years longer. Thus, the study findings may have been impacted by these time differences.

The difference in sample size (142 in South Africa to 454 in Singapore) also may have influenced the results, with the likelihood that some of the factors showing no statistical significance became significant with a larger sample. Relative advantage, and banking needs are most likely to have been affected. Once again, therefore, findings need to be interpreted keeping this in mind.

The large number of student responses is another major limitation, although this does not affect much the comparison between countries. In both studies, the number of students made up just over one-third of the responses (excluding MBA students in the South African sample). Besides, the profile of the typical Internet user in South Africa makes tertiary-level students the most likely potential adopters of innovations such as Internet banking. Tan and Teo (2000) furthermore state that the target group of their study was merely Internet users, regardless of their demographic profile. Thus, generalizability of results to the typical Internet user in each country may be limited. Generalizability of findings in South Africa is further limited by the fact that most respondents were residents of Cape Town, with respondents from other cities and regions not well represented.

In comparing income levels between the two studies, no account was taken of cost of living. However, given that Singapore has one of the highest GDP per capita in the world, it is fair to assume greater levels of affluence than South Africa, even though Internet users in South Africa are generally well off by South African standards.

Another potential limitation is the use of two different techniques for collecting data (online questionnaire and paper-based questionnaire). This changes the conditions under which data is gathered, which ultimately may have a potential impact on findings. Improvements to the instrument have also been suggested, which should be taken into account in any future replication of such a study.

Future research might also look at conducting studies simultaneously in different countries, with more attention given to sampling to get an accurate demographic profile, and incentives to bolster response rates.

In this study, Singapore and South Africa were compared. Other countries could also be included, and any differences explained in terms of the environmental context. Differences could be explained in terms of cultural values, as well as socioeconomic conditions, Internet diffusion, and government policy. A thorough and exhaustive analysis of the state of Internet diffusion in South Africa needs to be published, using the GDI framework as the basis. This would add greater weight to the assertions made in this study and allow for easy comparison to other countries. The impact of bank strategies on adoption has not been fully considered, but may be another important area for future research. Aggressive marketing by banks in one country, for example, may change people's perception of risk. The trend towards greater cooperation between banks and ISPs may also contribute to adoption of Internet banking.

The target group of the study was Internet users, and thus, an interesting extension would be to examine the potential for adoption among non-users, as suggested by Tan and Teo (2000), but with the view to comparing the adoption process with users. In the same vein, the dependent variable in this study was intentions to use Internet banking so as to allow those that have not as yet adopted the innovation an opportunity to respond. Future research might then compare the adoption process between non-adopters and current adopters of Internet banking, which, once again, is an extension to the suggestion by Tan and Teo (2000) for future studies to look at actual usage behavior, rather than just behavioral intentions.

CONCLUSION

Consumer Internet banking is offered by all the major retail banks in South Africa and Singapore. While initially hailed as the future of banking, it is now seen as just another channel (Leonard, 2002). Nevertheless, it is necessary, given the proportion of consumers that use it or intend to use it.

A framework of Internet banking adoption developed in Singapore identified attitudinal, social, and perceived behavioral control factors as possible influences on adoption. In the Singapore study and in a replicated study in South Africa, only attitudinal and perceived behavioral control factors, but no social factors, actually influenced adoption. In Singapore, furthermore, there was a wider array of influences than in South Africa. It was hypothesized that the national environment could explain such differences. National culture has been the most commonly used characteristic of the national environment, but is essentially only one element of a complex phenomenon. Thus, in this study, other dimensions of the environment were sought. Socioeconomic conditions, the state of Internet diffusion, and government ICT policies were the dimensions subsequently examined. These dimensions may have accounted for the fact that the attitudinal factors relative to advantage, banking needs, Internet experience, perceived risk, and government support were influences in Singapore but not in South Africa. Furthermore, they may have accounted for the greater influence of compatibility in South Africa as compared to Singapore.

The limitations in terms of sample size in South Africa, respondent profile, and timing of studies also may have affected the outcomes. Nevertheless, the highly successful government ICT policies in Singapore, which have resulted in an advanced state of Internet diffusion, help to promote the use of innovations such as Internet banking and provide valuable lessons for South Africa and other developing countries in their quest for similar success.

REFERENCES

Agarwal, R., & Prasad, J. (1997). The role of innovation characteristics, and perceived voluntariness in the acceptance of information technologies. *Decision Sciences, 28*(3), 557-582.

AISI: African Information Society Initiative (2000). *AISI connect national ICT profile, South Africa (ZA)*. Retrieved August 4, 2002, from *http://www2.sn.apc.org/africa/countdet.CFM?countries__ISO_Code=ZA*

Ajzen, I. (1985). From intention to actions. A theory of planned behavior. In J. Kuhl & J. Beckmann (Eds.), *Action control: From cognition to bBehavior* (pp. 11-39). New York: Springer-Verlag.

Aladwani, A. (2001). Online banking: A field study of drivers, development challenges, and expectations. *International Journal of Information Management, 21*, 213-225.

Anandarajan, M., Igbaria, M., & Anakwe, U. (2002). IT acceptance in a less-developed country: A motivational factor perspective. *International Journal of Information Management, 22*, 47-65.

BusinessDay (2004). Wind of change to howl through a sector shackled by a monopoly. *BusinessDay*. Retrieved September 10, 2004, from *http://www.bdfm.co.za*

CIA: Central Intelligence Agency (2001). *The World Fact Book 2001*. Retrieved January 28, 2002, from *http://www.odci.gov/cia/publications/factbook/*

CIA: Central Intelligence Agency (2004). *The World Fact Book 2004*. Retrieved September 10, 2004, from *http://www.odci.gov/cia/publications/factbook/*

CNETAsia (2002). Net banking on the rise in Singapore. *CNETAsia*. Retrieved August 4, 2002, *from http://asia.cnet.com/newstech/industry/0,39001143,39040200,00.htm*

Davis, F. (1989). Perceived usefulness, perceived ease of use and user acceptance of information technology. *MIS Quarterly, 13*(3), 319-340.

de Villiers, C., & van der Merwe, J. (2001). An investigation into the adoption of electronic commerce by South African consumers—Who they are and what they think. *Research Report*. Retrieved August 5, 2002, from *http://informatics.up.ac.za/tuksjvdm1.doc*

Goldstuck, A. (2001). South Africa—How many use Web sites and who are they? *Balancing Act News Update, 71*. Retrieved January 18, 2002, from *http://www.balancingact-africa.com/news/back/balancing-act_71.html*

Goldstuck, A. (2002a). The Goldstuck report: Internet access in South Africa, 2002. *World Wide Worx*. Retrieved August 5, 2002, from *http://www.theworx.biz/access02*

Goldstuck, A. (2002b). The Goldstuck Report: Online Retail in South Africa, 2002. *World Wide Worx*. Retrieved August 5, 2002, from *http://www.theworx.biz/retail02*

Goldstuck, A. (2004). Online banking in South Africa, 2004. *World Wide Worx*. Retrieved August 16, 2004, from *http://www.theworx.biz/bank04.htm*

Hair, J., Anderson, R., Tatham, R., & Black, W. (1992). *Multivariate data analysis with readings*. New York: MacMillan.

Hoppe, R., Newman, P., & Mugera, P. (2001). Factors affecting the adoption of Internet banking in South Africa: A comparative study. *Unpublished Honours Empirical Research Report*, 1-38.

IDA: Infocomm Development Authority (2001). Infocomm development authority of Singapore. Retrieved November 7, 2002, from *http://www.ida.gov.sg*

ISPA: Internet Service Providers Association (2002). About ISPA. ISPA. Retrieved August 5, 2002, from *http://www.ispa.org.za*

ITU: International Telecommunications Union (2001a). Singapore country profile. *ITU Telecommunication Indicators Update, 2*(4). Retrieved August 1, 2002, from *http://www.itu.int/ITU-D/ict/update/pdf/Update_2_01.pdf*

ITU: International Telecommunications Union (2001b). South Africa country profile. *ITU Telecommunication Indicators Update, 3*(4). Retrieved August 1, 2002, from *http://www.itu.int/ITU-D/ict/update/pdf/Update_3_01.pdf.pdf*

ITU: International Telecommunications Union (2001c). The e-city: Singapore Internet case study. *ITU Country Report,* 1-57. Retrieved August 1, 2002, from *http://www.itu.int/ITU-D/ict/cs/singapore/material/Singapore.pdf*

ITU: International Telecommunications Union (2002). ICT – Free Statistics. ITU. Retrieved August 1, 2002, from *http://www.itu.int/ITU-D/ict/statistics/*

James, T., Esselaar, P., Bowmaker-Falconer, A., Quansah, Y., Sibthorpe, C., Mosebi, J., & Mokoena, T. (2000). A survey of the IT industry and related jobs and skills in South Africa. *South African IT Industry Strategy (SAITIS) Baseline Studies*. Retrieved August 5, 2002, from *http://www.dti.gov.za/saitis/studies/jobs_skills/index.html*

Jensen, M. (2002). *The African Internet—A status report*. Retrieved August 4, 2002, from *http://www3.sn.apc.org/africa/afstat.htm*

Jiang, J., Hsu, M., Klein, G., & Lin, B. (2000). E-commerce user behavior model: An empirical study. *Human Systems Management, 19*, 265-276.

King, J., Gurbaxani, V., Kraemer, L., McFarlan, W., Raman, K., & Yap, S. (1994). Institutional factors in information technology innovation. *Information Systems Research, 5*(2), 139-169.

Kleinbaum, D., Kupper, L., & Muller, K. (1988). *Applied regression analysis and other multivariate methods*. Boston: PWS.

Leonard, R. (2002). Prepare for the future of online banking. *IT in Banking*. Retrieved November 7, 2002, from *http://www.itweb.co.za/sections/industryinsight/itinbanking/leonard020611.asp*

Liao, S., Shao, Y., Wang, H., & Chen, A. (1999). The adoption of virtual banking: An empirical study. *International Journal of Information Management, 19*, 63-74.

Licker, P., & Motts, N. (2000, June 11-13). Extending the benefits of e-commerce in Africa: Exploratory phase. *Proceedings, First Annual Global Information Technology Management World Conference*, Memphis.

Mbarika, V. (2002). Rethinking information and communication technology policy focus on Internet versus teledensity diffusion for Africa's least developed countries. *Electronic Journal of Information Systems in Developing Countries, 9*(1), 1-13. Retrieved August 1, 2002, from *http://www.ejisdc.org*

News24 (2002). Absa to charge for web banking. *News24*, Retrieved June 7, 2002, from *http://www.news24.com/News24/Finance/Companies/0,4186,24_1175849,00.html*

Nunally, J. (1967). *Psychometric theory*. New York: McGraw-Hill.

Rogers, E. (1983). *Diffusion of innovations* (3rd Ed.). New York: Free Press.

Rogers, E. (1995). *Diffusion of innovations* (4th Ed.). New York: Free Press.

Sa Voix (2002). About us: Internet user profile. *Sa Voix Magazine*. Retrieved November 7, 2002, from *http://www.savoixmagazine.com/aboutUs-NetUserProfile.htm*

Straub, D., Keil, M., & Brenner, W. (1997). Testing the technology acceptance model across cultures: A three-country study. *Information & Management, 33*, 1-11.

Tan, M., & Teo, T. (2000). Factors influencing the adoption of Internet banking. *Journal of the Association for Information Systems, 1*(5), 1-42.

Taylor, S., & Todd, P. (1995). Understanding information technology usage: A test of competing models. *Information Systems Research, 6*(2), 144-176.

The Star (2002, July 31). Controversial Internet bill becomes law. *The Star Newspaper*. Retrieved August 4, 2002, from *http://www.iol.co.za/index.php?set_id=1&click_id=13&art_id=ct20020731214902902M125812*

Wolcott, P., Press, L., McHenry, W., Goodman, S., & Foster, W. (2001). A framework for assessing the global diffusion of the Internet. *Journal of the Association for Information Systems, 2*(6), 1-50.

World Bank (2004). World development indicators database. Retrieved September 10, 2004, from *http://www.worldbank.org/data/countrydata/countrydata.html*

APPENDIX 1

Questionnaire Items (Phase 2 Extract)

1. *Internet Experience*

- EXP1-For how long have you been using the Internet?
- EXP2-On the average, how frequently do you use the Internet?
- EXP3-On the average working day, how much time is spent on the Internet?

- EXP4. Diversity of Internet usage (1 – Not at all, to 7 – A great extent)
- EXP4a-Gathering information.
- EXP4b-Communication
- EXP4c-Download SW
- EXP4d-Shopping
- EXP4e-Searching for Job
- EXP4f-Swapping/Selling

- SKILL. Internet skills (1 – Strongly Disagree, to 7 – Strongly Agree)
  SKILL1-I am very skilled at using the Internet
- SKILL2-I am knowledgeable about good search techniques
- SKILL3-I know more about the Internet than most
- SKILL4-I know how to find what I want on the Internet

2. *Banking needs*

Savings Account	Current Account
Housing Mortgage	Personal Investment Account
Share Margin Trading	Forex Margin Trading
Fixed Deposit	Personal Loan
Unit Trust	Pay Bills
Telephone Banking	Online Banking

3. *Attitudinal Factors (1 – Strongly Disagree, to 7 – Strongly Agree)*

Relative Advantage

- RADV1-Internet banking makes it easier to do banking
- RADV2-Internet banking gives me greater control over my finances

- RADV3-Internet banking allows me to manage my finances more efficiently
- RADV4-Internet banking is a convenient way to manage my finances
- RADV5-Internet banking allows me to manage my finances more effectively
- RADV6-I find Internet banking useful for managing my financial resources.

Compatibility
- CPAT1-Internet banking is compatible with my lifestyle
- CPAT2-Using Internet banking fits well with the way I like to manage my finances
- CPAT3-Using Internet banking fits into my working style

Complexity
- CMPX1-Using Internet banking requires a lot of mental effort
- CMPX2-Using Internet banking can be frustrating
- CMPX3-Internet banking is a difficult way to conduct banking transactions

Trialability
- TRI1-I want to be able to try Internet banking for at least one month
- TRI2-I want to be able to use Internet banking on a trial basis first

Perceived Risk
- RISK1-I am concerned over the security aspects of Internet banking
- RISK2-Information concerning my Internet banking transactions will be known to others
- RISK3-Information concerning my Internet banking transactions can be tampered with by others

4. *Social Factors*
   Subjective Norm (1 – Strongly Disagree, to 7 – Strongly Agree)

My decision to adopt Internet banking is influenced by my ....
- SN1-Friends
- SN2-Family
- SN3-Colleagues/Peers

5. *Perceived Behavioral Control (1 – Strongly Disagree, to 7 – Strongly Agree)*

Government Support
- GSUP1-The government supports Internet commerce
- GSUP2-The government is active in setting up facilities to enable Internet commerce
- GSUP3-The government promotes the use of Internet commerce

Technology Support

- TSUP1-Advances in Internet security provide for safer Internet banking
- TSUP2-Faster Internet access speed is important for Internet banking
- *TSUP3-Cell phone technology like WAP makes Internet banking more feasible

(Originally: Internet technology like the Singapore ONE network makes Internet banking more feasible.)*

Self-Efficacy

I am confident of using Internet banking....

- SEFF1-If I have only the online instructions for reference
- SEFF2-Even if there is no one around to show me how to use it
- SEFF3-Even if I have never used such a system before
- SEFF4-Even if I have just seen someone using it before trying it myself.
- SEFF5-If I have just the online help function for assistance

6. *Intentions to use*

To what extent would you be interested in using Internet banking if it is available?

	Not at all interested					Very interested
INT1	1	2	3	4	5	6	7

If Internet banking were available to you, how likely would you be using it in the next:

		Very Unlikely						Very Likely
INT2a	6 months	1	2	3	4	5	6	7
INT2b.	12 months	1	2	3	4	5	6	7
INT2c.	18 months	1	2	3	4	5	6	7

APPENDIX 2

Psychometric Properties

Table A1. Results of factor analysis

	Factor1	Factor2	Factor3	Factor4	Factor5	Factor6	Factor7	Factor8	Factor9
EXP1	0.04	-0.02	0.03	0.12	-0.28	***0.61***	-0.17	0.31	-0.06
EXP2	0.04	0.02	-0.01	-0.04	0.00	***0.73***	-0.01	-0.21	-0.17
EXP3	-0.01	0.11	0.09	0.07	0.05	***0.45***	0.30	-0.34	0.33
EXP4	0.08	0.10	-0.07	0.00	0.09	***0.69***	0.20	0.06	0.19
SKILL	0.18	0.30	-0.01	0.04	0.02	***0.67***	0.17	0.16	-0.03
RADV1	***0.84***	0.14	0.04	-0.05	-0.03	0.08	0.12	-0.07	-0.02
RADV2	***0.83***	0.18	0.03	0.06	0.11	0.05	0.13	0.16	0.10
RADV3	***0.88***	0.12	0.09	0.08	0.02	0.08	0.05	0.13	0.14
RADV4	***0.92***	0.19	0.03	0.05	-0.01	0.02	0.02	0.10	0.03
RADV5	***0.85***	0.18	0.05	0.08	0.12	-0.03	0.08	0.13	0.09
RADV6	***0.85***	0.15	0.11	0.10	0.01	-0.02	0.09	0.09	0.13
CPAT1	***0.79***	0.29	0.03	0.04	-0.02	0.15	0.01	-0.13	-0.09
CPAT2	***0.88***	0.23	-0.04	0.09	0.02	0.08	-0.03	0.04	-0.02
CPAT3	***0.87***	0.23	-0.05	0.05	0.02	0.07	0.03	-0.04	0.01
CMPLX1	-0.17	-0.22	0.09	0.13	0.03	-0.18	-0.07	***-0.69***	-0.05
CMPLX2	-0.06	-0.06	-0.13	-0.15	-0.06	0.08	-0.24	***-0.70***	0.02
TRI1	0.11	0.00	0.07	0.03	***0.96***	-0.04	-0.02	-0.01	0.02
TRI2	0.03	-0.08	0.06	0.07	***0.96***	0.02	-0.05	0.03	0.07
RISK2	-0.14	-0.15	-0.02	-0.04	0.03	-0.14	***-0.90***	-0.14	0.01
RISK3	-0.17	-0.13	-0.03	0.02	0.05	-0.10	***-0.89***	-0.13	0.03
GSUP1	0.09	0.11	-0.05	***0.82***	0.10	0.18	0.06	0.02	-0.06
GSUP2	0.11	-0.03	-0.04	***0.89***	0.01	-0.02	-0.04	0.05	0.12
GSUP3	0.12	-0.10	0.04	***0.83***	-0.01	-0.08	0.01	-0.04	0.17
TSUP2	0.25	0.02	-0.05	0.01	0.08	0.05	-0.17	-0.07	***0.67***
TSUP3	-0.02	0.09	0.15	0.22	0.00	-0.07	0.12	0.09	***0.69***
SEFF1	0.33	***0.80***	0.01	-0.05	-0.06	0.16	0.08	-0.01	0.09
SEFF2	0.34	***0.78***	-0.03	-0.01	-0.12	0.21	0.00	0.17	0.11
SEFF3	0.28	***0.86***	0.01	-0.01	-0.03	0.08	0.07	0.08	0.02
SEFF4	0.26	***0.81***	0.02	0.05	0.00	-0.09	0.16	0.03	-0.12
SEFF5	0.30	***0.79***	-0.02	-0.02	0.10	0.12	0.08	0.08	0.13
SN1	0.10	-0.04	***0.89***	0.02	-0.02	-0.03	-0.03	0.03	0.01
SN2	-0.02	0.01	***0.82***	-0.08	0.06	0.05	0.10	0.10	0.04
SN3	0.10	0.02	***0.89***	0.02	0.09	-0.06	-0.02	-0.11	0.04
Expl.Var	7.34	3.87	2.36	2.34	2.03	2.32	2.02	1.47	1.28
Prp.Totl	0.22	0.12	0.07	0.07	0.06	0.07	0.06	0.04	0.04

Several similarities emerged in the factor-analytic structure of the data, as with the Tan and Teo (2000) study as follows:

- The Relative advantage items (RADV1 to RADV6), and compatibility items (CPAT1 to CPAT 3) loaded on the same factor in both studies, thus indicating a lack of discrimination between the two constructs.
- The technology support item TSUP1 cross-loaded on several factors.
- In both studies, for the risk factor, the RISK1 item was slightly problematic. In the Singapore it loaded less than 0.4, but still highest on is own factor, and in this study it loaded on two factors, although still loading highest on its own factor.
- For the complexity factor, the CMPX3 item did not load on its expected factor, and in the Singapore study, it loaded less than 0.4 on its own factor.

Future studies that intend to employ these constructs therefore need to take account of these shortcomings by:

1. Developing measures that will show greater discrimination between relative advantage and compatibility.
2. Developing a better measure for technological support that is not specific to a country (Tan and Teo refer to the Singapore ONE network).
3. Developing better measures for complexity and risk where items that require reversal appear to be problematic.

Table A2. Reliability analysis

Construct	Number of Items	Cronbach alpha
Relative Advantage	6	0.96
Compatibility	3	0.94
*Complexity	2	0.52
Internet Experience	5	0.63
Trialability	2	0.96
Risk	2	0.94
Banking Needs	1	-
Subjective Norm	3	0.85
Self-efficacy	5	0.93
Government Support	3	0.83
*Technology Support	2	0.18

* *Below accepted level of 0.6 (Nunally, 1967)*

APPENDIX 3

Global Diffusion of the Internet (GDI) Framework

This GDI framework consists of six dimensions, defined and briefly summarised in this section from Wolcott et al. (2001, pp. 9-19).

- **Pervasiveness** is principally a function of the number of users per capita represented on a scale from 0 to 4. At the extremes, level 0 corresponds to the situation where the Internet does not exist in a viable form in a country, while level 4 is where the ratio of Internet users per capita is at least 10% or more.
- **Geographic Dispersion** describes the physical dispersion of the Internet within a country. Again, at the extremes, level 0 is where the Internet does not exist in any viable form, while level 4 is where Internet points-of-presence are located in essentially all first-tier political subdivisions (e.g., provinces) of a country, and rural access is widely available.
- **Sectoral Absorption** focuses on the extent to which organizations in the four major sectors of academic, commercial, health, and public have made a tangible commitment to Internet use. Within each sector, use is rated as non-existent (0 points), minimal (1 point), medium (2 points), or great majority (3 points). Adding these points together for each sector gives a value that can be used to get a rating (level 0 to 4) for sectoral absorption.
- **Connectivity Infrastructure** assesses the extent and robustness of the physical structure of the network and comprises four components: the aggregate bandwidth of the domestic backbone, the aggregate bandwidth of international IP links, the number and type of interconnection exchanges, and the type and sophistication of local access methods used. Ratings again are from level 0 to level 4, depending on the above factors.
- **Organizational Infrastructure** is centered on the number of Internet Service Providers (ISPs), and their competitive environment, with levels again from 0 to 4.
- **Sophistication of Use** reflects the fact that it is also necessary to know how the Internet is employed in order to characterizse diffusion. In some countries, the Internet may be used simply for e-mail, whereas in others, there may be greater levels of sophistication in use. This dimension is again rated from none (level 0) to innovating (level 4).

Internet Diffusion in South Africa (based on data available in 2002)

- **Pervasiveness:** South Africa is the largest Internet market in southern Africa, with approximately 3 million Internet users (Goldstuck, 2002a). Thus, the ratio of Internet users per capita is about 6-7%.
- **Geographic Dispersion:** In each of the nine provinces, all major cities and towns had Internet points-of-presence (POPs) (AISI, 2000). In many rural areas, however, there was insufficient infrastructure (i.e. Internet hosts) and demand for the Internet.
- **Sectoral Absorption:** Within the academic environment, the great majority of South Africa's tertiary institutions has Internet connectivity, but this is less so in schools, and especially formerly disadvantaged schools. In the commercial sector, most large corporations have Internet servers, with a great demand for bandwidth (Goldstuck, 2002a). This is less so in the small and medium enterprise sector, and even to a lesser extent in the informal sector (Licker & Motts, 2000). In the private health care sector, connectivity is more common than in the public health care system, and in the public sector, most government departments and parastatals

have a Web presence, with plans to implement Internet-based e-services to the public similar to those provided in Singapore (James et al., 2000).

- **Connectivity Infrastructure:** International bandwidth in South Africa totals about 380 Mbps (Jensen, 2002). The ISP Association manages two Internet exchanges, to which ISPs may join by becoming members (ISPA, 2002). Most domestic access is via dial-up modems, with higher capacity links via satellite and ADSL lines also now available. For businesses, it is mainly through 64 kbps leased lines (AISI, 2000) or ISDN lines (Jensen, 2002).
- **Organizational Infrastructure:** In 2001, there were approximately 80 main ISPs in South Africa (Jensen, 2002). The ISPA was formed in 1996 and currently has 44 members (ISPA, 2002). International bandwidth is provided by about four organizations, including the national telecom operator (Telkom). With lack of competition in fixed-line telephone operations, ISPs and customers continue to pay high charges for Internet access (Goldstuck, 2002a).
- **Sophistication of Use:** Web user behavior in South Africa "is characteristic of an Internet community that is building rapidly and has yet to reach maturity," according to Nielson//NetRatings (Goldstuck, 2001). Online retail sales in South Africa totalled ZAR162million (USD16 million) in 2001, just 0.1% of total retail sales in South Africa (Goldstuck, 2002b). The government is increasingly providing ways for people to use the Internet for transactions (i.e., online tax returns).

Internet Diffusion in Singapore (ITU, 2001c)

A recent ITU country study (ITU, 2001c) assessed Singapore according to the GDI framework, with the following summarized results:

- **Pervasiveness:** Singapore has about 43% of the population with Internet access, and government data states that Singapore has the highest household Internet access penetration in the world (42%).
- **Geographic Dispersion:** Given the small size of Singapore (647 square km) wide geographic dispersion has been easily achieved, with Internet access available to all telephone subscribers.
- **Sectoral Absorption:** All sectors of education, commerce, health, and public make use of the Internet.
- **Connectivity Infrastructure:** There is more than 805 Mbps of international bandwidth and a national backbone with speeds of up to 155 Mbps. Besides conventional dial-up for local access, broadband access is available to home subscribers via ISDN, ADSL, and cable modem.
- **Organizational Infrastructure:** Singapore had 37 licensed ISPs as of 2000, one Internet exchange, and free entry into the ISP market.
- **Sophistication of Use:** In Singapore, SGD 2.75 billion (USD 1.3 billion) was expected to be generated from business to consumer e-commerce in 2001. Of this, SGD 438.6 million (USD 220 million) is estimated as being spent on the Internet from home for consumer products (Sa Voix, 2002).

Chapter XV

An Integrated Model of E-Business Strategy: The Case of Haier from China

Peter Ping Li, California State University, Stanislaus, USA

Steven Tung-lung Chang, Long Island University, USA

ABSTRACT

This chapter proposes a conceptual framework of e-business strategy. It argues that such a framework must be holistic, dynamic, and dialectical. This framework will assist both researchers and practitioners regarding the key issues of e-business strategy. Further, a case study of Haier from China reveals that all firms need to learn how to design effective e-business strategies that should be built on a sound organizational form or business model. The case of Haier also suggests that the local firms from the developing countries need to be creative in formulating e-business strategies in order to operate effectively in the underdeveloped e-business environment. Strategic alliance is especially critical for the local firms with regard to their e-business success.

INTRODUCTION

Although many doubters regard it as a fad, we are convinced that e-business, as the convergence of digital technology, intellectual property, and customer supremacy, has the potential to transform our society as much as the Industrial Revolution did. This conviction is based on our belief that e-business is facilitating a new business paradigm—the paradigm of network capitalism, which emphasizes the central role of interconnectivity in both configuration and coordination of business activities between various partners in an open network across interlinked value-adding chains, thus transcending the traditional boundaries of space, time, individuals, and organizations (Afuah, 2003; Afuah & Tucci, 2003; Brews & Tucci, 2004; Dunning, 1995; Kalakota & Robinson, 2002; Li, 1998, 2001a, 2001b; Porter, 2001).

THE CONCEPTUAL FRAMEWORK OF E-BUSINESS

The Components of E-Business Framework

Four key developments occurred in the past few decades, and they seem to have been accelerating in recent years (Economist, 1999, 2000a, 2000b; Kalakota & Robinson, 2002; Li, 2001a, 2001b): (1) digital technologies (e.g., the Internet); (2) intellectual properties (e.g., information/knowledge); (3) consumer supremacy (e.g., customization and reduced cycle time); and (4) productivity growth (e.g., long-term economic growth without inflation). The first three elements can be viewed as the *antecedents* of e-business because they jointly define the conditions or drivers for the emergence and development of e-business, while the last can be viewed as the *consequence* of e-business because it defines the long-term effect and result of e-business. We argue that e-business (defined as the integration of digital technology, intellectual property, and customer supremacy with business activities, which alters and often replaces the conventional rules at both industrial and corporate levels) is one of the core components and the primary drivers of the so-called "new" economy (defined as the macro-economy undergoing a paradigm shift due to the convergence of digital technology, intellectual property, and customer supremacy). E-business seems to have the capability to change the competitive landscapes in the traditional industries, create new industries, and affect the configuration and coordination of firm-level business activities, especially such external activities as customer relationship management (CRM), which is about developing close cooperative relationship with key customers by applying new technologies and sticking to long-term goals, and supply-chain management (SCM), which is about developing close cooperative relationship with key suppliers by applying new technologies and sticking to long-term goals), and such internal activities as business process reengineering (BPR), which is about streamlining and integrating internal execution processes for more efficient and effective operations, and enterprise resource planning (ERP), which is about streamlining and integrating information flows for better coordination of all internal resources (Afuah & Tucci, 2003; Kalakota & Robinson, 2002; Li, 2001a, 2001b; Porter, 2001).

The Central Theme of E-Business

The central theme shared by the above issues lies in the core concept of network (Afuah, 2003; Afuah & Tucci, 2003; Brews & Tucci, 2004; Kalakota & Robinson, 2002; Li, 2001a, 2001b; Rayport, 1996; Shapiro & Varian, 1999; The Economist, 2000c). The digital network technology is at the core of digital technology, which is the enabler of e-business. The social network of intellectual contributors is critical for the creation and exchange of information and other intellectual properties, which constitute the content of e-business. The business network of suppliers, customers, and even competitors is shaping the new ecosystem of the global economy, which is the effect of e-business. All these phenomena invoke the notion of network (even Porter [2001] acknowledges connectivity as the core of e-business), so we can define the emerging theme of e-business as the network paradigm. Borrowing from Kuhn (1970), we define *paradigm* as a set of basic guiding principles for organizational form or business model in terms of the configuration and coordination of business activities at the micro or firm level as well as economic structure or market system in terms of the configuration and coordination of economic activities at the macro or national level (Dunning, 1995; Li, 1998). According to the above definition of paradigm, the concepts of organizational form and business model converge with the emerging notion of strategy—the so-called "organic" perspective of strategy (Farjoun, 2002; Li, 1995, 1997). In contrast to the traditional notion of strategy, whose epistemological basis is increasingly criticized as fragmented, static, and linear, the organic view of strategy is holistic, dynamic, and dialectical (Farjoun, 2002; Li, 1995, 1997, 1998, 2003; Parkhe, 1993). The broader view of strategy is particularly evident in the study of global strategy (Bartlett & Ghoshal, 1989; Li, 1995, 1997; Porter, 1990) and also in the study of e-business strategy (Afuah, 2003; Afuah & Tucci, 2003; Brews & Tucci, 2004; Kalakota & Robinson, 2002; Li, 2001a, 2001b). Such a view of strategy has been embraced by some scholars who study the generic strategy content (Hambrich & Fredrickson, 2001; Li, 1994, 1998, 2003).

According to the new views, organizational form, business model, and strategy content overlap with each other, each consisting of two key components: (1) business configuration and (2) business coordination. Business configuration refers to the scope of business coverage by a firm, which is the traditional conceptualization of strategy (Afuah, 2003; Porter, 1980). Business coordination refers to the style of business relationship within and without a company (Bartlett & Ghoshal, 1989; Porter, 1990), including the modes for internal control and external market entry (Li, 1994, 1998; Hambrick & Fredrickson, 2001). In short, focusing on the emerging phenomenon of highly specialized firms interconnected through various modes of strategic alliance for a proper balance between paradoxical forces (Li, 1998), the network paradigm seems to have the best prospect for addressing the paradoxes of e-business (paradox is defined as an interactive and interdependent pair of opposite elements [Li, 1995, 1998]), including growth-profit, customization-standardization, specialization-diversification, core-non-core, brick-click, competition-cooperation, control-trust, flexibility-continuity, and separation-integration (Li, 2001a, 2001b).

A Working Model of E-Business Strategy

According to the central theme of e-business and in light of the new perspective of strategy, we propose here a working model of e-business strategy. The proposed working model contains two components: (1) a four-dimension typology of strategy content and (2) a three-dimension typology of strategy process (Li, 2003). This model is consistent with the argument that the concept of strategy should cover both the spatial (i.e., state) aspect and the temporal (i.e., path) aspects of strategy (Farjoun, 2002). The working model is a specific application of the network paradigm that is holistic, dynamic, and dialectical in nature.

Four-Dimension Typology of E-Business Strategy Content

A four-dimension typology of strategy content (Li, 1994, 2003) can be used to measure the spatial content of e-business strategy. These four dimensions are (1) strategic target, (2) strategic thrust, (3) strategic posture, and 4) strategic mode. Similar to the prevailing view of strategic goal (Farjoun, 2002), strategic target refers to the means (i.e., to maximize volume size and/or margin rate) to achieve a firm's ultimate intent (i.e., to become the industry leader or the most admired in the world) (Bartlett & Ghoshal, 1989). Similar to the prevailing view of strategic focus (Hambrick & Fredrickson, 2001; Porter, 1980), strategic thrust refers to the means (i.e., to lower cost and/or improve value) to achieve strategic target. Similar to the prevailing view of strategic position (Bartlett & Ghoshal, 1989; Hambrick & Fredrickson, 2001; Porter, 1980), strategic posture refers to the means (i.e., to configure broad and/or narrow scope of market coverage, supply-chain coverage, function coverage, and location coverage) to achieve strategic thrust. Contrary to the prevailing view of strategy content, we incorporate the style or mode of coordination into our model of strategy content in accordance with the new organic perspective of strategy (Farjoun, 2002; Li, 1995, 1997, 1998, 2003). Strategic mode refers to the means (i.e., to coordinate business relationships both within and without the firm with various control modes regarding internal execution (e.g., business process reengineering) and external market entry (e.g., strategic alliance) to achieve strategic posture. Many scholars of global strategy have recognized the critical role of coordination in global strategy (e.g., Bartlett & Ghoshal, 1989; Porter, 1990), and even some scholars of generic strategy have accepted at least external coordination modes as the integral part of strategy content (e.g., Hambrick & Fredrickson, 2001). The above four dimensions make up a holistic and dialectical ends-and-means hierarchy (Farjoun, 2002; Li, 1994, 2003).

As it is generically applicable to all strategies, the above four-dimension typology can be applied to the special case of e-business strategy with necessary adaptations according to the unique characteristics of e-business. In light of the central theme of e-business (i.e., network paradigm), we can refer to the study of network form for the application of the findings to the case of e-business. The key feature of network form is the distinction between core and non-core business activities and their configuration along the value-adding chains (Li, 1998). This feature of network form is especially applicable to e-business in terms of business configuration scope (e.g., SCM design, brick and click configuration, etc.) as well as business coordination style (e.g., BPR, ERP, and CRM). The other key feature of network form is the concurrent pursuit of seemingly paradoxical ends and means, such as achieving both high revenue volume and high profit margin, both low cost and high value, both broad and narrow scope, and both tight and loose control (Li, 2001a, 2001b). Below are the propositions about e-business strategy content.

P1: The strategic target for e-business is the concurrent pursuit of volume size and margin rate, with the focus of core business on margin rate while the focus of non-core business is on volume size; thus, the expected effect of e-business strategy is an increase in both market share and profitability.

P2: The strategic thrust for e-business is the concurrent pursuit of low cost and high value, with the focus of core business on high value via customization while the focus of non-core business is on low cost via standardization; thus, the expected effect of e-business strategy is a reduced cost and an enhanced value.

P3: The strategic posture for e-business is the concurrent configuration of narrow and broad scopes of business coverage in terms of market, supply-chain, function, and location, with the focus of core business as in-house specialist with a narrow configuration at the firm level and the focus of non-core business as outsourcing generalist with a broad configuration at the network level.

P4: The strategic mode for e-business is the concurrent application of tight and loose coordination, with the focus of core business on tight coordination, both internal (e.g., ERP and BPR) and external (e.g., SCM and CRM), while the focus of non-core business on loose coordination (e.g., e-marketplace). The distinction between SCM and e-marketplace is that SCM is firm-specific and firm-centric for core business, while e-marketplace is community-generic for non-core business. A proper brick-and-click scheme should be one in which the bricks focus on the physical elements as their core while the clicks focus on the virtual elements as their core, but they collaborate as loosely coupled, strategic partners (i.e., two separate yet closely coordinated units). However, the real and virtual elements of global presence should be a closely coupled integration.

Three-Dimension Typology of E-Business Strategy Process

A three-dimension typology of strategy process (Li, 2003) can be used to measure the temporal process of e-business strategy. The temporal aspect of business in general and strategy in particular has been largely ignored for too long (Farjoun, 2002; Li, 1998, 2003). These three dimensions are (1) strategic simultaneity, (2) strategic directionality, and (3) strategic rhythm. Strategic simultaneity refers to the temporal synchronization of the strategic content events or activities that occur at a given point in time (Li, 2003). Strategic directionality refers to the temporal sequence of the strategic content events or activities that occur at different points in time (Li, 2003), such as the timing of market entry as a first-mover or a quick-follower and the phases of evolution (Hambrick & Fredrickson, 2001; Li, 2003). Strategic rhythm refers to the temporal variation of the strategic content events or activities that occur at different tempos at different times (Li, 2003), such as the tempo of activities (Hambrick & Federickson, 2001), especially for the latecomers committed to catching up (Li, 2003).

As it is generically applicable to all strategies, the above three-dimension typology can be applied to the special case of e-business strategy with necessary adaptations according to the unique characteristics of e-business. A fast tempo seems to be one of the key characteristics of e-business (Kalakota & Robinson, 2002; Li, 2001a, 2001b). A well-timed synchronization is another key feature of e-business in terms of instant access to information by all the members of an organization, a supply chain, or a marketplace. The third key feature is the sequencing of different activities at different stages. Hence, we developed the following propositions with regard to the temporal process of e-business strategy:

Figure 1. A network-based conceptual framework of e-business strategy

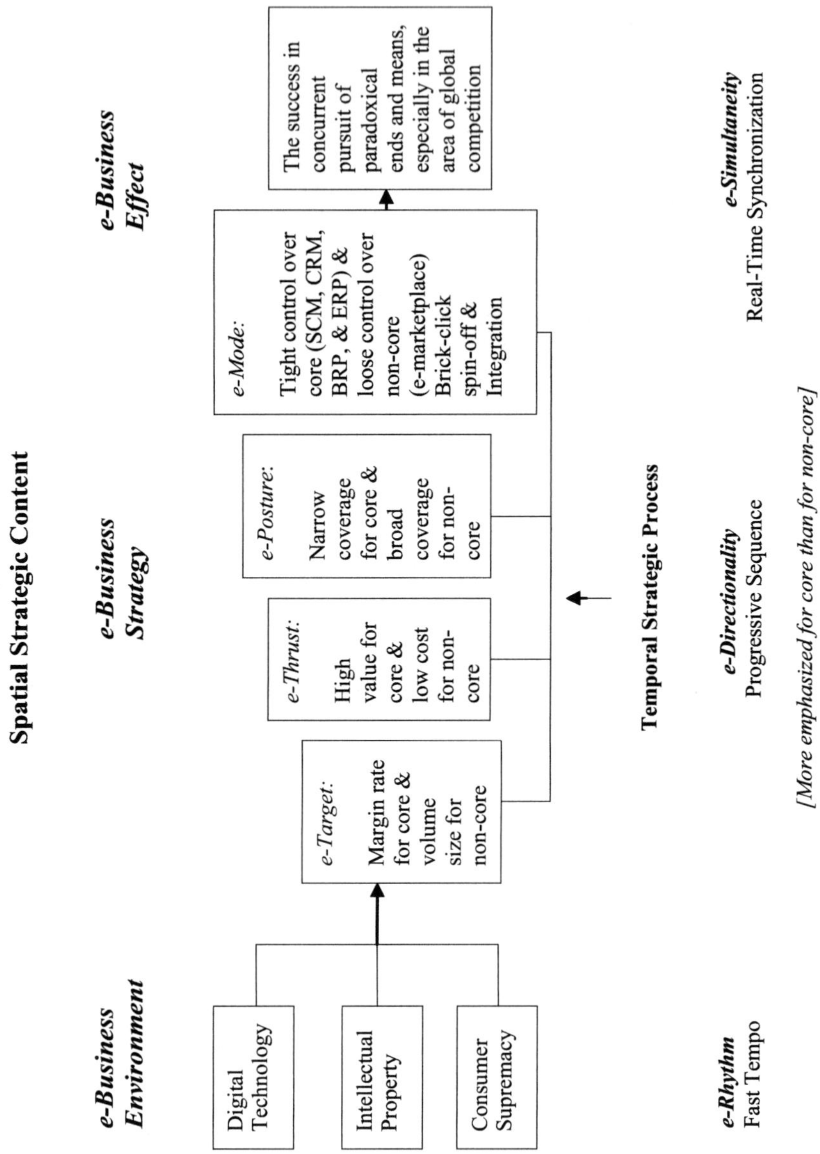

P5: The strategic rhythm for e-business in general is a fast tempo, especially for the latecomers to the global competition, with a faster pace for core business and a slower pace for non-core business.

P6: The strategic simultaneity for e-business is a real-time synchronization of e-business and other programs, with core business as the primary focus and non-core business as the secondary focus.

P7: The strategic directionality for e-business is a progressive sequence from a trial-and-error phase to a mature one, with an earlier entry timing for core business and a later entry timing for non-core business.

THE CASE EVIDENCE OF CHINA'S HAIER

Methodology

We chose to adopt the method of longitudinal case study because this method offers the best prospect for advancing our knowledge of strategic process (Eisenhardt, 1989; Numagami, 1998). We chose Haier for our case study of e-business in China due to four considerations. First, Haier is one of the most successful companies in China. Second, Haier is well prepared for e-business. Third, Haier has successfully formulated and implemented an e-business strategy. Fourth, Haier has a clear strategic intent of internationalizing itself, which is closely related to its e-business strategy. We used secondary data from corporate brochures, corporate annual reports, books, newspaper/magazine reports, others' case studies, corporate Web sites, and public information Web sites. Only those data that were corroborated from multiple sources were used.

China's E-Business Status

The prevailing view is that China is not yet ready for e-business due to the barriers of poor technical infrastructure, legal protection, payment system, and consumer acceptance (Bin & Chen, 2003; Stylianou, Robbins, & Jackson, 2003). However, China has great potential for e-business in the future (Stylianou, Robbins, & Jackson, 2003).

China's Haier Group

Restructured in 1985 as a small manufacturer of refrigerators burdened by a debt of RMB 1.47 million, Haier Group not only has survived a series of radical reforms, but also has become one of the most successful companies in China. As a multinational company with an annual revenue of RMB 80.6 billion (US $10.4 billion) in 2003, Haier had 22 overseas manufacturing facilities, and its international sales accounted for about 10% of its total revenue with a value of over US $1 billion, among which the U.S. market accounted for 25%. Haier produces a full line of electronic household appliances and other products in 42 major categories and more than 9,000 models. Haier ranks number one in China's household appliance industry as measured by brand recognition and sales revenues. Haier has gained an international reputation by exporting to over 100 countries. In the global survey by *Euromonitor,* Haier was ranked as the world's number-one refrigerator brand and also the world's fifth largest white appliance producer after Whirlpool, Electrolux, Bosch, Siemens, and GE (Xinhua News Agency, 2002). Haier has achieved a high growth for 17 consecutive years with an annual average growth rate of 78% (Xinhua News Agency, 2002). Haier has been undergoing three key stages of corporate development (Sun, Ji, & Wang, 2001):

1. Brand-building stage (1984–1991). It took Haier seven years to build a strong brand in refrigerators through a well planned total quality management (TQM) program.

2. Diversification stage (1992–1998): It took Haier another seven years to diversify and broaden its product offerings.
3. Internationalization stage (1998–Present). Haier has been undergoing an internationalization program. Its strategic intent is to become one of the Global Top 500 with a global brand.

Background of Haier's E-Business

The Strategic Intent of Haier's E-Business

The overall long-term strategic intent or vision of Haier is to be one of the world's top 500 companies with a global brand. Haier views e-business as the most effective means to achieve Haier's strategic intent of joining the Global 500 Club.

Haier's Readiness for E-Business

Haier appeared ready for e-business when it launched its e-business initiatives. First, Haier had a strong commitment from its top management team, who has led Haier in its e-business effort. Haier's leadership was aware of the imperatives of e-business and the implications of the new economy. Zhang Ruimin, the CEO of Haier, points out, "Do e-business or die." Zhang saw e-business as an effective tool to offer more value-added services through customization, shortened cycle time, and innovation to reach out to the worldwide market and to minimize overall cost (Yan & Hu, 2001). He also saw employees as the sources of self-innovation. According to him, each employee should become a boss of his or her own with enough flexibility (i.e., delegated decision-making power) and enough incentive (i.e., rewards closely linked to performance), as reflected by Haier's Employee-as-SBU-Boss program (Li & Liu, 2002).

Haier laid a solid foundation for the success of e-business with the implementation of two critical programs (Li & Liu, 2002). The first was Hiaer's ERP for building the technical infrastructure for information collecting and processing. Haier realized that digital information on various business functions was integrated for sharing by all parties, so a firm could adapt to the volatile market and survive in the new economy. The second was Haier's BPR. Haier implemented e-business programs together with its BPR programs, because it realized that e-business was not simply about setting up a Web site or an information system; rather, it was about redesigning the ways firms used to conduct its business.

The Content of Haier's E-Business Strategy

Haier's E-Target

The primary strategic targets for Haier's e-business were the revenue size, market share, and revenue growth rate. These targets were based on Haier's long-term intent and vision of becoming one of the top 500 global players. Toward that vision, Haier chose to target the highly competitive U.S. and European markets first before it moved to easier markets (Wu, 2003). However, Haier appeared to pay much less attention to profit growth and profit margin. Further, there was no clear differentiation between Haier's core business and non-core businesses with regard to the choice of e-target, thus lacking the evidence to support the view that the e-target for core business would be profit-oriented while the e-target for non-core business would be revenue-oriented. This may be a source of Haier's problems with regard to the effect of its e-business strategy.

Haier's E-Thrust

With the above strategic target in mind, Haier identified its strategic thrust as a concurrent pursuit of both low cost and high value. Haier wanted to provide high value-added services along with superior products at the lowest total cost possible. Haier wanted to reduce its total working inventory and the associated working capital cost. It also wanted to reduce the cost of transaction by streamlining the commercial interface between business partners in order to share information. Haier also wanted to improve the value of its products with enhanced services and customization. Haier wanted to achieve the above ends not only in its domestic market, but also in the global market. For example, Haier set up three key e-business objectives: (1) zero distance between customers and suppliers; (2) zero inventory; and (3) zero working capital. The first objective was directly related to the strategic thrust of high value by way of product and service customization as well as shortened cycle time in product development and logistic delivery, both of which required Haier's close cooperation with its customers and suppliers. Further, the objectives of zero inventory and zero working capital were directly related to the strategic thrust of low cost. However, there was no clear differentiation between Haier's core business and non-core businesses with regard to the choice of e-thrust, thus lacking the evidence to support the view that the e-thrust for core business would be value-oriented, while the e-thrust for non-core business would be cost-oriented. Again, this may be a source of Haier's problems with regard to the effect of its e-business strategy.

Haier's E-Posture

To accomplish its e-thrust, Haier chose both broad and narrow coverages for its configuration of a business portfolio. Within its core business of household appliance sector, Haier produced 42 types of household appliances. It also diversified into computer and cellular phone areas, which were somewhat related to its core business. Further, Haier had a certain level of vertical integration, and it had a fairly extensive supply chain. Nevertheless, the level of outsourcing of non-core operations in Haier was still rather limited. For example, Haier expanded its logistic unit to a great extent, so the logistic unit had an annual revenue of more than RMB 20 billion (US $2.4 billion) in 2002, or 27.7% of Haier's total annual sales. Further, Haier had a limited real global presence, for which Haier tried to compensate by establishing a strong virtual global presence. This implied that a real global presence could not be substituted with a virtual global presence; therefore, Haier would have to work hard to establish its real global presence. Further, there was no clear differentiation between Haier's core business and non-core businesses with regard to the choice of e-posture, thus lacking the evidence to support the view that the e-posture for core business would be narrow, while the e-posture for non-core business would be broad. This may another source of Haier's problems with regard to the effect of its e-business strategy.

Haier's E-Mode

The most interesting among the four dimensions of strategy content is e-mode. The strategic mode of Haier's e-business was clearly reflected by its e-business platform. Such a platform consisted of four elements. The first was Haier's SCM project, which included its distributor management, supplier management, and other corporate partnership management. The second was Haier's CRM project, which included the programs for both institutional and individual consumers. The third was Haier's ERP project. The last and the most important was Haier's BPR project. The first two projects were closely tied with Haier's B2B (business-to-business e-business transactions) and B2C (business-to-customer e-business transactions) external networks, while the last two projects were closely tied to Haier's internal networks. The above four projects constituted Haier's e-business platform as its e-mode. Although they were all related to Haier's global effort, its SCM and CRM projects were more closely tied to Haier's global expansion.

Haier's SCM system as its external B2B network was built on the platform of iHaier.com. With this platform, Haier was able to seek its best suppliers, set up close partner relationships with its suppliers, and reduce the purchase cost while improving product quality. Haier planned to use this system not only for itself, but also for other related companies. This platform had the functions of ordering, automated stock replenishment, payment processing, and production control processes. This platform was built on an open architecture where information was fully shared. Haier had not been a member of any neutral e-marketplaces yet. Further, iHaier.com limited itself only to the household appliance industry. Finally, Haier's platform only supported Haier's own procurement, even though Haier stated that it would also host third-party procurement in the future.

Haier's CRM system as its external B2C network was built on the platform of eHaier.com. This Web platform had the function of taking online orders, but online payment was limited to selective cities only (most consumers paid upon delivery). However, Haier's B2C platform still only handled Haier's own products. It would carry other producers' products in the future. To support its B2C platform, Haier built one of the best distribution networks in China. Haier had more than 30 call centers in China's major cities and more than 10 thousand distributors that reached more than 60 thousand rural areas in China. Further, Haier's B2C platform was able to process customized orders. Finally, Haier's B2C platform had a much weaker impact on global consumers than on domestic consumers. Haier had been working hard to establish a global logistic distribution and delivery system to support its B2C and B2B platforms as well.

What is really unique about Haier's e-mode is its creative adaptation of the B2C model from the U.S. to China's special situations. Since many Chinese individual consumers did not have direct access to the Internet at home, Haier adopted an innovative B2B2C model. Haier was trying to leverage its established distribution systems with its Web site system. Haier asked its distributors and retailers to provide Internet access for its consumers. As a result, in addition to being the point of delivery, service, and payment collection, Haier's distributors and retailers also served as a point of providing product information, offering consultation or advice and taking customized orders from the consumers who did not have Internet access or did not feel comfortable placing online orders by themselves. Such a B2C platform was the unique feature of Haier's CRM system.

As discussed earlier, Haier implemented an internal ERP system, which served as part of its IT system for its e-business strategy. The current problem with Haier's ERP system was that its internal logistic flows, capital flows, and information flows still were handled separately by different internal subsystems. This was due to the different times when various parts of the information systems were adopted. Haier would have to work hard to integrate its subsystems and complete its internal ERP system. Haier had worked with many global IT partners (e.g., Lucent and SAP) and some consulting firms (e.g., McKinsey) for technical and business advice. In a promising market with various entry barriers, such complementary alliances among local companies and foreign multinationals would be critical to the success of both types of companies (Chang, Chuang, & Jan, 1998).

Haier had also restructured its organization and management process toward an e-management system through its continuous BPR program. Haier's BPR program had two major goals. First, it wanted to implement a shift from a traditional function structure to a flat customer-driven, process-oriented, market-chain structure. Second, it wanted to instill a new culture into the company, where an employee was responsible for customers and responsive to market signals rather than to his or her boss.

The concrete examples of Haier's BRP system are its Market Chain program and its Employee-as-SBU-Boss program. The core value of the Market Chain program was to take customers (external and internal) as the focus of all the activities. To accomplish that, Haier reorganized itself from the traditional functional structure to a process flow structure with four basic divisions: (1) commercial flow centers for acquiring sales orders; (2) workflow centers for manufacturing products; (3) logistic flow centers for distribution and delivery; and (4) overseas centers for international activities. The central link between the four divisions was the flow of order information. In addition, the flows of capital and material were integrated with the flow of order information. Finally, the SCM and CRM systems were integrated with the above activities through the flow of order information. Together, the above activities constituted Haier's Market Chain program.

Besides the Market Chain program, Haier also implemented the Employee-as-SBU-Boss program. First, every unit in the company, regardless of its function or level, was granted the status of a profit center with full responsibilities of its own. Second, every employee, regardless of his or her function or level, was also granted the status of a profit center, with full responsibilities as though he or she was the boss of his or her own SBU. In short, the central theme of Haier's BRP was about the incorporation of external market competition into its internal operations in order to create an internal market network inside the company.

Further, the mode of Haier's entry into e-business was that of brick-click separation. In 2000, Haier set up a separate unit with a small staff of 45 (Haier E-Business Co.), which was responsible for the implementation of Haier's e-business initiatives. This indicated that e-business was not yet well integrated with the rest of Haier's business operations. Similarly, Haier's global presence was more virtual than real, although the trend was toward a more well balanced integration of its virtual and real elements.

Finally, there was no clear differentiation between Haier's core business and non-core business regarding the choice of e-mode, thus lacking the evidence to support the view that the e-mode for core business would be tight, while the e-mode for non-core business would be loose. This is the final source of Haier's problems with regard to the effect of its e-business strategy.

The Process of Haier's E-Business Strategy

Haier's E-Rhythm

Realizing that e-business was largely about the fast speed of information flow, capital flow, and logistic flow, Haier accelerated all of its programs and projects. For instance, within the past three years, Haier made various changes to its organizational structure more than 40 times. Further, as a latecomer to the global competition, Haier had to accelerate its process of internationalization with whatever means available, including e-business.

More importantly, Haier saw speed as the source of its new core competence. Haier launched a series of projects as part of its BPR program to shorten the cycle time for new product development, global market access, logistic delivery, and payment collection. The core of those projects was the application of e-business tools as the most effective in enhancing the speed of operations. For example, Haier ranked speed at the top of its three-pronged strategic initiative for 2002 (i.e., speed, innovation, and SBU), where innovation and SBU were regarded as the means to achieve the end of speed. However, Haier did not differentiate its core business from its non-core business with regard to the tempo of e-business process.

Haier's E-Directionality

Haier's e-business strategy underwent a progressive sequence from an initial phase of trial and error to a later phase of maturity. Haier started its e-business program in 1998. Initially, Haier copied the model from the West to develop Web sites, which did not work well in China. Later, Haier invented its unique and effective B2B2C model. Even later, Haier started to integrate its e-business effort with those of internationalization, (which started in 1999) and BPR restructuring (which started in 1998). Despite the remaining problems, Haier's e-business effort is becoming increasingly more effective.

There were two good examples of e-directionality. The first was Haier's internationalization program, which underwent four phases. In the first phase, Haier started to learn from the Western companies such as Dell for the Just-in-Time (JIT) system as well as from GE for the diversification program. In the second phase, Haier started to make strategic alliances with the Western companies. In the third phase, Haier started to localize its overseas operations, including local design, local production, and local service. In the current fourth phase, Haier has started to seriously build a global brand. The other example was Haier's logistic operation, which underwent three phases. In the first phase, Haier started to consolidate all the logistic operations into one business unit. In the second phase, Haier started to integrate its logistic unit with its SCM and CRM programs. In the third phase, Haier started to turn the logistic unit as a profit center that offers services to both internal and external customers (as the third-party logistic provider). However, Haier did not differentiate its core business from its non-core business with regard to the sequence of e-business process.

Haier's E-Simultaneity

As indicated, Haier did not integrate its e-business strategy with other initiatives at the early stages (e.g., the cases of brick-click separation and unbalanced virtual-real global presence), but Haier adjusted its approach later to incorporate them into a coherent stream of programs. The most significant integration occurred in the areas of its global program and restructuring program. In 2001 and 2002, there was a pattern of better synchronization with regard to e-business, internationalization, and BPR, which was much more pronounced in the years to come.

The best example of Haier's e-simultaneity was the link between the three e-business objectives (i.e., zero distance between customers and suppliers, zero inventory, and zero working capital). The objective of zero distance implied that e-business was largely about meeting the real need of the customers, which could be accomplished through product and service customization as well as shortened cycle time in product development and logistic delivery. Zero distance required well synchronized cooperation between customers and suppliers. The objective of zero inventory implied the need for effective JIT, SCM, and CRM systems, so the achievement of this objective relied heavily on the achievement of the objective of zero distance. Similarly, the achievement of the third objective (i.e., zero working capital) had to rely on the achievement of the first two objectives. Put differently, the three objectives were closely related, so a good synchronization among the three was critical. However, Haier did not differentiate its core business from its non-core business with regard to the synchronization of e-business process.

The Effect of Haier's E-Business Strategy

The overall effect of Haier's e-business strategy was generally positive. The effect of Haier's e-business strategy can be evaluated by three sets of measures. The first set was the measures of *reduced cycle time* and *enhanced customization* according to the objectives of zero distance and fast speed. The evidence strongly suggested that Haier's e-business strategy was very effective. For instance, the cycle time for developing new products was reduced from four to six months in 2000 to two to three months in 2002. Further, the cycle time for logistic distribution was reduced to a great extent. Haier could distribute its items in less than eight hours within major cities, less than 24 hours within the same province, and less than four days nationally inside China. This was also the case overseas. For instance, Haier reduced its delivery time in Europe by half during the past three years. The export delivery cycle time was reduced from more than 15 days in 2000 to less than seven days in 2002. Another example of the effect of e-business on Haier was that its cycle time from sale order to procurement order was reduced from more than seven days in 2000 to less than one hour in 2002. Also in 2002, Haier achieved a significant milestone by having a perfect 100% of ordering online. Finally, the procurement cycle was reduced from 10 days in 2000 to three days in 2002.

In terms of enhanced customization, Haier also achieved a lot. The customized orders accounted for more than 3% of the total orders, up from less than 1% in 2000. Further, newly innovated products accounted for more than 60% of its total sales in 2002. Haier was able to introduce new products with an average rate of 1.5 per day and apply new patents with an average rate of 2.6 per day. Let us take a closer look at one specific instance. When Haier announced its customized refrigerator in August 2000, within just one month, it received orders of more than 1 million customized refrigerators, which was about one-third of Haier's annual refrigerator sales. By 2002, about half of Haier's orders for refrigerators were customized. Another example is that it took Haier only five months to design a new wine storage cabinet for the U.S. and European markets, which would normally take other companies 18 months (Wu, 2003). The new product was so popular that it had more than 50% market share in the U.S. (Wu, 2003).

The second set was the measures of cost saving and cooperative value creation according to the objectives of zero inventory and zero working capital. The evidence strongly suggests that Haier's e-business strategy was largely successful. For instance, the size of procurement staff was cut by one-third from 2000 to 2002, and the cost of procurement was dropping at an annual rate of 5%. Further, the average inventory turnover time was reduced from more than 30 days in 2000 to less than 18 days in 2001 and further down to 10 days in 2002. With regard to the SCM program, the number of suppliers was reduced from more than 2300 in 2000 to 721 in 2002, among which were 59 of the world's top 500 companies. More than 70% of the suppliers participated in various joint R & D projects with Haier. There was also measured success in the CRM area where the goal was to help the customers succeed in their business ventures.

In terms of zero working capital, Haier also achieved a great deal. For instance, the turnover of its working capital for supply procurement was reduced from more than 36 days in 2000 to less than 10 days in 2002. Further, more than 80% of the online orders were paid online in 2002. During the past three years, the old inventory dropped by about 74%, the warehouse space utilization was improved by about 100%, and the working capital tied up with the inventory dropped by about 67%.

The last set was the measures of overseas sale, profit, and brand recognition according to the objective of internationalization. Despite the increasingly large revenue base, Haier still achieved a high export growth rate. For instance, Haier's export sales grew more than 100% in 2000, more than 50% in 2001, and 37% in 2002. Haier had 62 overseas distributors covering more than 30,000 overseas distribution terminals. Haier's products reached more than 100 countries. Haier's exports to the U.S. and Europe accounted for about 60% of its total exports. Haier was one of the largest exporters in China.

Haier enjoyed a high market share in many overseas markets. For instance, Haier's small refrigerators had a market share of 25% in the U.S.; its freezers had a market share of 28%. In Indonesia; its air conditioners had a market share of 60% in Uruguay; and its washing machines had a market share of 30% in Sri Lanka. In addition to the overseas markets, Haier's refrigerators had a domestic market share of 33.4%; its freezers had a domestic market share of 41.8%; its air conditioners had a domestic market share of 30.6%; and its washing machines had a domestic market share of 30.5%. In sum, the effect of Haier's e-business strategy was generally positive, but not without problems.

The Remaining Problems of Haier

Despite its apparent success, Haier still faced many challenges. First, there was no proven e-business model either in China or abroad for Haier to copy. However, Haier was good at learning from the successful practices of other companies such as GE and Dell, and Haier was creative in developing its own unique e-business models (e.g., the B2B2C model) in a trial-and-error process. Second, the e-business conditions in China limited Haier's e-business expansion. Many of Haier's suppliers did not have adequate in-house information systems, and even Haier's own in-house IT system had not been fully integrated. Third, Haier was fighting an uphill battle in its pursuit of global brand. Judged by revenue size, Haier was almost ready to join the Global Top 500 Club. However, Haier still lacked a strong global presence with a strong global brand.

Figure 2. Evidence of Haier's e-business strategy

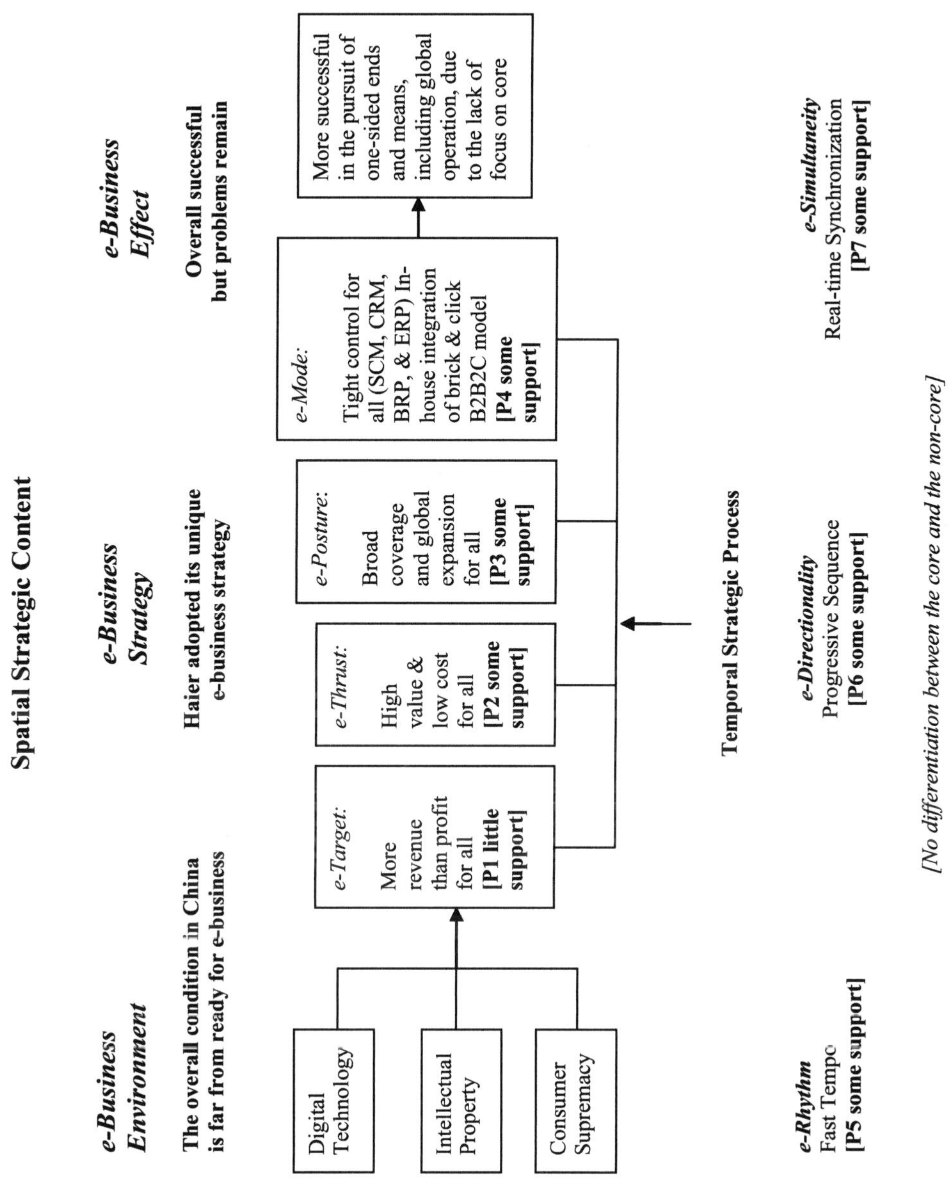

Finally, there was a concern that Haier might have made several wrong choices for its e-business strategy content. It was controversial for Haier to focus so much on joining the Global Top 500 Club as its strategic intent and achieving revenue volume as its strategic target. Further, Haier might have to reexamine its strategic posture of broad diversification, which was largely driven by its single-minded pursuit of the Global Top

500 Club membership. Haier might have expanded too far beyond the scope of its core competency and its managerial control. Haier has failed to differentiate its core business from its non-core business in both content and process of e-business strategy. This could be the most serious mistake for Haier, as it might have missed the biggest potential benefit of e-business—the concurrent pursuit of paradoxical ends via balanced means according to the holistic, dynamic, and dialectical network paradigm.

Discussion about the Case Evidence

The case evidence can be summarized in Figure 2 as analyzed against Figure 1.

In general, the case evidence of Haier provided partial support for most propositions. Specifically, the case evidence did not lend much support for Proposition 1 about the choice of strategic target. The case evidence provided partial support for the other six propositions. The partial support is largely due to the lack of Haier's differentiation between its core business from its non-core business regarding both the content and process of e-business strategy. The failure to make such distinction might have hindered Haier from benefiting fully from e-business, which could be the root problem of its current e-business strategy and would be the biggest obstacle to its future e-business success. The problem is related to such general problems in China as a pride-driven focus on large size, the underdevelopment of component supply industries, and the general lack of effective legal protection. The significant practical implication is that a company must clarify its strategic core in order to build its e-supply and e-demand chains around its core.

CONCLUSION

This chapter has attempted to make two basic contributions. First, it has proposed a holistic, dynamic, and dialectical conceptual framework of e-business. The proposed framework offers a working model by which a firm can map out its strategic choices with regard to the critical issues of e-business (e.g., the antecedents, content, process, and consequence of e-business, such as the paradoxes of growth-profit, customization-standardization, specialization-diversification, core-non-core, competition-cooperation, tight-loose, brick-click, and spatial-temporal). The chapter has conducted a case study to test the proposed model. Further research, both conceptual and empirical, is needed to expand the application of the proposed framework.

Second, the chapter has also provided practical lessons for local firms in the developing countries to avoid the major pitfalls of e-business. The local firms need to be creative in adapting the business models from the West to their unique local conditions (e.g., Haier's B2B2C Model) in order to overcome the unique barriers in the developing countries. Strategic alliance with foreign partners is imperative for the local firms to effectively implement e-business strategies. The local firms must build their in-house information systems before initiating their e-business programs. They need to be patient and persistent with their internationalization programs. They particularly need to avoid the pitfall of a single-minded pursuit of one-dimensional objectives (e.g., the overemphasis on growth target, cost thrust, broad posture, tight mode, fast tempo, non-synchronized coordination, and premature leapfrogging sequence).

REFERENCES

Afuah, A. (2003). Redefining firm boundaries in the face of the Internet: Are firms really shrinking? *Academy of Management Review*, *28*, 34-53.

Afuah, A., & Tucci, C.L. (2003). *Internet business models and strategies: Text and cases* (2nd Ed.). New York: McGraw-Hill.

Arndt, M. (2002, April 15). Can Haier freeze out Whirlpool and GE? *Business Week Online*. Retrieved October 24, 2002, from *http://www.businessweek.com/bwdaily/dnflash/apr2002/nf20020411_3930.htm*

Bartlett, C.A., & Ghoshal, S. (1989). *Transnational management*. New York: Free Press.

Bin, Q., & Chen, S. (2003). Cultural differences in e-commerce: A comparison between the U.S. and China. *Journal of Global Information Management*, *11*(2), 48-55.

Brews, P.J., & Tucci, C.L. (2004). Exploring the structural effects on internetworking. *Strategic Management Journal*, *25*, 429-451.

Chang, T, Chuang, C., & Jan, W. (1998). International collaboration of law firms: Modes, motives and advantages. *Journal of World Business*, *33*, 241-262.

Dunning, J.H. (1995). Reappraising the eclectic paradigm in an age of alliance capitalism. *Journal of International Business Studies*, *26*, 461-491.

Economist. (1999, June 26). Survey: Business and the Internet. *Economist*, s1-40.

Economist. (2000a, February 26). Survey: E-commerce. *Economist*, s1-54.

Economist. (2000b, September 23). Survey: The new economy. *Economist*, s1-45.

Economist. (2000c, November 9). Survey: E-management. *Economist*, s1-63.

Eisenhardt, K.M. (1989). Building theories from case study research. *Academy of Management Review*, *14*. 532-550.

Farjoun, M. (2002). Towards an organic perspective on strategy. *Strategic Management Journal*, *23*, 561-594.

Financial Times. (1999, December 7). World's most respected companies: Entrepreneurs win high admiration. *Financial Times* (special ed.). Retrieved October 23, 2002, from *http://specials.ft.com/ln/specials/sp41d2.htm*

Flannery, R. (2001, August 6). China goes global: The ambiguous Haier leads an international brand push by mainland multinationals. *Forbes*, 36-45.

GEO Award (2000). GEO Conference 2000 highlights. Retrieved October 23, 2002, from *http://atkearney.com/geoaward/conference.htm*

Hambrick, D.C., & Fredrickson, J.W. (2001). Are you sure you have a strategy? *Academy of Management Executive*, *15*(4), 48-59.

Kalakota, R., & Robinson, M.M. (2001). *E-business 2.0*. New York: Addison-Wesley.

Kelly, K. (1998). *New rules for the new economy: 10 radical strategies for a connected world*. New York: Penguin Group.

Kuhn, T.S. (1970). *The structure of scientific revolution*. Chicago: University of Chicago Press.

Li, M., & Liu, Y. (2002). The foggy Haier. *Chinese Entrepreneur*, *208*, 12-26.

Li, P.P. (1994). Strategic profile of indigenous MNEs from NIEs: Case of South Korea and Taiwan. *The International Executive*, *34*, 525-537.

Li, P.P. (1995, August 5-8). Inevitability, desirability and resolvability of paradox: A dialectic approach to global strategy. *Annual Meeting of the Academy of Management*, Vancouver, Canada.

Li, P.P. (1997). The unique nature of international joint venture in the People's Republic of China. In S. Stewart & A. Carver (Eds.), *Advances in Chinese industrial studies, 5* (pp. 13-34). Greenwich, CT: JAI.

Li, P.P. (1998). Towards a geocentric framework of organizational form. *Organization Studies, 19*, 829-861.

Li, P.P. (2001a). Towards a conceptual framework of e-business. *E-Business Review, 1*, 132-135.

Li, P.P. (2001b, May). The paradoxes of e-business and the network paradigm. *Decision Line*, 14-16.

Li, P.P. (2003). Toward a geocentric theory of multinational evolution: The implications from the Asian MNEs as latecomers. *Asian Pacific Journal of Management, 20*(2), 217-242.

Numagami, T. (1998). The infeasibility of invariant laws in management studies: A reflective dialogue in defense of case study. *Organization Science, 9*, 2-15.

Parkhe, A. (1993). "Messy" research, methodological redispositions, and theory development in international joint ventures. *Academy of Management Review, 18*, 227-268.

People's Daily. (2002a, June 13). Fortune selects most favored foreign firms in China. *People's Daily Online*. Retrieved October 24, 2002, from *http://english.peopledaily.com.cn/200206/13/eng20020613_97789.html*

People's Daily. (2002b, April 15). Haier takes lead in exporting electrical appliances. *People's Daily Online*. Retrieved from *http://english.peopledaily.com.cn/200204/15/eng20020415_94065.shtml*

Porter, M.E. (1980). *Competitive strategy*. New York: Free Press.

Porter, M.E. (1990). *The competitive advantage of nations*. New York: Free Press.

Porter, M.E. (2001). Strategy and the Internet. *Harvard Business Review, 79*(3), 63-78.

Rayport, J.F. (1996). The virus of marketing. *Fast Company, 6*, 68.

Rayport, J.F., & Jaworski, B.J. (2001). *E-commerce*. New York: McGraw-Hill/Irwin.

Roberts, D. (2002, July 17). Baby steps for a Chinese giant. *Business Week Online*. Retrieved October 23, 2002, from *http://www.businessweek.com/bwdaily/dnflash/jul2002/nf20020717_8966.htm*

Roberts, D, Arndt, M., & Zammert, A. (2002, April 1). Haier's tough trip from China. *Business Week Online*. Retrieved October 23, 2002, from *http://www.businessweek.com/magazine/content/02_13/b3776139.htm*

Shapiro, C., & Varian, H. (1999). *Information rules*. Cambridge, MA: Harvard Business School Press.

Sprague, J. (2002a, October 28). China's manufacturing beachhead. *Fortune* (Industrial Management Edition). Retrieved December 10, 2002, from *http://www.fortune.com/indexw.jhtml?channel=article.jhtml&doc_id=209794*

Sprague, J. (2002b, September 16). Haier reaches higher. *Fortune* (International Edition). Retrieved October 24, 2002, from *http://www.fortune.com/indext.jhtml?channel=article.jhtml&doc_id=209394*

Stylianou, A.C., Robbins, S.S., & Jackson, P. (2003). Perceptions and attitudes about e-commerce development in China: An exploratory study. *Journal of Global Information Management, 11*(2), 31-47.

Sun, J., Ji, J., & Wang, F. (2001). *Direct from Haier: The growth of a Chinese firm* (in Chinese), Beijing, China: Business Management Publishing House.
Wu, Y. (2003). China's refrigerator magnate. *The McKinsey Quarterly*, 3, 106-115.
Xinhua News Agency (2002, October 15). Haier rates as world's top refrigerator brand. *Xinhua News Agency*, 1, 1.
Yan, J., & Hu, Y. (2001). *Haier: Made in China* (in Chinese). Hannan, China: Hanan Publishing House.

Chapter XVI

Relationship of Information Infrastructures and Social Development Among the Visegrad-Four Countries of Central Europe at the Time of EU Accession

László A. Poók, Metropolitan State College of Denver, USA

Norman E. Pence, Metropolitan State College of Denver, USA

ABSTRACT

The following is an examination of the developmental status of four of the next candidate countries' information infrastructures for accession into the European Union (EU). It develops significant relationships between nations' levels of information infrastructure development and their economic and social developments and applies these relationships to evaluate investment needs for a select group of accession countries called the Visegrad Four. Using cross-country analysis, the candidate countries were compared to other countries and were classified into like groups using cluster analysis while their relative developments were evaluated using regression modeling. Developmental requirements and capital needs to promote growth in their information and communications industries were then identified. It is hoped that the chapter will offer a comparative glimpse of the information and communications infrastructures of some Central and East European countries as compared to other nations.

INTRODUCTION

In 1991 after the fall of communism and the disintegration of the USSR, four of the Eastern block nations institutionalized a policy of coordination and laid the foundations of their transition from totalitarian regimes to free, pluralistic, and democratic societies. Their agreement was formalized as the Visegrad Declaration (at the Hungarian city of Visegrad), and Visegrad-Four (or V4) is the name sometimes given the four Central European post-communist countries—the Czech Republic, the Republic of Hungary, the Republic of Poland, and the Slovak Republic (TASR, 2002). The Visegrad Declaration led to a continuation of free trade agreement among the four signatories, and trade increased with Western Europe; however, it decreased with other former communist countries and declined considerably with the former Soviet republics (Baylis, 1994). Also, agreements among the Visegrad-Four led to these four countries simultaneously and jointly submitting their applications for European Union (EU) membership (Shea & Stefes, 2002).

To date, the EU has had accession talks with 13 Eastern European nations and designated 10 countries as candidates for integration into the EU to join in a first wave in 2004. The 10 front-runners, named at a summit in Laeken, Belgium, in December 2001, are the original V4 (the Czech Republic, Hungary, Poland, and Slovakia), three Baltic nations (Latvia, Lithuania, and Estonia), two Mediterranean island nations (Malta and Cyprus), and the well-advanced ex-Yugoslav federation state of Slovenia [1]. Bulgaria, Turkey, and Romania were designated as second-wave candidates.

Motives of the V4 for joining the EU range from purely ideological to a desire for political, economic, and military stability, and for foreign direct investments in their economies. Conversely, the motives of the EU to expand eastward are morals, fear, and economics: morals, because there is no other region in the world closer to Central and Eastern Europe capable of defusing potential strife in a historically war-prone region; fear, because in case of strife, Western Europe first would have to receive migrants from the east fleeing potential conflicts and repressions; and economics because these ex-satellite states of the USSR constitute enormous markets as well as educated and underpaid work forces ready to be plugged into the West's production systems.

Accession into the EU is a lengthy process of meeting designated criteria on several factors (i.e., enabling laws, policies, and development on telecommunications and information). We will employ cross-country analyses to evaluate the information and telecommunication preparedness of the V4 and the extent to which their national information infrastructures are developed to contribute to their needed economic growth for accession to the EU and, once admitted, to match the information infrastructures of the advanced members of the present EU. Furthermore, the V4 as a group are of particular interest to the EU because of their relative political stability and proximity to present union boundaries. As part of this investigation, we also will examine the practice by the World Bank of assessing information penetration by using telephone mainline counts as surrogate for the penetration of information technologies.

Research Model

In the ensuing discussion, by *social development* we mean the extent to which major sources of social distress and instability for the family and for society have been eliminated (United Nations, 1995). Sources of distress are traditionally described by a collection of factors measuring social well being; accordingly, a Social Development.

Indicator (SDI) was described, though not constructed as such, by the World Bank (1997) and by the United Nations (1989, 1996) to include measures of social condition. The indicator was constructed and verified by Meso (1999, 2000) for African countries and by Pook and Pence (2001) in a 210-country analysis. National Information Infrastructure (NII) is a robust collection of widely used media for the creation, dissemination, and use of basic information. NII forms the foundation of sophisticated information services often disseminated via high technology equipment such as computers. Intuitively one may wish to use PCs and Internet hosts as indicators of NII. However, NII does not include computing technology itself; computing, Internet hosts, and so forth are included in the technology infrastructure. While NII was described by the Information Infrastructure Task Force (Lehman et al., 1995), an index was not actually constructed. Technology infrastructure is used here to indicate the collection of computer and telecommunication hardware, software, data, storage technology, networks, and other related foundation industries that form the delivery system for the NII (Laudon, 2001).

Expanding on Peter Meso's model (Meso et al., 2000), we explicitly include technology as part of the model in Figure 1. This is to accommodate the influence of technology on the development of NII, social development, investment, economic growth, and economic activities.

Even in famously market-based economies such as the U.S., markets do not satisfy all public needs and certainly do not work perfectly and do require outside intervention (Stiglitz, 2002). For the purposes of our topic, governments serve to gratify public needs and wants by providing access to education, fresh water, health care, welfare, highways, defense, information services, and so forth, whereas markets are incapable of providing satisfactory results (Summer, 1980, Meso et al. 2000). Here, governments and markets are shown to govern all aspects of model components in Figure 1.

Meso et al. effectively argue that abundant information impacts the quality and magnitude of investments in the economy; it indirectly fosters the growth of the economy; and, while improving "... awareness and knowledge of the nation's citizens," it will result in improvement in the social development of the nation; hence, the importance of the NII. UNESCO (United Nations Education Science and Cultural Organization) was founded on the basic premise that information is not only a necessity to all human beings, but also a human right. Furthermore, the ITU (International Telecommunication Union) stresses that communication and communication infrastructures are preconditions for economic and social development (Ferguson, 2000). In other words, UNESCO implies and the ITU states direct causality between economic and social development and information infrastructures, and, by implication, technology as well.

It is further argued here that economic growth and investment influence technology, which then supports the foundation of the NII. Conversely, national developments of the technology industry will attract further investments and spur economic development. For electronic commerce to develop in an economy, financial institutions and a certain level of technology are needed to be in place, which then indirectly influence education and other aspects of social development. Furthermore, as Meso et al suggest, "... access to quality information depends largely on the nature of the nation's NII," which in the model presented here is influenced by available technology.

The dashed line in Figure 1 delineates the extent of the model tested on 210 countries as reported in Pook and Pence (2001). This chapter will examine the effects of social

Figure 1. Research model for relationship of social development, NII, and technological development

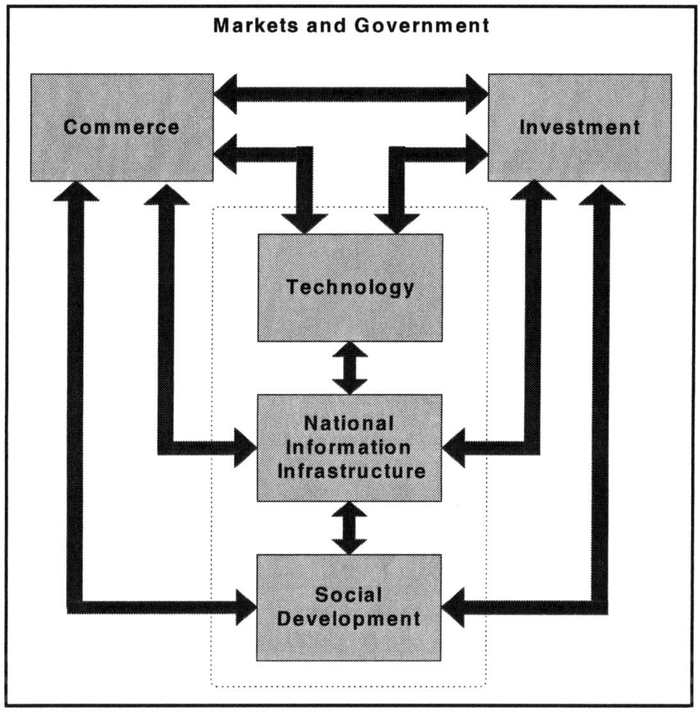

development, national information infrastructures, and investments in technology and information infrastructures on the information and telecommunications preparedness of the accession countries of the Czech Republic, Hungary, Poland, and Slovakia. This will be accomplished by comparing development of these countries' social and economic as well as NII indicators to those of current member EU nations, and by identifying development needs and resources likely to contribute to achieving the NII goals of the V4. Analysis will be performed on the macro level of national policy.

Hypotheses

V4's telecommunications and information infrastructures had only 12 years to catch up to Western European standards set by their prospective peer countries in the EU. Many of their industries benefited from the influx of foreign direct investment (FDI). Are the V4 competitive at present with their future EU peers, or would foreign direct investment in their information technologies also raise their per capita GDPs? As Nicholas Stern (2002) observed, information is a necessary ingredient for economic development; hence, we also need to examine if there is an "information gap" between the V4 and their future EU peers in terms of the availability of modern information services.

H1 a. It is hypothesized that the V4 will need large amounts of investment in information technology to reach parity levels of GDP with EU countries in order to help them move into Rapidly Developing and Developed Information Infrastructure nations clusters.

H1 b. It is further hypothesized that the V4 will need to improve information penetration to reach parity levels of GDP with EU countries in order to help them move into Rapidly Developing and Developed Information Infrastructure Nations clusters.

The main focus of information and communication policies is the development of information infrastructure (Ferguson, 2000). In order to finance infrastructure growth, a nation needs investment capital. Since the V4 have not had a track record of being able to accumulate private capital, would they be able to attract capital from abroad, and what are the preconditions for attracting foreign capital?

H2. It is hypothesized that the V4 will need to accelerate development of their information infrastructures in order to attract foreign investment capital to finance growth.

National governments' responsibility is first to fulfill basic social needs (e.g., hygiene, health care, availability of safe drinking water, minimum purchasing power, life expectancy, etc.) before higher-level conveniences are targeted. However, once basic needs are satisfied, it may be assumed that investments may be channeled to achieve higher level needs such as those associated with communications and information.

H3. In order for the V4 to reach levels of NII comparable to EU standards, they will need to further improve on the component factors of their respective social development indicators (SDI).

Proper functioning of democratic institutions depends on the public's access to available information. Information is not only power; it is also a prerequisite for development (Pfiser, 2000). In order for the V4 to become information societies and to be able to utilize needed information and associated technologies in order to improve their quality of life and working conditions and to control their own evolution, they will need increased levels of wealth as measured by GDP per capita (EC, 2003). That is, the penetration of information into society as measured by NII increases as individual wealth increases.

H4. Development of National Information Infrastructure of nations is directly related to GDP per capita, indicating that information penetration is positively related to GDP per capita.

The National Information Infrastructure index contains indicators of basic communications industries consisting of radios, television, newspapers, and telephones. Beyond the basic communication media, what other factors may impact the wealth of a developing nation as measured by GDP per capita? Investment capital is fungible—it moves toward expected highest profits. Should we use investments in telecommunications rather than the four basic communications industries to gauge development impact?

H5. Information delivery systems and investment in those systems are crucial ingredients of per capita GDP in the accession countries.

Measuring information penetration and the digital divide, the World Bank (Navas-Sabater et al., 2002) uses telephone mainlines per 1,000 population. While the model is parsimonious, it can be argued that it may miss other, equally important factors such as social or cultural preferences, or unique local information infrastructures. We therefore propose the following.

H6. Using telephone mainlines per 1,000 population is a less effective measure of basic information penetration overall than using the broader NII-SDI relationship or NII-GDP per capita models.

METHODOLOGY AND DATA USED

It is argued that countries in similar circumstances display similar social, economic, and technical characteristics. Where these characteristics are numerically measurable, they can be used to identify and group countries that have similar development histories and predict from their collective situations their short-term economic development and technological futures.

In order to locate the V4 among peers that are most alike on social development as well as NII indicators, cluster analysis was used. Cluster analysis identifies relatively homogeneous groups of countries based on selected variables. Using F statistics, the procedure finds groups that are significantly different statistically on a given set of variables. Further, Kendall's W test was used to determine if similarity of membership exists among clusters formed using variables representing social development and NII, as well as World Bank classifications by income.

In order to match the World Bank's grouping of countries by income classes, four clusters were defined. Clusters of the social development indicator variables just described resulted in variable descriptors for Underdeveloped Nations labeled cluster 1, Developing Nations labeled cluster 2, Rapidly Developing Nations labeled cluster 3, and Developed Nations labeled cluster 4. The four countries targeted for accession into the European Union are all located in cluster 3, Rapidly Developing Nations, when considering social development indicator variables only.

In order to summarize apparent relationships among descriptor variables characterizing the countries in question, regression modeling was used. Results of regression analysis allowed for the determination of preferred NII growth policies among the four nations.

Following World Bank and United Nations practice, as discussed previously, social development was measured by primary education participation rate, GDP per capita, Purchasing Power Parity (PPP) (in constant 1987 international dollars), literacy rate, infant mortality rate, health care (% of population with access), safe water (% of population with access), and life expectancy at birth. The Social Development Index used here contains these factors with equal weight on a scale of zero to 100, where the value 100 represents the highest possible social development.

Components of the National Information Infrastructure indicator were defined by Lehman et al. to consist ideally of "[t]elephones, television, radios, computers and fax machines ... to receive, store, process, perform, display and transmit data, text, voice, sound, and images in homes and businesses" (Lehman et al., 1995, p.4). In order to include voice, image, and print media in the NII index, a merging of conventional information industries suggested by Tarjanne (1994), Meso (1999, 2000) defined the NII to include telephones, radio, television, and newspapers. Therefore, National Information Infrastructure is measured by the saturation rates of newspapers per 1,000 population, radios per 1,000 population, television sets per 1,000 population, and telephone mainlines per 1,000 population. The National Information Infrastructure Index employed here uses these factors with equal weight on a scale from zero to 100, which represents complete saturation.

Data items in this set are measured on different scales. To avoid erroneous results, where appropriate, variables were standardized into z-scores, using their means and standard deviations resulting essentially in the same distributions without certain variables overpowering others. In order to measure the impact of foreign and domestic investments on the economy, investments in telecommunications and foreign direct investments will be used.

Data sources were the World Bank, World Development Indicators 1998, UNESCO 1999, United Nations, Women's Indicators and Statistics Database 2001, the International Telecommunications Union (ITU) 1996/97, Telcordia Technologies Web Site 2002, Industrial Development: Global Report (UNIDO) 1997, CIA World Factbook 2001, and the International Labor Organization (ILO) databases 1999. All of the above databases are available to the public either free of charge or for a nominal fee.

Graph 1. World Bank income class and cluster relationships

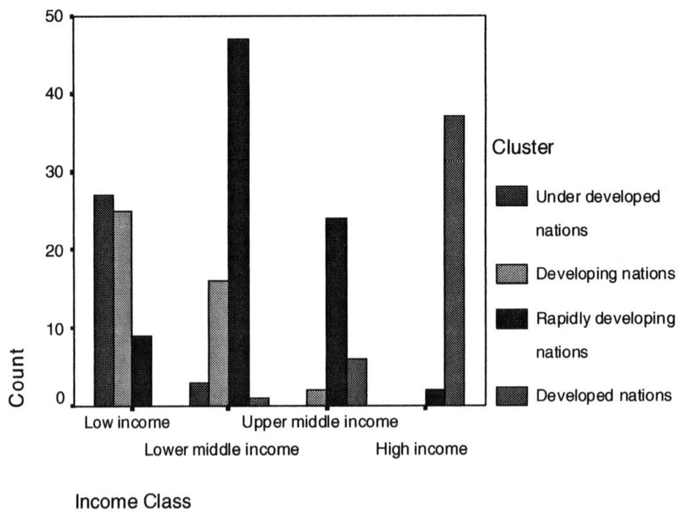

Table 1. Descriptive social statistics for rapidly developing nations

Factors	N	Minimum	Maximum	Mean	Std. Deviation
Literacy rate, adult total (% of people 15+)	74	61.60	99.00	90.86	8.91
Primary education particip. rate. (%)	59	69.00	133.00	102.08	13.64
Infant survival (%) (100-mortality)	85	93.48	99.35	97.65	1.26
Health care (% of population with access)	45	44.50	100.00	93.20	10.72
Safe water (% of population with access)	61	35.65	100.00	82.28	14.02
Life expectancy at birth, total (years)	86	59.19	76.61	70.49	3.23
GDP per capita, PPP (constant 1987 international $)	85	800.00	9924.03	4257.29	2372.75

Sources: (CIA, 2001; ILO Bureau of Statistics, 2002; World Bank, Social Indicators of Development, 1997)

Table 2. Descriptive social statistics for V4 accession countries

Factors	Czech Republic	Hungary	Poland	Slovakia
Literacy rate, adult total (% of people 15+)	99.00	99.00	99.00	98.00
Primary education particip. rate. (%)	104.00	103.00	96.00	102.00
Infant survival (%) (100-mortality)	66.20	98.81	98.62	98.87
Health care (% of population with access)	100.00	100.00	100.00	100.00
Safe water (% of population with access)	89.35	98.00	78.00	92.00
Life expectancy at birth, total (years)	72.95	69.42	71.67	72.29
GDP per capita, PPP (constant 1987 international $)	7474.85	5095.16	4103.25	5137.22

Sources: (CIA, 2001; ILO Bureau of Statistics, 2002; World Bank, Social Indicators of Development, 1997)

RESULTS

Results of Cluster Analysis

Upon examining clusters plotted against the four income classifications of the World Bank (WB) in Figure 1, it can be demonstrated visually that our clustering matches that of the WB income groups (Pook & Pence, 2001). However, coefficients of concordance for Kendall's test ranged from 0.527 to 0.739 with a significance of $p<0.001$,

Table 3. Descriptive social statistics for developed nations

Factors	N	Minimum	Maximum	Mean	Std. Deviation
Literacy rate, adult total (% of people 15+)	31	78.60	100.00	94.90	6.40
Primary education particip. rate. (%)	27	77.00	116.00	100.15	7.06
Infant survival (%) (100-mortality)	47	94.40	99.57	99.04	0.82
Health care (% of population with access)	29	94.00	100.00	99.29	1.62
Safe water (% of population with access)	33	90.00	100.00	98.54	2.47
Life expectancy at birth, total (years)	47	62.95	83.46	76.09	3.09
GDP per capita, PPP (constant 1987 international $)	48	9147.22	25000.00	15831.97	3780.73

Sources: (CIA, 2001; ILO Bureau of Statistics, 2002; World Bank, Social Indicators of Development, 1997)

Table 4. Descriptive statistics for developing information infrastructure nations

Factors	N	Minimum	Maximum	Mean	Std. Deviation
Telephone mainlines (per 1,000 people)	60	3.53	517.57	248.27	122.44
Television sets (per 1,000 people)	59	73.06	644.18	326.80	117.50
Radios (per 1,000 people)	60	62.02	1461.19	530.59	251.79
Newspapers (per 1,000 people)	49	2.00	297.00	133.16	76.97

Sources: (ITU, Yearbook of Statistics, 2000; Telcordia, 2001; World Bank, Social Indicators of Development, 1997)

indicating a statistically close match between our clusters on social development indicator variables and those of the WB income classifications. One can observe that 85.3% of "low income" countries are "underdeveloped" and "developing" according to our social development index; 94% of "lower middle-income" countries are "developing" or "rapidly developing" countries; 75% of "upper middle-income" countries are "rapidly developing"; and 94.9% of "high-income" nations are "developed."

According to our clusters, none of the Underdeveloped or Developing Nations is in Europe. However, of the Rapidly Developing nations, 23.3% are in Europe; an overwhelming majority of them are ex-USSR states or liberated satellite states, including the Czech Republic, Hungary, Poland, and Slovakia, which are of interest here. The rest

Table 5. Descriptive statistics of information infrastructures for Visegrad-Four countries

Factors	Czech Republic	Hungary	Poland	Slovakia
Telephone mainlines (per 1,000 people)	273.12	261.12	169.06	231.94
Television sets (per 1,000 people)	406.35	443.51	417.97	384.26
Radios (per 1,000 people)	279.75	613.29	534.58	952.59
Newspapers (per 1,000 people)	219.00	228.00	141.00	256.00

Sources: (ITU, Yearbook of Statistics, 2000; Telcordia, 2001; World Bank, Social Indicators of Development, 1997)Finding identical members is not an objective; hence, the variations in the ranges of the cluster member data. Table 5 depicts NII descriptive statistics for the V4 accession countries.

of Europe belongs to the group of Developed Nations. For purposes of comparison, South America clustered primarily into Rapidly Developing, and North America into the Developed Nation cluster. Descriptive statistics that characterize Rapidly Developing Nations are displayed in Table 1. The list of factors shown in the table was used to represent and differentiate social development of nations, classifying them into the four resulting clusters. Because of data availability, not all nations are represented on all seven factors; however, the list is dense enough to draw relevant conclusions.

It should be noted that "Primary education participation rate (%)" in some cases exceeds 100% in the table. This is related to the manner in which the World Bank accounts for this variable. While many countries consider "primary school age" to be six to 11 years, others do not. For some countries with universal primary education, the gross enrollment ratios may exceed 100% because some pupils are younger or older than the country's standard primary school age (World Bank, 1997).

Table 2 provides descriptive statistics for the four accession countries of the V4 accession countries.

Nations with developing information infrastructures are characterized by factor values as shown in Table 4 representing the V4 nations' current peer group.

It should be noted that Tables 2 and 3 represent bounded data with limited minimum and maximum values resulting in apparently small standard deviations. On the other hand, data in Tables 4 through 6 represent absolute values without predetermined upper limits resulting in large potential variations represented by large standard deviations displayed in these tables. Also, the process of clustering finds cluster members which are relatively similar by excluding candidates that are, in general, statistically dissimilar.

Table 6 reports descriptive statistics for rapidly developing nations' information infrastructures. This is the group the V4 nations are attempting to join.

Table 6. Descriptive statistics for rapidly developing information infrastructure nations

Factors	N	Minimum	Maximum	Mean	Std. Deviation
Telephone mainlines (per 1,000 people)	23	232.27	683.22	479.72	121.13
Television sets (per 1,000 people)	23	215.09	699.74	417.68	117.02
Radios (per 1,000 people)	22	184.52	1207.60	738.46	271.69
Newspapers (per 1,000 people)	19	250.00	757.00	472.74	149.39

Sources: (ITU, Yearbook of Statistics, 2000; Telcordia, 2001; World Bank, Social Indicators of Development, 1997)

Analysis of the Status Quo

In order to evaluate national policies on the development of information infrastructures, graphical analysis and regression modeling will be used for the testing of the six hypotheses offered previously.

Testing Hypothesis 1a

GDP per capita for all 210 countries was regressed on investments per capita in telecommunications. Resulting R-square = 0.73 indicates that investments in telecommunications indeed have a close relationship to per capita GDP. Standardized beta coefficient of telecommunications investment is 0.855 with $p<0.0001$.

The unstandardized regression equation is:

GDP per capita (1987 US $) =
2432+74.47*telecommunications investment

The model is shown in Figure 2.

Examination of the data in Figure 2 suggests that investments in telecommunications grow asymptotically rather than linearly, indicating that a saturation point is reached in traditional telecommunications formats; therefore, a quadratic model may better describe the data as in Figure 3.

GDP per capita =
19554.6*(1-exp[-.01*telecommunications investment]),
with R-square = 0.79.

With 79% of variability in GDP per capita explained by telecommunications investment, and with V4 nations at the lower end of the investment spectrum, hypothesis H1a has been confirmed.

Figure 2. GDP per capita vs. investment in telecommunications (with 95% confidence interval)

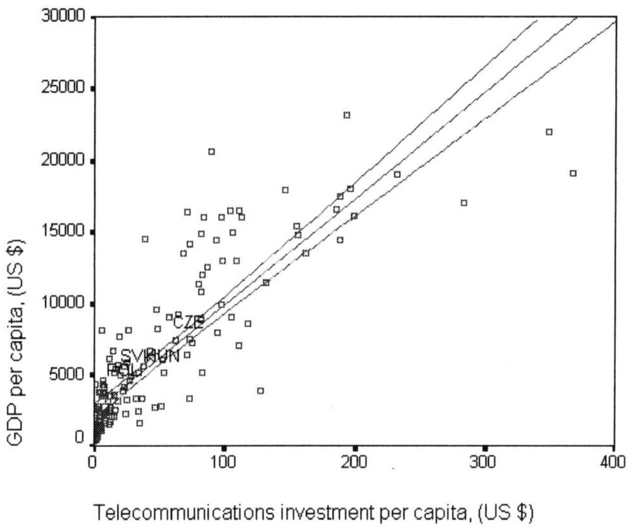

Figure 3. GDP per capita vs. investment in telecommunications (with 95% confidence interval)

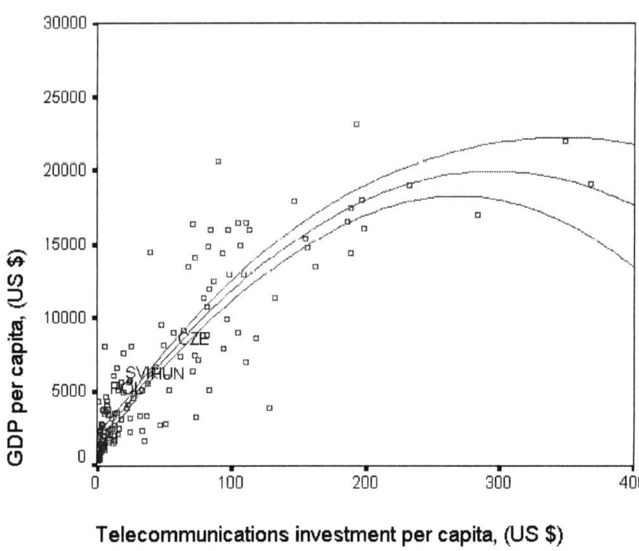

Figure 4. PCs vs. GDP per capita for all countries (information technology penetration) (with 95% confidence interval)

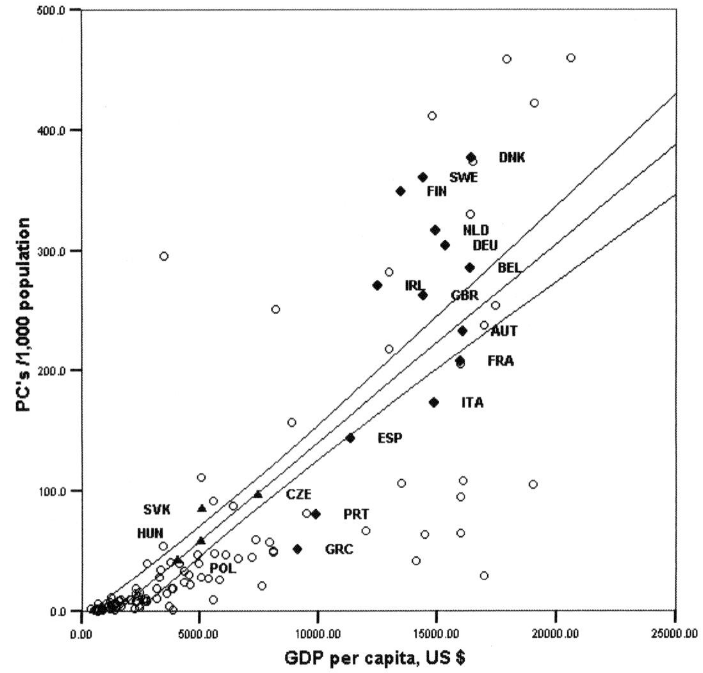

Testing Hypothesis 1b

Information technology diffusion is the penetration of information technology as measured by the number of PCs per capita and per capita GDP. Regressing PCs per 1,000 population on per capita GDP, R-square was found to be 0.646 with a beta coefficient of 0.803 at $p<0.0001$.

PCs/1,000 pop. =
-23.99 + 0.02*GDP per capita, with R-square of 0.65

The model is shown in Figure 4. Sixty-five percent of variation in PCs/1,000 population is explained by GDP per capita, indicating a strong relationship. Since V4 nations cluster at the lower end of information technology penetration as compared to their EU peers, hypothesis H1b is confirmed.

Testing Hypothesis 2

Examining foreign direct investments (FDI) in the Czech Republic, Hungary, and Poland, it is reported to be 4.65%, 4.32%, and 4.73% of GDP, respectively (WB, 1997), which are above the developing country group mean of 3.76% of GDP. Slovakia is below that mean at 2.96%. The data indicate that with the exception of Slovakia, V4 accession countries seem to be on track, attracting foreign direct investment to bolster their economies.

Figure 5. Foreign direct investment (FDI) as % of GDP vs. national information infrastructure index (NII)

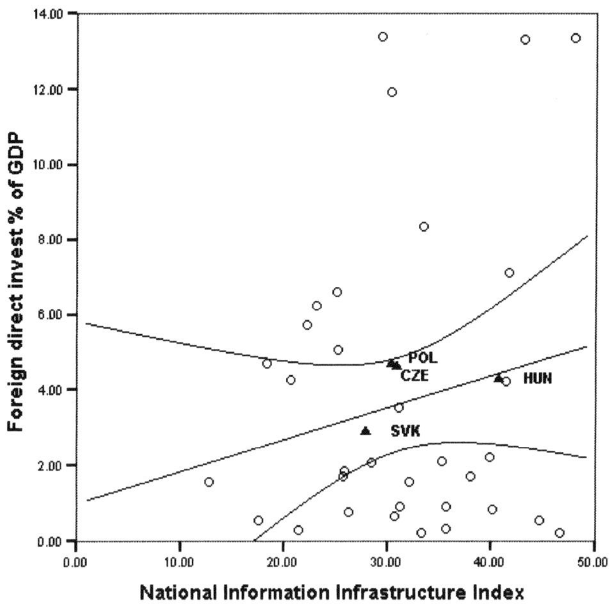

Figure 6. NII vs. SDI for all countries (Information penetration) (With 95% confidence interval)

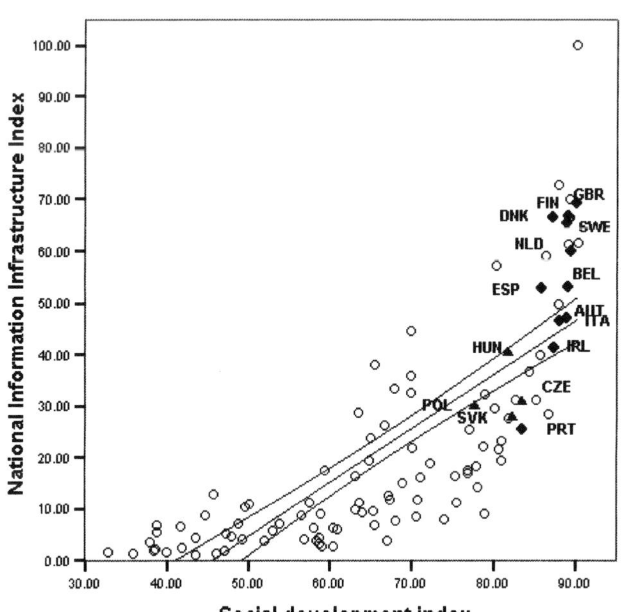

Figure 7. NII vs. SDI for all countries (information penetration) (With 95% confidence interval)

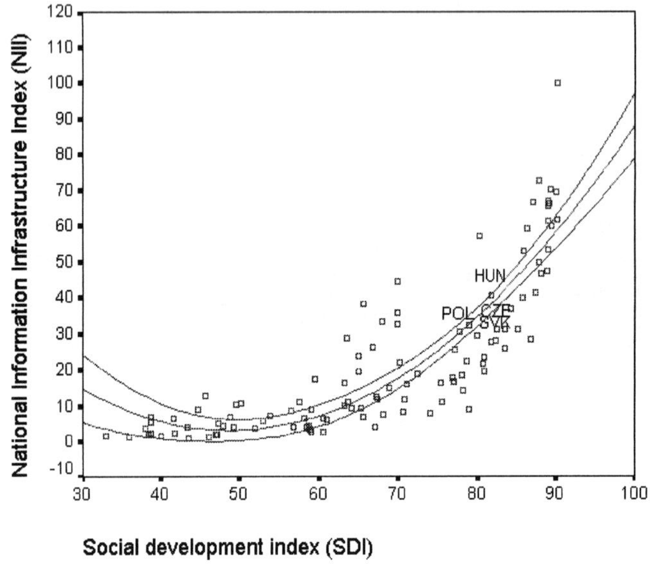

Figure 8. NII vs. GDP per capita for all countries (information penetration) (With 95% confidence interval)

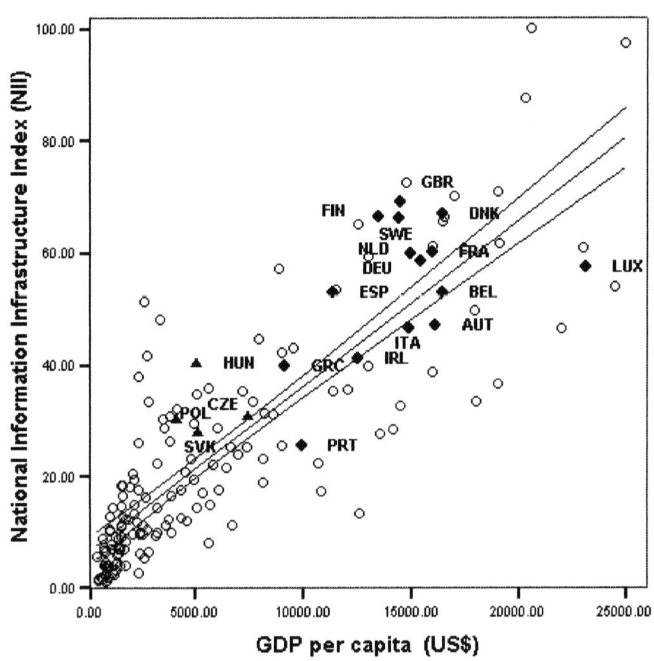

Figure 9. Telephone mainlines vs. GDP per capita for all countries (information penetration) (With 95% confidence intervals)

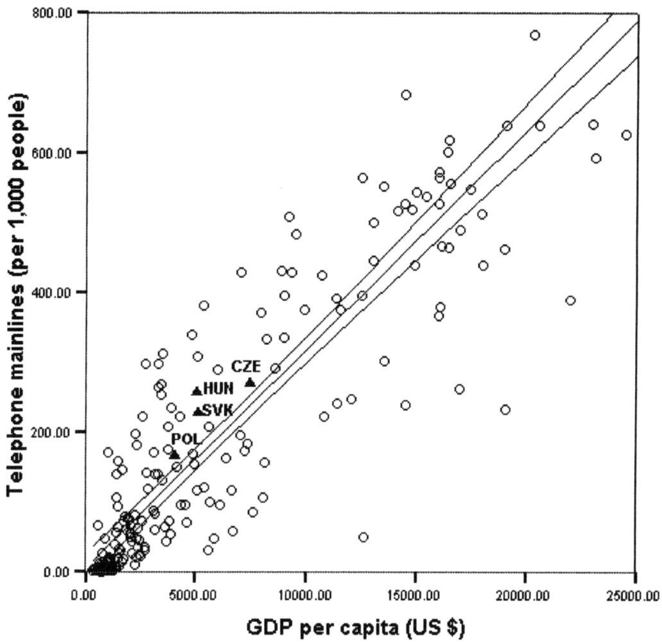

Further analysis of the World Bank data shows that poor countries' average FDI rate is 2.83% of GDP; rapidly developing countries' average FDI rate is 3.1% of GDP; and developed countries' average FDI rate is 3.57% of GDP. It is significant that mean FDI rates are the highest for developing countries at 3.76% of GDP, if one considers that these countries offer the greatest potential returns, all else being equal. However, Figure 5 illustrates the relative position of the V4 countries' ability to attract FDI with respect to their cluster peers to date, indicating that the V4 group is not doing as well as they could be and indicating the validity of Hypothesis 2.

Testing Hypothesis 3

We next inquire if there are some beneficial trade-offs to be gained between social development and information infrastructures developments by V4 countries. When regressing the overall National Information Infrastructure Index (NII) for all countries on the Social Development Index (SDI) as illustrated in Figure 6, the standardized beta coefficient was found to be 0.792 at $p<0.0001$ and the unstandardized regression equitation was:

$$NII = -48.04 + 1.05 * SDI, \text{ with R-square of } 0.628$$

As observed with telecommunications investments, when social needs become fulfilled with each unit of social need satisfied, NII penetration increases at a growing rate. Using a quadratic model to represent increasing rates of NII growth we offer the quadratic model.

NII = 0.1*exp (0.07* SDI), with R-square of 0.762.

R-square increased from 0.628 for the linear model to 0.762 for the quadratic model; in addition, the quadratic model does make logical sense, since it suggests that as capital earmarked for social projects is freed up, it can move into national information infrastructure development and improve it at increasing rates, confirming Hypothesis 3. Figure 7 represents this second order relationship.

Testing Hypothesis 4

We examine information penetration as the function of per capita GDP by regressing NII on per capita GDP. The liner model is:

NII = 6.64 + 0.003*per capita GDP, with R-square of 0.730.

The standardized beta coefficient of GDP per capita is 0.854 significant at $p<0.0001$. The model and Figure 8 indicate that per capita GDP is a strong (73%) predictor of NII and that information penetration increases with increasing per capita GDP, confirming Hypothesis 4.

When attempting a quadratic regression model on the above data, R-square remained essentially the same at 0.7288, indicating that there are not likely to be significant rates of diminishing returns.

Testing Hypothesis 5

When we tested the apparent relationship of telephone, television, radios, newspapers, and investment in telecommunications (i.e., basic information industries and investment in those industries on GDP per capita), the following regression model using standardized beta coefficients resulted:

GDP per capita = 0.243*telephone mainlines (per 1,000 people)
+ 0.613*telecommunications investment (per capita)
-0.358*radios (per 1,000 people)
+ 0.228*television sets (per 1,000 people), with R-square of 0.693.

Significance levels for the coefficients were: for telephone mainlines (per 1,000 people) < 0.1; for telecommunications investment (per capita) < 0.0001; for radios (per 1,000 people) < 0.01; and for television sets (per 1,000 people) < 0.05. Newspapers is not a significant contributing variable of the above relationship. With a predictive power of 69%, basic information delivery systems are crucial ingredients of per capita GDP, confirming Hypothesis 5.

Testing Hypothesis 6

The World Bank (Navas-Sabater et al., 2002) traditionally used the number of telephone mainlines vs. GDP per capita to measure information penetration. To test the validity of using single variables, we compared our results of using the NII in lieu of using the number of telephone mainlines and the social development index instead of just GDP per capita. Figure 9 depicts using telephone mainlines vs. GDP per capita with an R-square of 0.75. The beta coefficient for the model is 0.866 with $p<0.0001$. The unnormalized model is:

Telephone mainlines (per 1,000 people) =
4.115 + 0.0314*GDP per capita, with R-square of 0.748.

Conversely, the model in Figure 8 uses NII vs. GDP per capita with an R-square of 0.73, and Figure 7 uses the quadratic model for NII vs. SDI with a resulting R-square of 0.762. Correlations of the three models are remarkably close, indeed; Hypothesis 6 cannot be confirmed.

DISCUSSION

All four accession countries are located within the cluster 3 ranges (rapidly developing nations) on the seven factor variables for social indicators of Table 1. Furthermore, when comparing social indicator statistics of the accession countries to those of the developed countries in Table 3, it can be seen that the accession countries exceed or at least match mean values on most factors of the developed countries, with the exception of access to safe water and life expectancy in some of the V4 countries. All of the V4 countries trail developed countries on GDP per capita in all cases by more than half.

All four accession countries are located within the Developing NII Nations ranges on the four factor variables of Table 4. We can conclude that the four accession countries match mean television densities but fall well behind on the rest of the factors of rapidly developing information infrastructure nations. Of the present EU member countries, Greece, Ireland, Italy, and Portugal possess NII characteristics similar to the accession countries, Austria, Belgium, Germany, Netherlands, Norway, and

Sweden falls among the rapidly developing information infrastructure group, while Denmark, Finland, France, Spain, United Kingdom, and Luxembourg belong in the group called well-developed information infrastructure nations consisting of nations with the best developed NIIs to date. Canada, Australia, and the US are also members of the developed NII cluster.

In examining Figure 2, it should be noted that the V4 countries lag far behind developed Western European countries, as corroborated by EC (2003), in the amount invested in telecommunications, and they have a long way to go to catch up. ITU (2000) data indicate that mean per capita investment in telecommunications by developed countries is $126, whereas accession country investments are $73, $53, $23, and $34 per capita in the Czech Republic, Hungary, Poland, and Slovakia, respectively. These data indicate large discrepancies, indeed, suggesting a need to attract investments into the telecommunications sectors of the V4 countries. To attract the required capital, Poland, Hungary, and the Czech Republic are actively moving to an open market position in telecommunications, though only the mobile market has been partially opened to competition in Slovakia (Navas-Sabater, 2002).

A strong, apparent relationship can be observed between telecommunications investment and GDP per capita, suggesting that impressive productivity gains are achievable through telecommunications and information technology investments. However, once the saturation point in traditional telecommunications is reached (on Figure 3, around $300 per capita), one should not expect significant GDP growth from further investments in traditional telecommunications technologies, but rather alternative formats or multiple communications technologies need be considered. Clearly, the V4 countries are well below this saturation point and, at least for the time being, need additional investment in their traditional information and telecommunications infrastructures to impact on their respective per capita GDP.

On Figure 4, EU countries appear on the upper right hand of the figure, indicating that the V4 countries are well below EU standards in information technology penetration. Among developing countries in the NII cluster, data show that the average number of PCs per 1,000 population is 74, while among developed countries there are 314 PCs per 1,000 population. Among rapidly developing countries on the SDI scale, the data indicate that the average number of PCs per 1,000 population is 47 and among developed countries there are on the average 228 PCs per 1,000 population.

Clearly, with per capita GDPs of the V4 countries (Table 2) on the average 50% below developed Western countries, the purchase of personal computers is not readily made as part of discretionary spending, yet they will need significant investments in information technology to reach parity levels of information penetration with existing EU countries and thus close the information gap.

H2: Examining Figure 5 for V4 countries' ability to attract foreign direct investment, Slovakia appears to lag the group. Early extremism by the Meciar government may be responsible for Slovakia's slow performance to date. With recent establishment of political stability in Slovakia, foreign investments are predicted to resume. In fact Slovakia's 2001 FDI was 6.3%, while the Czech Republic's five-year average was 7.8%, Hungary's five-year average was 4.3%, and Poland's was 4.2%, all above the mean for developing countries. During the same period, real per capita GDP growth rates have nearly doubled in all four countries, but rates still are well below the EU average.

Further examination of data in Figure 5 indicates that there are several countries in the developing country cluster that attract FDI at rates better than double the rates for the V4 nations, as well as countries with better developed NII structures that attract FDI at lesser rates. As Nicholas Stern (2002) observed, in addition to having an up-to-date information infrastructure, an attractive investment climate also must be established with appropriate institutions, governance, policies, electric power, water, and transportation systems. In Figure 5, countries that lead in attracting FDI are Malta, Lithuania, and Estonia, all of them EU candidate countries with attractive investment climates. Among the countries slow to attract foreign direct investment are Argentina, Italy, and Belarus with their lackluster economies; Turkey and Lebanon with unstable political and social environments; and Russia, Ukraine, and Greece with notably corrupt business environments (TI, 2002).

V4 nations have newly formed market economies; they lack sufficient domestic capital and, therefore, need to increase the inflow of foreign investment funds. They further need to open up these economies, privatize still remaining government monopolies, strengthen institutions, and formulate the investment climate needed for foreign capital. While all of the V4 countries instituted pro-competitive telecom policies, formed independent regulatory agencies, and privatized their fixed line phone companies, only Poland and the Czech Republic were expected to have all telecommunications services liberalized by the end of 2003 (World Bank, 2001). Time is of the essence here; a well-developed information industry will promote the influx of foreign direct investment (Navas-Sabater et al., 2002) and will contribute to economic growth preventing these countries from becoming a drag on an enlarged EU. The disparity for the V4 countries offers an opportunity to make their investment environments attractive to potential foreign as well as domestic investors by improving, among other things, their national information infrastructures (Stern, p. 165).

V4 countries' social development indices (SDI) are all around 80 with a cluster mean of 74, while their national information infrastructure indices (NII) are around 40 with a cluster mean of 34, indicating that the accession countries lead their groups in both SDI and NII. However EU nations are in the socially developed cluster and the developed NII cluster, with mean SDI and NII of 88 and 69, respectively. These results suggest that while some development in social standards is warranted, V4 countries need considerable improvement in their information infrastructure.

However, the non-linear nature of the relationship displayed in Figures 6 and 7 suggests that the V4 countries should be able to reach parity with EU nations at an ever-increasing rate, since smaller and smaller amounts of improvements in social factors will release capital resources to improve information infrastructure benefits at increasing rates.

NII and per capita GDP play important roles in a nation's welfare, both as preconditions for and as integral components of its economic and social development (Ferguson, 2000). It can be seen from Figure 8 that the accession countries lag considerably behind developed countries of the EU (except Portugal, Ireland, and Greece) in information penetration. Lack of information penetration leads to the public's inability to make informed decisions and to be productive participants in a democratic society. Information penetration will be secured only by economic well being and delivered by a secure information infrastructure. It is no coincidence that the world's richest democracies are also wealthy nations with developed information infrastructures.

It was discovered that *newspapers* is not a significant contributing variable of the predictive relationship, possibly indicating that the availability of electronic information delivery systems is gaining in importance as countries advance toward the developed NII cluster. Newspapers may need to be replaced by PCs in the model for developed and rapidly developing NII nations once adequate data on PCs by country becomes available. Also significant is that radios play a negative role in the model suggesting that high numbers of radios are indicative of lower levels of GDP per capita, whereas telephone lines and television are signs of advanced NII. Investment in telecommunications proved to have significant predictive ability in the model suggesting that harnessing communications for development should be a necessary ingredient of government policies in these countries

Examining the World Bank model (Navas-Sabater et al., 2002) used to measure digital divide, it was learned that the use of single variable prediction model with telephone mainlines per 1,000 population provided technically equivalent predictive capability as using a broader NII-SDI based model. However, one can argue that while simplicity in modeling may be preferable, the indices presented here provide a more all-encompassing picture of a nation's social development, cultural preferences, and information infrastructure, which is a logically defensible model and, therefore, should be favored over simpler models.

CONCLUSION

The V4 countries, of all other ex-satellite countries as well as Slovenia, have liberalized their economies and opened their markets to foreign investments with favorable incentives, tax laws, and business practices. The Commission of the European Communities (CEC) reported that the Czech Republic (CEC, Czech, 2002), Hungary (CEC, Hungary, 2002), Poland (CEC, Poland, 2002), and Slovakia (CEC, Slovakia, 2002) have made adequate and, in some instances, remarkable strides to meet the Copenhagen Criteria for accession into the EU. It should be noted that GDP growth rates in all four countries have been erratic over the past five years and averaged from a low of 1.1% in the Czech Republic to a high of 4.5% in Hungary. Further, per capita GDP among the V4 countries ranged from a low of around 40% to just about half of the EU average. Slovakia and Poland's relatively large current account balances impede the inflow of foreign investment, and high unemployment in all four countries drains government funds needed for improvements of infrastructures. Other EU candidate countries have succeeded at creating investment climates that attract foreign direct investments at rates several times those of the V4 countries, suggesting the importance of quality of infrastructures, government institutions, and a behavioral environment conducive to trust (Stern, 2002). On the other hand, high skill levels of workers in the V4 countries should be incentive for FDI (Pook, 1999; Pook, Fustos, & Marian, 2003).

It was demonstrated that both telecommunications and information technologies show a close relationship to national economic output measured here by per capita GDP. It was further demonstrated that telecommunications offers only diminishing returns with respect to GDP, while information technology, measured by numbers of PCs per capita, displays more linear effects on GDP; that is, PC penetration sustains continuing growth of outputs, suggesting that PCs do, indeed, contribute to information generation, while telecommunications can reach saturation. However, none of the V4 countries has reached the telecommunications saturation point; they are far below EU levels, and their economies can benefit for the time being from investments in these sectors.

As indicated in Figure 1, national social development, measured by the Social Development Index (SDI), forms the foundation of national information and technology infrastructures. That is, without appropriate level of social development, it would not make any sense to talk about developments in telecommunications and technology. We demonstrated through cross-national analysis that not only is there a high correlation between the NII and SDI, but also the relationship is non-linear, indicating that as a nation fulfills its social obligations, less and less investment in SDI-related projects will result in ever increasing payoffs in NII. Since the V4 countries are on the steep right hand side of the curve, having well developed social infrastructures, they can look forward to high rates of NII developments as they catch up with their prospective EU peers.

Examining NII penetration, we noted that the higher the per capita GDP a country enjoys, the greater national investment is likely to be in basic information delivery systems and the greater the availability of basic information to the population to utilize for their economic and political gain (Arnbak, 1996). Accelerating V4 countries toward EU parity policies, which result in the increase of per capita GDP, would benefit the national information infrastructure. Information leads to knowledge, and knowledge leads to empowerment; that is, to the empowerment of the population to participate in economic growth (Stern, 2002). Such a primer we identified earlier as that of investments in telecommunications, with foreign direct investment as the source. Recent positive outcome of the Irish plebiscite on EU expansion should have a further positive effect on making available foreign capital for the V4 countries.

The linear model being the best fit of the NII-GDP data in Figure 7 indicates that these country populations have a need for information in whatever format information can be delivered. Our conclusion, therefore, is that a national information infrastructure index (NII), which consists of several technologies, is a better measure of information penetration than a single technology (e.g., number of telephone mainlines) as a predictor variable by itself. This is confirmed by Howkins and Valantin when they argue that "each country has a clear priority to create an information society that reflects its culture and needs," and that "creating an information society is more important than using any specific technology" (1997, p. 45). On the other hand, the EU has raised cultural hackles before through its efforts at standardization (e.g., import-export quotas, proposed controls on French cheese, etc.) and is proposing an effort to reflect the "European social model" in social protection, information and communications infrastructures, European Internet, and content (EC, 2002).

Examining information technologies that most significantly affect national economic output, we conclude that electronic technologies (e.g., televisions, telephones, and PCs) and investments in electronic technologies have the greatest impact on per capita GDP, while radios may be indicative of backwardness, and newspapers do not play a significant role in these economies anymore. Underscoring this observation is the effort by the EU to build an information society because such a society "... has the potential to improve the quality of life of Europe's citizens, the efficiency of ... social and economic organization and to reinforce cohesion" (EU, 2000). The EU plan involves the establishment of an EU-wide ISDN broadband network, interconnecting telecom, cable television, and satellite networks; the provision for trans-European basic services, including electronic mail, file transfer, and video services; and the creation of a European Basic Services Forum. In addition, one minister from each country is to be designated responsible for the establishment and management of their respective "information society."

In view of the results highlighted by the last hypothesis and the effort by the European Union to build a broad-based information society, it is our conclusion that indices such as the NII index constructed and used here would offer a better mirror of information and communications technology developments. We further propose that using single technology indicators to gauge information technology developments even may be culturally misleading.

It is a common belief that information and telecommunication systems transform societies; influence economic performance, health care and education; and, in general, improve the welfare of nations. Economically advanced Western nations are often referred to as "information societies" (Arnbak, 1996). Some observers have noted that these information societies have experienced changes in language, culture, and methods of communication; a formation of relationships, executing business processes and transactions; and have generally enjoyed spectacular economic growth as the result of embracing the "information revolution." By making large local, regional, and international markets accessible to all producers, even small entrepreneurs can contribute to the national welfare, while government services can be electronically delivered in order to make social services available to the masses, opening alternative avenues to increase their productivity. At the least, "telecommunications and information services are tools for access to and processing of information and keys to poverty reduction" (Navas-Sabater et al, 2002). Howkins and Valantin (1997) concluded that the keys to national information and communication development are the global community and appropriate national responses to challenges to its economic growth.

While not directly addressed in this chapter, decision-makers in these nations must address the issue of state capture and administrative corruption, which, while rated as being less than severe among the V4 nations, extracts heavy economic and social costs (World Bank, 2000).

RECOMMENDATIONS

In the review and analysis of the V4 countries' information infrastructures, the term *digital divide* was purposefully avoided due to its tendency to suggest an oversimplified view of a void to be filled with hardware, software, computers, and Internet, which are already available in the V4 countries. The issue for accession is more complex; it is one of political, economic, social, and technological inclusion. For the V4 countries, successful achievement of appropriate e-commerce and information infrastructures on par with their EU counterparts also involves establishing the necessary basis of social development. Viewing the SDI index previously, it was shown that all of the V4 countries are on par with Western Europe in social development. However, in order to achieve full inclusion, emphasis must be placed on the effective integration of information and e-commerce technologies into all levels of their societies, governments, commerce, and institutions. Simply stated, government services, laws and parliamentary processes, banking services, remote transactions, and institutional services need to be made available using modern communications and information technologies seven days a week, 24 hours a day (Fletcher, 2002). To implement this, some EU member-state assistance is made available in the amount of EUR 1.5 billion through the Phare program, which is currently the main channel for the EU's financial and technical cooperation with the candidate countries. The rest of the financing for information access must be internally budgeted by governments or private businesses, or acquired through private direct investment from abroad.

In a recent survey, Eastern European countries indicated that while the information infrastructure concept and building of an information society were important, activities such as increasing telephone penetration, digitizing the network, and decreasing waiting times for services were their immediate objectives. Completing privatization is at the top of the agenda in the V4 countries; the predominant view in the region is that the private sector should bear the cost of the information infrastructures of the future. For the present, however, since capital formation is slow, at present these countries need to reduce barriers that slow the inflow of foreign capital and direct investments into the information and telecommunications industries.

Poland adopted its National Information Policy in 2001 (CEC-Poland, 2002), while the Czech Republic adopted its version and updated it in March 2002 (CEC-Czech, 2002). As of 2002, Hungary was still developing one (CEC-Hungary, 2002) at the behest of a group of professionals who were active in the national information technology industry. They prepared a study and submitted it to the Hungarian government for consideration. The Slovak Republic has not approved a national information policy to date (CEC-Slovak, 2002). These national information policies emphasize the importance of information infrastructure on all sectors of the economy and the importance of developing a national strategy for information technology (EU, 2000). Other candidate states may similarly benefit from such a grass-roots approach to information integration into their societies.

Ultimately, when the V4 countries reach the accession date in 2004, penetration of NII into their societies is expected to have increased. At that time, it will be interesting to confirm the above conclusion using a longitudinal study.

While this study employs a broad spectrum of data sources, certain desirable, topic-specific data (e.g., imports and exports of IT sector, Internet hosts, IT graduates, and IT workers) per country are sparse in the total international database to date. Therefore, in order to preserve the robustness of the analysis, some of the indices employed surrogate components. Over the coming years, as more data become available, further analysis will be recommended with topic-specific data components. Since not all data are available, conclusions drawn must take into account consequent inaccuracies. Also, statistical modeling relies as much on common sense as on tools and data. The objective was not to select large models with high R-squares; such models would offer little useful information. Rather, rationality and statistical parsimony were our guides.

Since EU accession requirements concentrate on national level policy implications, the authors did not examine industry and sector level data. While such analysis promises to be interesting, but also extensive and lengthy, we chose to defer such analysis to others.

One last issue needs to be considered by EU and national policymakers, as well as investors: differences in developmental starting points (e.g., size, level of economic and social development, and administrative capacity) and developmental priorities related to differences in cultures (EC, 2003). Though the study here addressed the comparative picture in the aggregate country level, detail needs to be studied further by researchers in order to determine optimal developmental paths best suited to each country's unique situation.

Results of this study have important consequences to national policy research, as well as to practitioners in information technology and international finance. On the theoretical level, we presented a research model for the study of the interactions of national policy and social, economic, and technological factors. Specifically, we offered significant operational relationships between economic and information policy variables. We learned that the V4 countries are in need of foreign capital; are in the process of creating attractive investment climates to attract foreign capital; and have the need to further develop their information infrastructures in order to ensure social, political, and economic security for their people. Furthermore, we pointed out that the V4 countries are socially developed and prepared to make an effort to catch up with Western European information technology in order to secure their place in western information society; they only need the capital to do it. Finally, there is a strong implication for national policymakers to create a regulatory environment in these countries that will foster the growth of their respective information industries; we offered strong rationale why EU decision-makers need to support the development of information societies among these new members.

REFERENCES

Arnbak, J. (1996, January 25-27). Foreword. *Conference on National and International Initiatives for Information Infrastructure, special Session on Information Infrastructures, OECD*, Paris.

Baylis, T. A. (1994). *The West and Eastern Europe. Economic statecraft and political change*. Westport, CT: Praeger.

CEC-Czech. (2002). Commission of the European communities. Regular report on Czech Republic's progress towards accession. Brussels, Belgium, SEC, 1402.

CEC-Hungary. (2002). Commission of the European communities. Regular report on Hungary's progress towards accession. Brussels, Belgium, SEC, 1404.

CEC-Poland. (2002). Commission of the European communities. Regular report on Poland's progress towards accession. Brussels, Belgium, SEC, 1408.

CEC-Slovakia. (2002). Commission of the European communities. Regular report on Slovakia's progress towards accession. Brussels, Belgium, SEC, 1410.

EU (2000). The European Union's information society. Retrieved January 25, 2000, from *http://www.ispo.cec.be/*

EC, European Commission. (2002). Towards a knowledge based Europe: The European Union and the information society. Retrieved March 19, 2003, from *http://europa.eu.int/information_society/newsroom/documents/catalogue_en.pdf*

EC, European Commission. (2003). Financial assistance: PHARE. Retrieved April 4, 2003, from *http://europa.eu.int/comm/enlargement/pas/phare/struct_funds.htm*

Ferguson, K. (2000). World information flows and the impact of new technology: Is there a need for international communications policy and regulation? In G. Garson (Ed.), *Social dimensions of information technology*. Hershey, PA: Idea Group Publishing.

Fletcher, P.D. (2002). Electronic government: New models of service delivery. *Journal of Global Information Management, 10*(4).

Howkins, J., & Valantin, R. (1997). Development and the information age. *International Development Research Centre*, Ottawa, Canada.

ILO Bureau of Statistics (2002). LABORSTA Database. Retrieved December 13, 2002, from *http://laborsta.ilo.org/*

ITU (2000). *Yearbook of statistics*. Geneva, Switzerland: International Telecommunications Union.

Laudon, K.C., & Laudon, J.P. (2001). *Essentials of management information systems*. Upper Saddle River, NJ: Prentice-Hall.

Lehman, B., & Information Infrastructure Task Force (1995). *Intellectual property and the national information infrastructure*. Washington, DC: Office of Legislative and International Affairs.

Meso, P. (1999, November). Relating social development to information infrastructure in the least developed countries. *Proceedings of the Decision Sciences Institute*, New Orleans, LA.

Meso, P. (2000). Can national information infrastructures enhance social development in the least developed countries? An empirical investigation. *Journal of Global Information Management, 8*(4).

Navas-Sabater, J., Dymond, A., & Juntunen, N. (2002). *Telecommunications and information services for the poor: Toward a strategy for universal access.* WB discussion paper no. 432. Washington, DC: The World Bank.

Pfiser, R. (2000). Information in and on Africa: Past, present, and future. In G. Garson (Ed.), *Social dimensions of information technologye.* Hershey, PA: Idea Group Publishing.

Pook, L.A. (1999, May 5-7). Information culture and management: Strategic information and strategic information systems. *Proceedings: Management at the Millenium: Managing Strategy, Proceedings of Production, Logistics and Quality*, Veszprem, Hungary.

Pook, L.A., Füstös, J., & Marian, L. (2003). The impact of gender bias on job satisfaction: Components of job satisfaction and advancement in post-liberation Hungary, Poland, and Romania. *Journal of Human Systems Management, 22*(1), 37-50.

Pook, L.A., & Pence, N. (2001, June). Relationship between technology and social development: A worldwide assessment with conclusions for Eastern Europe. *Proceedings of Informing Sciences*, Krakow, Poland.

Shea, J., & Stefes, C. (2002). *EU integration of the Visegrad countries.* Retrieved October 20, 2002, from *http://www.wws.princeton.edu/~jpia/July96/shea.html*

Stern, N. (2002). *A strategy for development.* Washington, DC: The World Bank.

Stiglitz, J.E. (2002). *Globalization and its discontents.* New York: W.W. Norton and Company.

Summer, C.E. (1980). *Strategic behavior in business and government.* Boston, MA: Little, Brown and Company.

Tarjanne, P. (1994). *Regulating the international information infrastructure.* Retrieved December 28, 2001, from *http://www.interesting-people.org/archive/0870.html*

TI, Transparency International. (2002). 2002 corruption perceptions index. Retrieved March 28, 2003, from *http://www.gwdg.de/~uwvw/2002.html*

UNIDO (1997). Industrial development: Global report 1997. UNIDO, ID/IDRG/1997.

United Nations. (1989). *Handbook on social indicators.* United Nations Publication, Series F, No. 49.

United Nations. (1995, March 6-12). Report of the world summit for social development. Retrieved July 21, 2001, from *http://www.un.org/documents/ga/conf166/aconf166-9.htm*

United Nations. (1996). *Social statistics: Follow-up to the world summit for social development.* Report to the Economic and Social Council, by Working Group on International Statistical Programmes and Coordination, Report no. E/CN.3/AC.1/1996/R.4.

TASR. (2002). *History of visegrad group: The visegrad-four.* Retrieved October 20, 2002, from *http://www.v-4.sk/may/history-of-visegrad.html*

World Bank. (1997). *Social indicators of development.* Retrieved October 28, 2002, from *http://www.ciesin.org/IC/wbank/sid-home.html*

World Bank. (2000). *Anticorruption in transition: A contribution to the policy debate.* Washington, DC.

World Bank. (2001). *Sector strategy paper: Information and communications technologies.* Washington, DC.

ENDNOTE

[1] As of May 2004, these 10 countries have been admitted into the EU as full members.

Chapter XVII

An Empirical Study of the Relationship of IT Intensity and Organizational Absorptive Capacity on CRM Performance

Ja-Shen Chen, Yuan-Ze University, Taiwan

Russell K.H. Ching, California State University, Sacramento, USA

ABSTRACT

In recent years, e-business has emerged as a mainstream business practice for engaging in global markets. To gain a competitive advantage in these highly competitive markets, many business organizations have turned to customer relationship management (CRM), an integrated system that draws upon the strengths of IT, to allow them to gain greater insights into their customers' needs. This study examines the relationship of information technology (IT) intensity and organizational absorptive capacity to CRM practices and performance. Data collected through a survey of Taiwanese financial service companies generally suggest that CRM practices mediate the effects of IT intensity and organizational absorptive capacity on CRM performance. Thus, it behooves organizations that seek to compete in global markets to invest in developing both their IT infrastructure and organizational absorptive capacity, and apply these resources toward building their marketing intelligence and innovating products and services that meet their customers' needs and expectations.

INTRODUCTION

Continual advances in information technology (IT) and the push for global marketing have led to the rapid expansion of global marketplaces and electronic business (e-business). In 1999, estimates placed business-to-business (B2B) e-business spending between $92 and $142 billion (Ah-Wong et al., 2001; Kelly, 2000), and business-to-customer (B2C) at $33 billion (Ah-Wong et al., 2001) in the U.S. Today, U.S. and worldwide B2B transactions have reached $2.4 trillion and $3.9 trillion, respectively, while U.S. B2C spending is forecasted to approach $95 billion (Mullaney et al., 2003). Future B2C spending can be expected to shadow worldwide B2B spending, estimated between $6 and $7.3 trillion for 2005. Gartner (2002) estimates the current 216.7 million adult worldwide Internet users to double in 2005. Thus, many opportunities avail themselves to businesses that pursue global electronic markets.

Yet, as more businesses transition to e-business, competition in the Internet-enabled marketplace becomes keener. Many have turned to information technology (IT) for solutions that provide a strong competitive advantage. One such IT-based solution that has gained popularity in recent years is customer relationship management (CRM), frequently described as an information system to assist the customer retention process or a methodology that extensively employs information technology, particularly database and Internet technologies, to enhance the effectiveness of relationship marketing practices. Generally, greater investments in IT provide CRM with greater capabilities. With its broader acceptance, the focus of CRM research has shifted from the examination of development and application issues to the discussion of management-oriented issues, particularly those concerned with its successful adoption, implementation and deployment.

As a competitive tool, CRM tunes the organization into listening to its customers, and allows the organization to develop customized products and services that cannot be easily duplicated, substituted or imitated by its competitors, and subsequently more precisely match its customers' needs (Peppard, 2000; Winer, 2001). Given this context, the primary objectives of CRM involve attracting, developing and maintaining successful customer relationships over time (Berry and Parasuraman, 1991; Day, 2000), and building customer loyalty (Kohli et al., 2001) through efficient and effective two-way dialogues (Peppers et al., 1999). As the customer-business relationship flourishes, both the customers and organization benefit (Yim and Kannan, 1999). Establishing these relationships becomes even more critical in e-business as the world becomes the marketplace.

Previous CRM studies have tended to focus on the virtues of using IT to improve organizational productivity and quality (Crosby and Johnson, 2001). However, as use became more pervasive, organizations discovered differences in their outcomes of CRM due to factors inherent to them. Recent studies now focus on identifying organizational factors that lead to the successful use of CRM, such as corporate culture, and process and technology improvement (Yu, 2001), and the alignment of people, processes and technology (Battista and Verhun, 2000; Crosby, 2002; Crosby and Johnson, 2000; Simpson, 2002). The results of these studies indicate that variations in these factors will affect the effectiveness of CRM. Therefore, CRM performance may vary among organizations due to organizational factors (all applications of IT being equal).

An important factor that may critically affect CRM performance lies in the organization's ability to leverage and exploit its knowledge toward innovating new products and services that benefit its customers. Previous studies (Boynton et al., 1994; Cohen and Levinthal, 1990; Hurley and Hult, 1998) suggest that an organization's ability to link its knowledge to its innovativeness (i.e., ability to innovate) depends upon its absorptive capacity, the ability to recognize and assimilate new information, and apply the ensuing knowledge to commercial ends (i.e., exploitation) (Cohen and Levinthal, 1990). Because absorptive capacity largely depends on precursory learning (within the organization) and the dissemination and integration of subsequent knowledge, and is facilitated through resource availability, its level will vary among organizations. Generally, greater investments devoted toward developing organizational learning often through research and development (R&D) lead to higher levels of absorptive capacity, which in turn often leads to more successful results with CRM.

The purpose of this paper is to examine the relationship among IT intensity, organizational absorptive capacity, CRM practices and CRM performance.

IT INTENSITY

CRM can be seen as an IT extension of relationship marketing (RM), a new marketing paradigm that focuses on developing close personal relationships, interactions and social exchange between an organization and its customers and business parties over time to enhance the organization's competitive response to continually changing markets (Zineldin, 2000). CRM expands upon this concept through its emphasis on information management (Peppard, 2000), and further draws upon IT for its strength and capabilities.

As an IT-enabled enterprise system, CRM's performance hinges on the resources and investments an organization commits to it. IT intensity refers to the IT infrastructure and applications that allow the organization to benefit from its IT investments and apply them toward its best interests. An IT infrastructure forms the foundation on which CRM systems are built, and specifies the extent of its capabilities since the success of CRM depends in part on organization-wide information and knowledge sharing as it draws upon the innovative and creative thoughts of its people. Thus, CRM promotes an enterprise-wide solution rather than a narrowly focused functional area solution.

Henderson and Venkatraman (1994) suggest that an organization's IT infrastructure has two components: (1) a technical IT infrastructure, and (2) a human IT infrastruc-

Figure 1. Research model

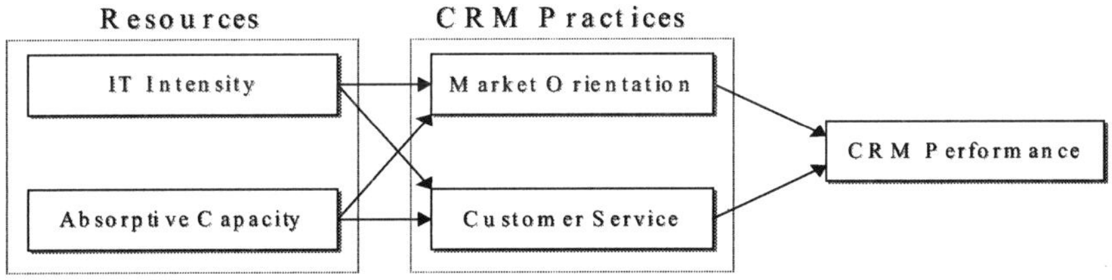

ture. Duncan (1995) views the technical IT infrastructure as a set of tangible, shared, physical IT resources, including hardware and operating systems, network and telecommunications technologies, data, and core software applications. The technical IT infrastructure should integrate and interconnect the organization in such a way to efficiently and effortlessly route information through its telecommunication network (Rockart et al., 1996).

In contrast, the human IT infrastructure addresses the necessary individual skills and knowledge required to develop, maintain, manipulate and support end-users in their abilities to leverage the technical infrastructure. Osterman (1995) discusses the importance of developing and acquiring individual skills and roles to enable an organization's investments in IT. Without an adequate human IT infrastructure, the organization will realize very few benefits from its IT infrastructure and investments.

Investments in IT intensity have opened new marketing opportunities to organizations through efficiency gains (i.e., marketing process automation) and improved intelligence (Applegate et al., 1996). While marketing process automation helps link marketing activities to support information sharing (i.e., efficiencies), marketing intelligence aims to enhance decision making through tools, such as data mining and knowledge management techniques, that provide greater insights and lead to the discovery of new information. Although the continual advances in IT have led to more sophisticated marketing applications of IT (Stone and Good, 2001), these investments alone are not sufficient to sustain competitive advantages in global markets.

ORGANIZATIONAL ABSORPTIVE CAPACITY

Another organization factor that may have a significant influence on CRM performance is organizational absorptive capacity. The absorptive capacity of an organization results from the cumulative learning activities of its individuals and the transfer of knowledge within the organization through a common language (Cohen and Levinthal, 1990). Learning activities occur with new experiences directed toward exploration (i.e., research), routine experiences and training. These activities help develop knowledge that can be used to recognize, acquire and assimilate new information and apply the ensuing knowledge. The more frequent learning occurs, the greater the accumulation process, which in turn reinforces prior knowledge, increases the capacity to retain new knowledge and yields the application of knowledge to new scenarios (Bower and Hilgrad, 1981). As a result, the greater the knowledge possessed and shared throughout the organization, the more the organization will be inclined to *absorb* new knowledge, and apply it toward innovative, creative and effective products and services. Leveraging becomes greater as knowledge becomes pervasive in the organization.

In developing and maintaining their relationships, organizations must be capable of applying their knowledge toward understanding the reasons that make their customers unique, and tailoring their products and services to accommodate their uniqueness (Massey et al., 2001; Morgan and Hunt, 1994). Thus, like IT intensity, absorptive capacity enables the organization, but through knowledge.

CRM PRACTICES: MARKET ORIENTATION AND CUSTOMER SERVICE

Market Orientation

Market orientation can be defined as the organization-wide generation, dissemination and responsiveness to market intelligence, and involves information sharing among multiple departments engaged in activities directed toward meeting customer needs. In contrast to product-driven marketing, which focuses on pushing end products into markets while promoting lower prices and good quality, market-orientation concentrates on continuously detecting customer needs and quickly filling them. As a means for promoting customer-centric values, market-orientation practices often have positive effects on the organization's performance and new products. However, given the underlying need to share information and knowledge, market-orientation becomes dependent on the organization's IT-intensity for its success. Previous studies have focused on performance implications (e.g., Matsuno and Mentzer, 2000), measurement (e.g., Homburg and Pflesser, 2000), and antecedents and performance outcomes (e.g., Jaworski and Kohli, 1993).

Customer Service: Customization and Loyalty Programs

Two fundamental components to customer service are customization and loyalty programs. The objective of customization is to tailor products and services that more precisely fit the individual customer's needs. Pine et al. (1995) describe customization as a learning relationship between the organization and its customers that results in delivering products or services in direct response to a particular customer's needs and preferences over time. Dialogs that increasingly lead to collaboration and loyalty, and further cultivate the relationship, underlie the success of customization. In contrast to mass-marketed items, customization adds value that cannot be easily obtained elsewhere. Gaining this knowledge requires eliciting customers for ideas and integrating this information into the production processes such that it provides the organization with its greatest competitive advantage (Pitta, 1998).

Mass customization attempts to customize products and services for the individual customers to reach a one-to-one marketing level. Compared to traditional product differentiation which strives to change the product's characteristics to competitively distinguish it from another's offering, mass customization achieves differentiation through targeting the product's or service's benefits toward satisfying the customer's specific needs. The advances in IT make mass customization more feasible as they allow individual customer behavior and characteristics to be traced and analyzed with greater detail through data warehouses and data mining techniques, all of which make customer service easier and solution-oriented (Kalakota and Robinson, 2001). The organization can then quickly identify its customers' needs and preferences, and translate them to products and services.

Loyalty programs link organizations to future opportunities. The development of loyalty programs frequently lead to increases in repeat-purchases and usage frequency, and raise barriers of entry into markets by making it difficult for new entrants to court customers away from existing businesses (Sharp and Sharp, 1997). The market research studies of Hughes (2001), and Reichheld and Sasser (1990) strongly suggest that loyalty programs can increase business revenue and total customer market share.

RESEARCH MODEL AND TEST OF HYPOTHESES

This study proposes that an organization's investments in IT and absorptive capacity will indirectly benefit its CRM performance. Both will have greater effects on how it applies CRM through its practices, and its practices will be a greater determinant of CRM performance success. Figure 1 illustrates this study's research model. The model suggests that CRM practices will mediate the effects of IT intensity and organizational absorptive capacity on CRM performance. IT intensity and organizational absorptive capacity represent the independent variables, while CRM practices a mediating variable, and CRM performance the dependent variable. The following hypotheses and sub-hypotheses are presented to test the model.

IT intensity reflects an organization's commitment to technology. The greater the investments, the more reasonable it is to expect greater achievements in other areas. IT has been attributed as the enabler of many opportunities, particularly as a means for increasing efficiencies and a tool for securing competitive advantages. Methods, such as online analytical processing (OLAP) and data mining, have allowed organizations to improve their intelligence gathering abilities to discover new information and knowledge about their markets and customers. IT has also enabled them to quickly and efficiently disseminate and share information across organizational boundaries, thereby empowering employees to make decisions that are consistent with the organization's goals and objects, and benefit the interests of both the organization and customer. In maintaining a competitive posture, organizations must also be able to anticipate competitive threats and adjust accordingly to maintain their customer base with novel and customized products and services that aptly meet their (customers') needs and expectations, and cannot be readily duplicated, imitated or substituted. IT can be used as the means to ensure customer feedback is integrated into the design and production of future products and services. Thus, IT intensity should facilitate the organization's market orientation and customer service, such that greater investments in IT will lead to more sophisticated and enriching CRM practices.

H1: IT intensity has a positive relationship with CRM practices.

H1a: IT intensity has a positive relationship with market orientation.

H1b: IT intensity has a positive relationship with customer service.

Absorptive capacity underlies organizational knowledge. It allows the organization to recognize and assimilate new information into its existing body of knowledge. As a result, the more information that can be absorbed into the organization and shared among its members, the greater the organization's knowledge will be, and potentially, the more information it can absorb in the future. Organizations that have amassed knowledge of their market requirements and customers' needs will be inclined toward practicing marketing intelligence and information sharing to produce knowledge, and leveraging information and knowledge to garner even greater levels of customer loyalty. Essentially, knowledge will affect the extent to which an organization focuses its market orientation and customer service.

H2: Absorptive capacity has a positive relationship with CRM practices.

H2a: Absorptive capacity has a positive relationship with market orientation.

H2b: Absorptive capacity has a positive relationship with customer service.

In highly competitive, IT-enabled marketplaces, all business organizations must stay in tune with the needs and expectations of their customers. Many have turned to CRM not only to retain their existing customers and win new ones, but to profit through both efficiency gains and new opportunities. Assuming the organization has made firm commitments to adopting critical IT and developing its knowledge and absorptive capacity, the enabled CRM practices should increase CRM performance, in particular organization and customer benefits. Essentially, understanding the customers' needs and tailoring products and services that more precisely meet their needs and expectations will lead to benefits for the organization and customer. Organization benefits involve achievements, such as cost reductions, increased profits resulting from repeat purchases (customer loyalty) and greater efficiencies (information sharing, empowerment), and one-to-one marketing success.

Gwinner et al. (1998) have identified four perceived benefits customers receive beyond the core service attributes, including social (personal recognition, friendship and fraternization), psychological (confidence, trust and reduced anxiety), economic (monetary and time savings) and customization (preferential treatment, special services or consideration, and history development). Of the four, psychological benefits remain invariant over service types and therefore provide more frequent benefits. Strong market orientation lends greater support towards ensuring the customers' needs and expectations are understood and acknowledged. Thus, greater achievements in CRM practices should be associated with greater CRM performance.

H3: CRM practices have a positive relationship with CRM performance.

H3a: Market orientation has a positive relationship with CRM performance.

H3b: Customer service has a positive relationship with CRM performance.

Organizations will eventually derive benefits from their investments in IT and R&D. IT intensity and absorptive capacity represent the fundamental enablers of CRM. Because CRM is IT-enabled and involves the leveraging and exploitation of the organization's knowledge, both are necessary (but not sufficient) conditions to CRM success. However, they alone will have little direct effect on CRM performance. The organization must learn to apply them in its practices to achieve desired outcomes. Yet, boosts to IT-intensity and absorptive capacity will eventually enhance CRM performance. IT-intensity and absorptive capacity allow an organization to develop its CRM practices, especially its market orientation. This in turn enhances CRM performance. Thus, market orientation should mediate the effects of IT intensity and absorptive capacity on CRM performance.

H4: Market orientation mediates the relationship of IT intensity and absorptive capacity on CRM performance.

Customer service will also mediate the effects of IT-intensity and absorptive capacity on CRM performance. As in the case of market orientation, customer service practices help ensure customer needs and expectations are more precisely met through the benefit of investments in IT and the development of the organization's absorptive capacity. Greater investments in both will provide the means to bolster customer service, which in turn allows the organization and customer to reap greater benefits. Thus, the effects of both (IT intensity and absorptive capacity) on organization and customer benefits will be mediated by customer service practices.

H5: Customer service mediates the relationship of IT intensity and absorptive capacity on CRM performance.

The importance of this model lies in understanding the effects of fundamental investments in IT and absorptive capacity on CRM success. Both enable the organization to gear itself to compete more effectively in electronic global markets.

RESEARCH METHODOLOGY

Data Collection and Sample

This study involved a survey of 542 Taiwanese financial service companies and was funded by the National Science Council (NSC) of Taiwan. The companies appeared on a list published by the Taiwan Joint Credit Information Center. Questionnaires were mailed with accompanying cover letters that explained the purpose of this research. Financial service companies were chosen because the highly competitive nature of their markets and their need to capitalize on customer-centric opportunities (i.e., highly customized services) make them prime adopters of CRM (Peppard, 2000). They are more motivated to differentiate and customize their products and services to better fit their customers' needs and expectations since many are engaged in Internet marketing (i.e., e-business) (Peterson et al., 1997). For example, to address e-business trends, many banks have looked into ways to build customer loyalty through customer segmentation, product differentiation and long-term relationships (Tomiuk and Pinsonneault, 2001). Because they represent a large segment of CRM users (Foss and Stone, 2002), their representatives tend to be more familiar with CRM features, applications and organizational objectives. The experiences of these companies would be representative of mature CRM users.

Table 1. Operational definitions

Variables	Operational definition	References
IT intensity	IT infrastructure IT applications	Sacha, 1993; Michael, 1996; Kalakota and Whinston, 1996
Absorptive capacity	Individuals cumulative learning activities Organization knowledge transfer Management climate	Cohen and Levinthal, 1990; Boynton et al., 1994; Bower and Hilgrad, 1981; Massey et al., 2001
Market orientation	Customer focus Competitor focus Cross functional integration	Narver and Slates, 1990; Han et al., 1998; Slater and Narver, 2000
Customer service		
Customization	Customized services Customized capability	Silveira et al., 2001; Gilmore and Pine, 1997; Kotha, 1995; Pine 1993
Loyalty program	Marketing campaigns Customer profitability Strategic alliance	Sharp and Sharp, 1997; Barnes, 2001; Winer, 2001; Griffin, 1995; Hughes, 2001
CRM performance		
Organization benefits	Profit increase, Cost down, New opportunities	Storey and Easigwood, 1999; Swith, 2001; Winer, 2001
Customer benefits	Social benefits, Psychological benefits, Economic benefits, Customized benefits	Gwinner et al., 1998

The recipients of the surveys were limited to CRM and marketing managers, and customer service department heads. Two weeks after the initial mailing, 99 responses were received. Follow-up telephone calls made a week later increased the total to 173 responses, for a response rate of 32 percent. Among the returned surveys, nine were incomplete and therefore discarded; this reduced the sample size to 164 (30.3 percent return rate). Missing values among nine of the surveys further reduced the analysis to 155 observations. The final sample covers a broad cross-section of companies engaged in domestic and foreign banking, insurance and securities trading as well as many others.

Measures

The items in the survey instrument were adopted from various sources (Table 1) to measure the five constructs: market orientation (1-8), IT intensity (9-15), customer service (16-20), absorptive capacity (21-28) and CRM performance (29-40). A standard psychometric scale development procedure (Gerbing and Anderson, 1988) was followed to generate multiple-item scales based on a review of the literature and interviews with IT and marketing professionals. The questionnaire was pre-tested and refined following the comments of five IT and marketing managers. All items were operationalized with five-point Likert-type scales ranging from (1) strongly disagree to (5) strongly agree. Table 1 provides the operational definitions and their sources (i.e., references) for the five constructs, and Table 2 reflects the instrument's items. The high Cronbach alphas support each construct's internal reliability (Table 2).

A factor analysis with a varimax rotation (using SAS 8.2) was performed on five factors to correspond to the five hypothesized constructs. Because a customer service variable (products/services satisfy customer needs) loaded onto the market orientation construct, it was dropped and a second analysis was performed. Table 2 shows the rotated factors with all variables properly loading onto their respective constructs. A Kaiser's measure of sampling adequacy (MSA) of .887 strongly indicates the appropriateness of the factor analysis, given the sample size (164 observations less 9 due to missing values). Composite scores were created for each factor from the summated values of its items to measure each construct.

Demographic data (number of employees, capital, and age of the organization) were also collected and compared to the population for potential non-response bias. T-tests of the sample and population means derived from the Joint Credit Information Center data suggest that none was statistically significant at a 0.05 level. Moreover, no significant differences between earlier and later responses for each item were detected. The absence of significant differences supports the contention that no response bias is present in the sample (Armstrong & Overton, 1977).

Table 2. Factor analysis

Kaiser's Measure of Sampling Adequacy: Overall MSA = 0.887

	Factor 1 CRM Performance	**Factor 2** Absorptive Capacity	**Factor 3** IT Intensity	**Factor 4** Market Orientation	**Factor 5** Customer Service
Cronbach Alpha	.948	.926	.915	.900	.782
Eigenvalue	13.742	4.837	2.904	2.33	1.596
Percent of Total Variance	.352	.124	.075	.057	.041
Cumulative Percent of Total Variance	.352	.476	.551	.608	.649
Factor Loadings:					
40. Value-added products and service	0.818				
37. Increased customer trust	0.813				
34. Enhanced image	0.806				
36. Reduced customer anxiety and uncertainty	0.801				
38. Reduced customer service time	0.796				
35. Customer service	0.795				
33. Increased service quality	0.780				
39. One-stop shopping features	0.752				
32. New markets	0.710				
29. Increased revenue	0.699				
31. Reduced new customer acquisition costs	0.696				
30. Marketing sales cost reduction	0.643				
25. CRM absorptive capacity		0.865			
23. CRM employee knowledge		0.850			
24. Employee business knowledge		0.835			
21. CRM training		0.756			
26. Cross-functional CRM involvement		0.730			
22. CRM training quality		0.728			
27. Help sources		0.715			
28. Relationships with IT staff/consultants		0.524			
12. IT hardware and software investments			0.808		
14. IT management			0.804		
13. IT system integration			0.781		
9. IT priority			0.744		
10. IT competitive advantage			0.716		
15. IT recruitment and training			0.698		
11. IT information features			0.642		
3. After sales service				0.759	
4. Predict customer needs				0.753	
2. Care and evaluate customer satisfaction				0.753	
6. Response to competitors				0.721	
5. Share information				0.686	
1. Customer satisfaction objective				0.671	
8. Cross-functional collaboration				0.607	
7. Understand competitors				0.571	
17. Add value to products/services					0.750
20. Promotions to customer segments					0.645
19. Products/services to customer segments					0.633
18. Incentive programs					0.586

Values greater than 0.4 shown.

Table 2. Factor analysis (cont.)

3. After sales service	0.759	
4. Predict customer needs	0.753	
2. Care and evaluate customer satisfaction	0.753	
6. Response to competitors	0.721	
5. Share information	0.686	
1. Customer satisfaction objective	0.671	
8. Cross-functional collaboration	0.607	
7. Understand competitors	0.571	
17. Add value to products/services		0.750
20. Promotions to customer segments		0.645
19. Products/services to customer segments		0.633
18. Incentive programs		0.586

Values greater than 0.4 shown.

ANALYSIS

Hierarchical regression models were developed to test each set of hypotheses. To study the effects of business capital and the number of employees on the dependent variables, both were included as control variables; this controls their effects to ensure that neither biases the results, and affords greater generalizbility of the findings (Pedhazur and Pedhazur Schmelkin, 1991). Tables 3 through 5 provide summaries of the statistical results. Scatter diagrams of the residuals for each model revealed no violations of homoscedasticity.

The regression models support all but two of the research hypotheses. Model I (Table 3) suggests that IT intensity and organizational absorptive capacity are positively related to market orientation, one of the elements of CRM practices. The p-values of their standardized coefficients indicate both are significant to the model. The low values of the variance inflation factor (VIF) reveal no colinearity problems. These results support H1a and H2a. Similar results in Model II (Table 3) suggest that IT-intensity and absorptive capacity are positively related to customer service, and lends support to H1b and H2b. Thus, it may be concluded that IT intensity and absorptive capacity are positively related to CRM practices. Increases to both will have similar effects on CRM practices.

Model III (Table 4) tested the effects of CRM practices on CRM performance. The results support the relationships, and suggest that increases to market orientation and customer service have positive effects to CRM performance.

Two models (IV and V) were developed to test the mediating effects of CRM practices. Table 5 summarizes the results. A mediating effect will occur when variations in the independent variables account for variations in the mediator, variations in the mediator account for variations in the dependent variable, and when controlled the independent variables have no effect on the dependent variable. The significance of models I and II provides strong evidence to suggest a positive relationship exists between the independent variables, IT intensity and absorptive capacity, and the mediators, market orientation and customer service (i.e., variations in the independent variable account for variations in the mediator).

Model III suggests a positive relationship exists between the mediators (market orientation and customer service) and the dependent variable, CRM performance.

Table 3. Regression analysis results for H_1 and H_2

		Dependent Variable					
		Model I			Model II		
		Market Orientation			***Customer Service***		
		Standardized Coefficient	T value	VIF	Standardized Coefficient	T value	VIF
Control Variables	Business Capital	-.089	-.82	2.881	-.779	-.66	2.881
	Number of Employees	-.104	-.95	2.873	.004	.04	2.942
Predictors	IT Intensity	.435***	5.70	1.426	.333***	4.00	1.426
	Absorptive Capacity	.271***	3.60	1.385	.267***	3.25	1.385
R^2		.387			.273		
F Value		23.67***			14.07***		
n		155			155		

*** $p < 0.01$, ** $p < 0.05$

Table 4. Regression analysis results for H_3

		Dependent Variable		
		Model III		
		CRM Performance		
		Standardized Coefficient	T value	VIF
Control Variables	Business Capital	-.034	-0.28	2.883
	Number of Employees	.144	1.20	2.872
Predictors	Market Orientation	.211**	2.53	1.374
	Customer Service	.336***	4.05	1.364
R^2		.244		
F Value		12.11***		
n		155		

*** $p < 0.01$, ** $p < 0.05$

However, the results of models IV and V (Table 5) reveal only a partial mediating effect. The significance of the mediators (market orientation in model IV and customer service in model V) and absorptive capacity, and the non-significance of IT intensity do not lend support to H4 and H5, respectively. This suggests that absorptive capacity has a direct positive relationship with CRM performance; increases to absorptive capacity will have similar effects on CRM performance. It can be concluded that market orientation and customer service mediate the effects of IT intensity on CRM performance.

Generally, the data analysis partially supports the proposed research model. The partial mediating effects of CRM practices on the relationship between the independent variables and CRM performance suggest the existence of a direct effect of absorptive capacity on CRM performance. However, the low R-squares of the models indicate other factors not included in the model may account for a greater portion of the variations.

Table 5. Regression analysis results for testing mediating effects

		Dependent Variable *CRM Performance*					
		Model IV			Model V		
		Standardized Coefficients	T value	VIF	Standardized Coefficient	T value	VIF
Control Variables	Business Capital	-.064	-0.52	2.896	-.063	-0.52	2.889
	Number of Employees	.149	1.19	2.954	.127	1.04	2.942
Predictors	IT Intensity	.048	0.51	1.729	.044	0.50	1.578
	Absorptive Capacity	.231**	2.61	1.487	.205**	2.37	1.495
	Market Orientation	.256**	2.67	1.592			
	Customer Service				.334***	4.02	1.375
R^2		.209			.253		
F Value		7.89***			10.07***		
n		155			155		

*** $p < 0.01$, ** $p < 0.05$

Future studies need to explore more precise measurements to capture these variations without increasing the number of items on the survey instrument (increasing the length of the instrument will result in a lower response rate).

DISCUSSION

Advances in technology have facilitated significant changes to business practices. Cases, such as American Airlines and Sabre (Copeland and McKenny, 1988), demonstrate the power of IT to secure unprecedented competitive advantages and change the competitive landscape of various industries. Today's IT and the Internet have enabled businesses to electronically broaden their reach to include global markets and increase their competitiveness in an unprecedented manner. Yet, the same IT that bestows a competitive advantage can also equalize the presence of all competitors in these markets. Thus, the challenge of today's global e-markets lies in an organization's ability to leverage and exploit its information to overcome IT's leveling factor and ensure a competitive advantage. The focus of marketing has shifted from mass marketing to long-term one-to-one learning relationships between an organization and its customers. The primary objective is to retain customers by understanding and meeting their needs and expectations. In global markets, this becomes a greater challenge as differences in cultures, societies and government regulations (as well as many others) increase uncertainty and complexity, yet (the differences) must be accommodated. E-business no longer confines business to geographical or political boundaries, and adopting a global marketing strategy may be the key to long-term survival. CRM has become a means to manage engagements in global markets as it incorporates the principles of RM and IT to establish, develop and maintain one-to-one relationships. Technology alone will not guarantee success; other organizational investments must also be made.

This study examined the relationship among IT intensity, absorptive capacity, customer service, market orientation and CRM performance in the context of the financial service industry in Taiwan. Because financial service companies face global marketing challenges and global marketing focuses on developing product markets (i.e., selling a product or service to customer seeking shared benefits) rather than country or regional markets (Muhlbacher et al., 1999), the results might be viewed as precursory. However, future studies can expand upon the results of this study to determine whether the same effects apply to other industries and industries of other national origins. As Markus and Soh (2002) point out, structural conditions (i.e., financial, legal, regulatory and telecommunications infrastructures, national policies, business practices, language and education, etc.) influence global e-commerce activities in different countries.

The investments an organization places in its resources will contribute to the profits it will reap. Both IT and absorptive capacity represent vital resources that are necessary to compete in highly competitive electronic marketplaces with global reaches. IT enables organizations to achieve greater levels of success through efficiency gains and new opportunities. Larger investments directed toward critical IT (linked to the achievement of their organizational goals and objectives) and developing a comprehensive IT infrastructure allow its members to share knowledge and information both within the organization and between the organization and others, and become innovative. They also help empower them in making decisions that are consistent with the organization's goals and objectives. Yet, these advantages tend to be offset by IT's equalizing effect. Consequently, more organizations have turned to CRM to (re)gain and maintain their competitive edge. However, for CRM to succeed, organizations must build and develop their organizational knowledge (R&D) and applications of that knowledge (i.e., practices) through absorptive capacity. When coupled with IT, the applications often lead to greater benefits.

The results of this study suggest that organizations will profit from their investments in IT and absorptive capacity through their ability to apply them. Simply investing in IT and R&D does not automatically lead to success; both are fundamental building blocks. Using IT and absorptive capacity to discover information of the environment and promote customer service activities increases the utility of CRM to capture certain benefits. The mediating effect of the organization's CRM practices supports this contention. Investments in IT and absorptive capacity facilitate the development of CRM practices. To a lesser degree, absorptive capacity was found to have a direct effect on CRM performance. Advanced CRM practices keep the organization in tune with the environment (i.e., market intelligence) and its customers' needs (i.e., customer service). As the learning relationship flourishes, both the customers and organization benefit. The customer not only receives products and services that add value and fit his/her needs, but also gains psychological benefits (i.e., confidence, trust and reduced anxiety), the most important in maintaining the longevity of a relationship. The organization benefits particularly from increased revenues and service quality, and an enhanced image. Hence, the organization profits from its long-term relationships with its customers. Its return on its investments may be measured in terms of the benefits it will receive from caring for its long-term relationships.

CONCLUSION

This paper examined the relationship of IT intensity and absorptive capacity to CRM practices and CRM performance, and the mediating effects of CRM practices. Based on the data collected through a survey of financial service companies in Taiwan, the study generally concludes that organizations will benefit from their investments in IT and absorptive capacity through their CRM practices. Organizational absorptive capacity may also directly affect CRM performance. Greater investments in IT and absorptive capacity allow the organization to use CRM to extend greater benefits to its customers and receive greater benefits from their relationships. The conclusions may be particularly applicable to e-business and global marketing as operating in the global marketplace comes with many challenges and opportunities that fall into the realm of CRM. With the current trend toward RM and the forecasted global opportunities, CRM can play a major role in helping build and maintain one-to-one relationships in B2B e-business.

REFERENCES

Ah-Wong, J., Gandhi, P., Patel, H., & Shah, U. (2001). E-commerce progress: Enablers, inhibitors and the short-term future. *European Business Journal*, 13, 98-107.

Amstrong & Overton (1997). Estimating non-response bias in mail surveys. *Journal of Marketing Research*, 14, 396-402.

Applegate, L.M., McFarlan, F.W., & McKenney, J.L. (1996). *Corporate information systems management: The issue facing senior executives* (4th ed). Chicago, IL: Irwin.

Barnes, J.G. (2001). *The secrets of customer relationship management*. New York: McGraw-Hill.

Battista, P. & Verhun, D. (2000). Customer relation management. *CMA Management*, 74, 34-37.

Berry, L.L. & Parasuraman, A. (1991). *Marketing service: Competing through quality*. New York: The Free Press.

Bower, G.H. & Hilgrad, E.R. (1981). *Theories of learning*. Englewood Cliffs, NJ: Prentice-Hall.

Boynton, A.C., Zmud, R.W., & Jacobs, G.C. (1994). The influence of IT management practice on IT use in large organizations. *MIS Quarterly*, 18, 299-318.

Cohen, W.M., & Levinthal, D. (1990). Absorptive capacity: A new perspective on learning and innovation. *Administrative Science Quarterly*, 35, 128-152.

Copeland, D.G., & McKenny, J.L. (1988). Airline reservations systems: Lessons from history. *MIS Quarterly*, 12, 353-370.

Crosby, L.A. (2002). Exploding some myths about customer relationship management. *Managing Service Quality*, 12, 271-277.

Crosby, L.A., & Johnson, S.L. (2000). Customer relationship management. *Marketing Management*, 9, 4-5.

Crosby, L.A., & Johnson, S.L. (2001). Friend or foe to customer relationships? *Marketing Management*, 10, 10-11.

Day, G.S. (2000), Managing marketing relationships. *Journal of the Academy of Marketing Science*, 28, 24-31.

Duncan, N.B. (1995). Capturing flexibility of information technology infrastructure: A study of resource characteristics and their measure. *Journal of Management Information Systems*, 12, 37-57.

Foss, B., & Stone, M. (2002). *CRM in financial services: A practical guide to making customer relationship management work*. London: Kogan Page.

Gartner, Inc. (2002). Gartner predicts 2002: Top 10 predictions. *www4.gartner. com/DisplayDocument?doc_cd=103726*

Gerbing, D.W., & Anderson, J.C. (1988). An updated paradigm for scale development incorporating unidimensionality and its assessment. *Journal of Marketing Research*, 25, 186-192.

Gilmore, J. & Pine, J. (1997). The four faces of mass customization. *Harvard Business Review*, 75, 91-101.

Gwinner, K.P., Gremler, D.D. & Bitner, M.J. (1998). Relational benefits in services industries: The customer's perspective. *Journal of the Academy of Marketing Science*, 26, 101-114.

Han, J.K., Kim, N. & Srivastava, R.K. (1998). Market orientation and organizational performance: Is innovation a missing link? *Journal of Marketing*, 62, 30-45.

Henderson, J.C., & Venkatraman, N. (1994). Strategic alignment: A model for organizational transformation via information technology. In T. J. Allen, & M. S. S. Morton (Eds.), *Information technology and the corporation of the 1990's* (pp. 202-220). Oxford, UK: Oxford University Press.

Homburg, C., & Pflesser, C. (2000). A multiple-layer model of market-oriented organizational culture: Measurement issues and performance outcomes. *Journal of Marketing Research*, 37, 449-462.

Hughes, A. M. (2001). *Strategic database marketing: The master plan for starting and managing a profitable, customer-based marketing program*. New York: McGraw-Hill.

Hurley, R.F., & Hult, G.T.M. (1998). Innovation, market orientation, and organizational learning: An integration and empirical examination. *Journal of Marketing*, 62, 42-54.

Jaworski, B. J., & Kohli, A. (1993). Market orientation: Antecedents and consequences. *Journal of Marketing*, 57, 53-71.

Kalakota, R. & Robinson, M. (2001). *e-Business: Roadmap for success*. Reading, MA: Addison-Wesley.

Kelly, S. (2000). An e-commerce explosion. *Communications News*, 37, 8.

Kohli, R., Piontek, F., Ellington, T., VanOsdol, T., Shepard, M., & Brazel, G. (2001). Managing customer relationships through e-business decision support applications: A case of hospital-physician collaboration. *Decision Support Systems*, 32, 171-187.

Kotha, S. (1995). Mass customization: Implementing the emerging paradigm for competitive advantage. *Strategic Management Journal*, 16, 21-42.

Li, E.Y., McLeod, R., Jr., & Rogers, J.C. (2001). Marketing information systems in Fortune 500 companies: A longitudinal analysis of 1980, 1990, 2000. *Information & Management*, 38, 307-322.

Markus, M.L., & Soh, C. (2002). Structural influences on global e-commerce activity. *Journal of Global Information Management,* 10, 5-12.

Massey, A.P., Montoya-Weiss, M.M., & Holcom, K. (2001). Re-engineering the customer relationship: leveraging knowledge assets at IBM. *Decision Support Systems*, 32, 155-170.

Matsuno, K., & Mentzer, J.T. (2000). The effects of strategy type on the market orientation-performance relationship. *Journal of Marketing*, 64, 1-16.

Michael, M. (1996). Practical advice for implementing corporate intranets. *Telecommunications*, April, 34-37.

Morgan, R.M., & Hunt, S.D. (1994). The commitment-trust theory of relationship marketing. *Journal of Marketing*, 58, 20-38.

Muhlbacher, H., Dahringer, L., & Leihs, H. (1999). *International marketing: A global perspective*. London: International Thomson Business Press.

Mullaney, T.J., Green, H., Arndt, M., Hoff, R.D., & Himelstein, L. (2003). The e-biz surprise. *Business Week,* (May 12), 60-68.

Narver, J. C. & Slater, S. F. (1990). The effect of a market orientation on business profitability. *Journal of Marketing*, 54, 20-35.

Osterman, P. (1995). Impact of IT on jobs and skill. In M. Scott Morton (Ed.), *The corporations of the 1990s*. New York: Oxford University Press.

Pedhazur, E.J., & Pedhazur Schmelkin, L. (1991). *Measurement, design, and analysis: An integrated approach*. Hillsdale, NJ: Lawrence Erlbaum Associates.

Peppard, J. (2000). Customer relationship management (CRM) in financial services. *European Management Journal*, 18, 312-327.

Peppers, D., Rogers, M., & Dorf, R. (1999). *The one-to-one fieldbook*. New York: Currency and Doubleday.

Peterson, R.A., Balasubramanian, S, & Bronnenberg, B.J. (1997). Exploring the implications of the Internet for consumer marketing. *Journal of the Academy of Marketing Science*, 25, 329-346.

Pine, B.J. (1993). *Mass customization: The new frontiers in business competition*. Boston: Harvard Business School Press.

Pine, B.J., Peppers, D., & Rogers, D. (1995). Do you want to keep your customers forever? *Harvard Business Review*, March-April, 103-114.

Pitta, D.A. (1998). Marketing one-to-one and its dependence on knowledge discovery in databases. *Journal of Consumer Marketing*, 15, 468-480.

Reichheld, F.F., & Sasser, Jr., W.E. (1990). Zero defections: Quality comes to services. *Harvard Business Review*, September-October, 105-111.

Reichheld, F.F., & Schefter, P. (2000). E-Loyalty: Your secret weapon on the Web. *Harvard Business Review*, 78(4), 105-113.

Rockart, J.F., Earl, M.J., & Ross, J.W. (1996). Eight imperatives for the new it organization. *Sloan Management Review*, 38, 43-55.

Sacha, C. (1996). Inside job: A guild to intranets. *Training and Development*, October, 63-65.

Sharp, B., & Sharp, A. (1997). Loyalty programs and their impact on repeat-purchase loyalty patterns. *International Journal of Research Marketing*, 14, 473-486.

Silveira, G.D., Borenstein, D. & Fogliatto, F.S. (2001). Mass customization: Literature review and research directions. *Journal of Personal Selling & Sales Management*, 72, 1-13.

Simpson, L (2002). The real reason why CRM initiatives fail. *Training,* (May), 50-54.

Slater, S.F., & Narver, J.C. (2000). The positive effect of a market orientation on business profitability: A balanced replication. *Journal of Business Research*, 48, 69-73.

Stone, R.W., & Good, D.J. (2001). The assimilation of computer-aided marketing activities. *Information & Management*, 38, 437-447.

Storey, C., & Easingwood, C.J. (1999). Types of new product performance: Evidence form the consumer financial services sectors. *Journal of Business Research*, 46, 193-203.

Swift, R.S. (2001). *Accelerating customer relationships*. Upper Saddle River, NJ: Prentice Hall.

Tomiuk, D., & Pinsonneault, A. (2001). Customer loyalty and electronic-banking: A conceptual framework. *Journal of Global Information Management,* 9, 4-14.

Winer, R.S. (2001). A framework for customer relationship management. *California Management Review*, 43, 89-105.

Yim, C.K., & Kannan, P.K. (1999). Consumer behavior loyalty: A segmentation model and analysis. *Journal of Business Research*, 44, 75-92.

Yu, L. (2001). Successful customer-relationship management. *Sloan Management Review*, 42, 18-19.

Zineldin, M. (2000). Beyond relationship marketing: Technologicalship marketing. *Marketing Intelligence and Planning*, 18, 9-23.

ENDNOTE

1 This study was partially supported by a Taiwan National Science Council grant (NSC90-2416-H-155-014).

Previously published in the Journal of Global Information Management, 12(1), 1-17, January - March 2004.

Chapter XVIII

Privacy, Trust, and Behavioral Intentions in Electronic Commerce: A Comparison of American and Taiwanese Perceptions

Chang Liu, Northern Illinois University, USA

Jack T. Marchewka, Northern Illinois University, USA

Catherine Ku, National Dong Hwa University, Taiwan

ABSTRACT

Although the Internet has opened the door for businesses and consumers around the globe, many issues and challenges remain in order to take full advantage of this unprecedented opportunity. More specifically, businesses engaged in electronic commerce can take advantage of current technologies that allow them to collect vast amounts of information about their customers. Although many companies use this information to tailor products and services to improve their relationship with their customers, many people simply do not trust most organizations that require the exchange of personal information. Recent studies suggest that many people either decline to provide this information over the Internet or simply provide false information. As electronic commerce companies extend their reach globally, the issues concerning privacy and trust must be extended beyond a single national culture. This study proposes and tests a Privacy-Trust-Behavioral Intention Model from a global perspective. More specifically, this study compares American and Taiwanese perceptions concerning online privacy and how it relates to the level of trust with a company's electronic commerce Web site. In turn, the model suggests that trust is an important intermediary variable that influences behavioral intentions for online transactions.

INTRODUCTION

The Internet has opened the door for global e-commerce. Although the United States has the lion's share of Internet users, it is estimated that Asia will have more than 188 million Internet users in 2004. In particular, China is expected to have 51 million users, while India and Indonesia are expected to have 10 million and 5.1 million users, respectively (CIO Metrics, 2001). Moreover, Latin American e-commerce revenues are expected to grow from $3.6B in 2000 to $67B in 2004 (Direct Marketing, 2001). There is no doubt that the number of potential global buyers is dramatically increasing.

E-commerce provides a considerable amount of product-related information to customers so they can make better-informed purchasing decisions. On the other hand, a vast amount of personal information about customers is being created and used by many companies. Customer information can be collected both explicitly through registration forms, order forms, online contests, and/or survey forms; and implicitly by using tracking software and/or cookies that allow businesses to follow customers' online activities and gather information about their personal interests and preferences. This data has become extremely valuable to online companies, because it not only enables them to sell products and services that are tailored to customers' demands, but it also provides an opportunity to boost their revenues by selling advertising space on their Web sites at a premium (Gilbert, 1999). This premium on advertising space may result from an opportunity to use customers' personal information to help advertisers better target consumers.

Many e-commerce customers, however, are concerned about privacy issues regarding the personal information that is being collected (Alderman & Kennedy, 2000; Whiting, 2000). In fact, a US study by Hoffman and Novak (1999) revealed that almost 95% of the Web users surveyed declined to provide personal information over the Internet. Moreover, 40% responded that they tend to make up personal information when online. It appears that many customers simply do not trust most Web sites enough to engage in "relationship exchanges" that involve the exchange of personal information.

Internet-based e-commerce activities on a global scale compress time and space and permit the duplication and sharing of scarce corporate resources. In recognition of the international marketplace afforded by e-commerce, organizations have begun to produce a number of products that assist in customizing sites for use outside of their native language audience. Moreover, Forrester Research predicts a growth rate of 50% a year for those companies who provide multilingual versions and other interpretation tools (Engler, 1999).

Customized language alone, however, is not enough. Web site builders also must consider political, cultural, social, and legal differences. For example, China has comparatively more lenient rules regarding the use of customer data (D'Amico, 2001). Yet in other countries, laws may be broken if business firms transfer customer data internationally. For example, the European Union (EU) Directive on the protection of personal data and on the free movement of such data prohibits the transfer of personal data to non-EU nations that do not meet the adequate protection standard set forth in European data privacy laws (Goldstein, Roth, &Young, 2001).

Culture is defined as a collective programming of the mind that distinguishes a group or category of people from another (Hofstede, 1991). Culture provides people with a sense of identity and an understanding of acceptable behavior within society. This includes values, shared beliefs, or group norms that have been internalized by individuals with some modification (Hofstede, 1980). In addition, culture influences people's perceptions and their interpretation of the world. It influences their expectations, values, beliefs, attitudes, and ultimately their behaviors in everyday life (Adler, 1991). Groups of people have unique backgrounds because of their shared history, economy, geography, religion, and demographics. Therefore, it is important for global e-commerce organizations to understand how culture may influence the Internet-based electronic commerce activities on the Internet.

Hofstede (1980) proposed a model of national culture. This model has provided a theoretical foundation in a number of information systems studies that focused on cultural differences (Early, 1993; Garfield & Watson, 1997; Katz & Townsend, 2000; Kim & Petersen, 2003). Hofstede's model includes four dimensions: (1) Individualism – Collectivism, (2) Power – Distance, (3) Masculinity – Femininity, and (4) Uncertainty Avoidance.

Individualism – Collectivism is the degree to which people in a country prefer to act on their own or as members of groups (Hofstede, 1980). More specifically, individualism is the degree to which an individual pursues self-interests and individual expression. Power – Distance, on the other hand, focuses on the extent to which members of a particular group accept unequal power distributions or specific superior-subordinate roles. Masculinity – Femininity concentrates on the degree to which individuals compete, value achievement, and resolve conflict. Lastly, Uncertainty Avoidance focuses on the degree to which individuals accept uncertainty of future events and the acceptance of rules, measures, or guidelines to reduce the anxiety or threat of uncertainty.

The US is traditionally recognized as having a strong culture based on individualism that stresses independence of each individual within a society. On the other hand, Asian culture (e.g., Taiwan) focuses more on collectivism, emphasizing the goals of the group over personal goals, stressing conformity and in-group harmony, and defining the self in relation to the group (Lu, Rose, & Blodgett, 1999; Triandis, 1995).

A study conducted by Kim and Peterson (2003) used Hofstede's cultural dimensions to compare the perceived importance of IS development strategies by developers from the US and Korea. Kim and Peterson characterized the US culture as being low in uncertainty avoidance and high in masculinity, while characterizing the Korean culture as being more risk averse and less competitive and aggressive in pursuing personal goals.

It follows, then, that individuals from the US and Taiwan may also differ in terms of their perceptions, beliefs, and use of e-commerce. For example, in order to mitigate uncertainty avoidance, individuals from Taiwan may require more assurances in terms of privacy and security to increase their level of trust when making an online purchase than individuals from the US. However, no studies that focus on privacy and levels of trust in e-commerce activities across cultural boundaries currently exist.

Therefore, the main research questions addressed in this study are as follows:

- How does ensuring the privacy of personal information influence an individual's level of trust?
- What role does trust play in influencing a customer's behavior intentions? This includes making an online purchase or visiting a Web site again.
- Does culture play a role in the privacy-trust-behavioral intention relationship?

This study proposes and tests a theoretical model from a global perspective. More specifically, this study compares American and Taiwanese perceptions concerning online privacy and how it relates to their level of trust with a company's electronic commerce Web site. In turn, the model suggests that trust is an important intermediary variable that influences behavioral intention for online transactions.

The results of this study should be of interest to both academics and business practitioners. For academics, a theoretical model for the relatively new and growing area of global electronic commerce is proposed and tested. For practitioners, the results can provide important guidelines for developing an electronic commerce Web site that will improve customer service and that supports the nurturing of critical relationships with customers across cultural boundaries. Hopefully, this will allow an organization conducting electronic commerce to improve their competitive position.

A PRIVACY-TRUST-BEHAVIORAL INTENTION MODEL

Concerns about privacy are not new. Privacy has long been defined as the right of an individual to be left alone and to be able to control the flow of personal information (Warren & Brandeis, 1890). However, privacy concerns often arise when new information technologies with enhanced capabilities for collection, storage, use, and communication of personal information come into play (Clark, 1988; Culnan, 1993; Mason, 1986; Milberg, Burke, Smith, & Kallman, 1995; Webster, 1998).

As business marketers place greater emphasis on building long-term relationships with their customers, trust has assumed a central role (Viega, Kohna, & Potter, 2001). Subsequently, a successful relationship between buyers and sellers requires a level of trust in order to create and sustain long-term relationships. Business enterprises forthcoming in their information collection practices may have stronger customer relationships. Customers, in turn, who are more willing to provide personal information enable businesses to advance the customer relationship through improved offerings and targeted communications (Milne & Boze, 1999). Therefore, it would make sense that addressing an individuals' privacy concerns or at least influencing their perceptions is an antecedent for gaining a potential electronic commerce customer's trust.

The theory of reasoned action (Albarracin, Johnson, Fishbein, & Muellerleile, 2001; Fishbein & Ajzen, 1975; Madden, Ellen, & Ajzen, 1992;) has been used extensively as a theoretical basis for predicting behavioral intentions and/or behaviors. The theory of reasoned action contends that behavioral intentions are antecedents to specific behaviors that an individual will perform. More specifically, an individual's attitudes, perceptions, and subjective norms will influence individuals' actions in a particular manner when they believe that their behavior will result in or be linked to a specific outcome. Furthermore, subjective norms that influence behavioral intentions can be determined by an individual's positive or negative evaluation of a certain behavior or by social pressures to perform or not perform a particular behavior. The relative strength of these beliefs may vary from person to person based upon their experiences (Ajzen & Fishbein, 1980) or even possibly their culture. Subsequently, a customer's perceptions and attitudes regarding privacy and trust should influence their behavioral intentions in a global electronic commerce environment.

Several studies address privacy and trust in the electronic commerce field (Cheung & Lee, 2000; Garbarino & Johnson, 1999); however, no studies include privacy as a major antecedent for trust. In addition, there is a lack of empirical research that examines the relationship among the privacy, trust, and behavioral intentions constructs from a global perspective. Therefore, the model in Figure 1 attempts to explain how privacy influences trust and how trust, in turn, influences consumer behavioral intention for online transactions.

Although the model in Figure 1 is simple, it is not simplistic. The relationship between privacy and trust and, in turn, trust and behavioral intentions is a complex relationship that requires careful study. Due to the complex nature of human behavior and various interactions, a model that attempts to explain everything is not feasible. Therefore, by focusing on a small component of this complex relationship, academics and practitioners may begin to explore and understand a complex relationship between consumer and firms conducting business online.

In addition, an important question posed by this study focuses on how culture may influence an individual's behavioral intention to purchase online. By comparing American and Taiwanese perceptions, a cultural perspective can be introduced and compared indirectly using the proposed model.

RESEARCH HYPOTHESES

The Federal Trade Commission (FTC) has proposed and advocated that fair information practices include four dimensions (FTC Congress Report, 2000):

- *Notice* – providing people notice that personal information is being collected prior to the collection of that information.
- *Access* – providing people with access to the data that is collected about them.
- *Choice* – providing people with a choice to allow an organization to use or share information collected about them.
- *Security* – providing reasonable assurance that personal information is kept secure.

The FTC privacy dimensions provide both a practical and theoretical foundation for the privacy construct of the model. The FTC promotes adherence to these principles to insure effective privacy protection. However, a recent study conducted by Liu and Arnett (2002) suggests that just slightly more than 50% of large business Web sites provide privacy policies or appropriate links to policies on their home pages. In addition, comprehensive privacy policies that address the four privacy dimensions (notice, choice, access, and security) are less common.

Trust has many shades of meaning in the social sciences and in common business usages (Houston, 2001). It is a complex social phenomenon that reflects technological, behavioral, social, psychological, and organizational aspects of interactions among various human and non-human agents. For example, trust is required when one party becomes vulnerable to the actions of another when that party expects the other to perform a particular action without the ability to monitor or control the other party (Mayer, Davis, & Shoorman, 1995). On the other hand, trust may be part of a person's personality or nature (Bowlby, 1982) that becomes forged by his or her experiences or simply one's faith in humanity (Ricer, 1971). Trust may also be associated with an individual's belief in an organization based upon the organization's norms, regulations, policies, and procedures (Lewis & Weigert, 1985; Sarker, Valacich, & Sarker, 2003; Scott, 1992; Shapiro, 1987).

All business transactions require an element of trust, especially those conducted in the uncertain environment of electronic commerce (Lee, 1998). In this context, trust becomes especially important for customers to engage in a relationship exchange with a company (Garbarino & Johnson, 1999). In electronic commerce, trust can be viewed as a perceptual belief or level of confidence that someone respects the intentions, actions, and integrity of another party during an online transaction (Jarvenpaa, Knoll, & Leidner, 1998). Therefore, privacy may be an important antecedent to trust because a customer first must believe that an online transaction in electronic commerce will occur in a manner consistent with their expectations.

The model in Figure 1 also suggests that the level or degree of trust an individual has will influence his or her behavioral intentions. In addition to behavioral intentions viewed under the theory of reasoned actions, Mehta (1994) presents a research model that suggests behavioral intention (i.e., buying interest or buying intention) is a suitable, dependable variable that can be used to measure the effectiveness of advertising strategy. Moreover, customer behavioral intention is an important dimension compared

Figure 1. Privacy - trust - behavioral intention model

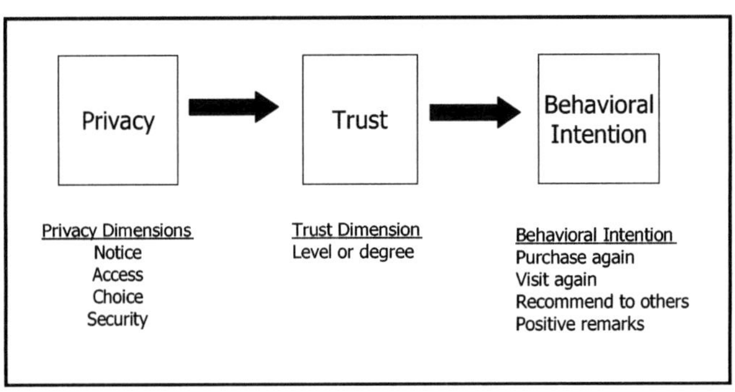

with measuring customer satisfaction or customer attitude as the dependent variables. As Zeuthaml, Berry, and Parasuraman (1996) suggest, customer behavioral intention variables are the best indication of customer satisfaction and business service quality and are defined as follows:

- Repeat visits to the Web site
- Recommendation of Web site to others
- Positive remarks or comments about the Web site
- Repeat purchases

It is not known for sure at this point whether people with cultural differences have the same needs and expectations for privacy and trust when making online purchases. Therefore, the following hypotheses are written in the form of null hypotheses to test the proposed model.

H1.0 There is a positive relationship between privacy and the degree or level of trust an individual has when making an online transaction, regardless of different cultural backgrounds.

H1.1 Notification of how a business will use personal information will be positively related to its overall degree of trust, regardless of different cultural backgrounds.

H1.2 Allowing access to personal information collected about an individual will be positively related to the overall degree of trust, regardless of different cultural backgrounds.

H1.3 Allowing an individual a choice to opt in or opt out will be positively related to the individual's overall degree of trust, regardless of different cultural backgrounds.

H1.4 Making reasonable attempts to keep data secure will be positively related to the overall degree of trust, regardless of different cultural backgrounds.

H2.0 There is a positive relationship between the level or degree of trust an individual has with an online business and the individual's behavioral intentions, regardless of different cultural backgrounds.

H2.1 The degree or level of trust an individual has will be positively related to whether the individual makes positive comments about the online business's Web site, regardless of different cultural backgrounds.

H2.2 The degree or level of trust an individual has will be positively related to whether the individual will visit the online business's Web site again, regardless of different cultural backgrounds.

H2.3 The degree or level of trust an individual has will be positively related to whether the individual will recommend the online business's Web site to others, regardless of different cultural backgrounds.

H2.4 The degree or level of trust an individual has will be positively related to whether the individual will make an online purchase again with the online business, regardless of different cultural backgrounds.

RESEARCH METHODOLOGY

This study employed a laboratory experiment as a research strategy. The goal was to achieve a high degree of internal validity in order to test the model and reduce the influence of confounding variables. Research subjects included undergraduate and graduate students from a large Midwestern university in the US and undergraduate and graduate students from a large university in Taiwan. Over 500 subjects participated in the experiment; however, only 212 data points were usable in the US sample, and 224 data points were usable in the Taiwan sample. Although the use of students is largely criticized in academic research, the use of students as subjects not only may be acceptable, but appropriate when studying certain patterns of relationships (Dickson, 1989).

Experimental Site Design and Treatment Groups

Subjects for the experiment were recruited from several junior- or senior-level undergraduate Information System courses in the US and Taiwan, and a Masters of Science in MIS program in the US. Subjects from both the US and Taiwan were randomly assigned to one of two treatment groups—High Privacy or Low Privacy—and then asked to browse the assigned Husky Virtual Bookstore (HVBS) site, view textbook requirements and descriptions, and then purchase one or more books online. The sites were written in English. However, subjects from Taiwan received copies of Chinese language interpretation for each page they visited.

Two Web sites were developed for this research study. One Web site was created to include the four privacy dimensions—notice, access, choice, and security—proposed by the FTC and depicted in the proposed model in Figure 1. Subjects assigned to this Web site were considered to be part of the High Privacy treatment group. On the other hand, the other Web site did not include any of these dimensions, and subjects assigned to this Web site were considered the Low Privacy treatment group.

The content of both Web sites was a fictitious e-commerce bookstore (the HVBS), that sells a variety of textbooks online. The first page for each storefront site included a consent form that confirmed that the subject voluntarily agreed to participate in the study. By agreeing to participate in the study, the subject would visit an experimental e-commerce Web site and then complete a questionnaire as part of the study. Keeping within the researchers' university guidelines on studies that involve human subjects, the subject was assured that any information gathered about the subject would remain confidential and that participation was voluntary and may be withdrawn without penalty or prejudice. A radio button gave the subject the choice to either agree or not agree to participate in the study. If the subject agreed and clicked the submit button, the subject then could continue with the study.

Next, subjects in both treatment groups were presented with an overview of the three types of information that could be collected by a Web site. The intent was to ensure that each subject had an understanding of how various types of information could be collected about them. This included the following:

- *Personally identifiable information* – Information collected through online forms and includes such personal information as name, address, phone number, income, gender, age, and so forth. A person is no longer anonymous once he or she provides this information.
- *Anonymous Information* – Information that cannot be tied specifically to a person. However, information about IP address, domain names, type of browser, hardware platform, and the address of the last Web site visited can be determined easily by the host server.
- *Passive information* – Usually in the form of a cookie or a small text file that may be sent from a server to the subject's computer. A cookie contains a unique number so the subject's computer can be identified; however, cookies cannot search for information on a subject's computer, transmit information to anyone else, or erase important files on a computer.

After reading this information, the subject clicked a button to begin the experiment. Subjects from both the High Privacy and Low Privacy treatment groups followed the same process. This included browsing the HVBS site, viewing textbook requirements and descriptions, and purchasing one or more books with a fake credit card number provided by the researchers.

The following provides a brief explanation of the site that included the four privacy dimensions – notice, choice, access, and security.

(1) *HVBS Privacy Statement* – Provided a detailed explanation of the information that would be collected and how it would be used by the HVBS. The intent of this page was to satisfy and support the notice dimension.

(2) *Main Page With Privacy Sea* – Contained general information about the HVBS. This page also included a prominent privacy seal that provided a means of telling the high privacy subject treatment group that certain standards were in place in order to increase the subjects' trust perceptions. In addition, a scrolling message was displayed at the bottom of the screen that assured the subject that the HVBS had taken reasonable measures to ensure data security and privacy protection.

(3) *Textbook Category Page.* – Contained available textbooks that included database applications, project management, operating systems, and system analysis and design. When a subject visited this page, a pop-up window informed the subject that an electronic cookie was being downloaded to the subject's computer. The intent of the pop-up window was to satisfy and support the notice dimension.

(4) *Textbook Description Page* – Contained information specifically related to the book selected by the subject and included the book's description and price. After viewing this information, the subject could then add the book to an electronic shopping cart. In addition, a scrolling message at the bottom of the screen explained and enforced the perception of how security was ensured at the HVBS site. The intent was to satisfy and support the security element of the privacy principles.

(5) *Shopping Cart Page* – Contained a list of books that were selected by the subject for purchase. The subject could alter the quantity or cancel the order and return to make another selection.
(6) *Order Registration Page* – Collected personal information such as name, address, telephone number, e-mail, and so forth.
(7) *Verification Page* – Used by the HVBS site to allow subjects to view and update their personal information. Because the site was integrated with a database management system, any changes to personal information would be updated immediately. The intention of this page was to satisfy and support the access dimension.
(8) *Order Processing Page* – Supported the subject's perception that the HVBS site was secure. The subject was able to view a message that the HVBS Web site uses Secure Socket Layer technology (a firewall) and stringent auditing procedures.
(9) *Choice Page* – Before subjects left the HVBS Web site, they were given the opportunity to allow the HVBS to share the personal information they provided with external secondary sources. By allowing the subject to opt out, the choice dimension was supported.

Online Questionnaire

After making an online purchase, each subject was asked to complete an online questionnaire. The data were stored in a relational database. The entire process took about 15 minutes to complete. Perceptions were measured using a Likert scale ranging from 1 to 7 with 1 for strongly disagree and 7 for strongly agree. Again, Taiwanese subjects received Chinese language interpretation for the survey questionnaire. A copy of the questionnaire can be found in Appendix 1.

DATA ANALYSIS AND FINDINGS

A total of 514 subjects participated in the experiment and answered the questionnaire survey. Of these responses, 16 and 29 were rejected from the US and Taiwanese samples, respectively. These responses were unusable because the subjects browsed both experimental sites and answered both surveys. In order to have an equal number of participants in the two treatment groups for both samples, an additional 35 surveys were randomly selected and dropped. Therefore, the final data analysis included 436 data points. This included 212 data points from the US sample, and 224 data points from the Taiwanese sample. Subjects were equally divided into the High Privacy and Low Privacy treatment groups for both samples. Table 1 presents the demographics of the subjects who participated in the study.

Reliability of the Constructs

Prior to conducting this study, a pretest of the experimental design and survey questionnaire was conducted using over 200 US subjects. The intent was to test and hone the two Web sites and to ensure the reliability and validity of the questionnaire. Several minor modifications were made, and the current study was conducted. None of the original pretest data is included in this study.

Table 1. Characteristics of respondents

		US Number (N=212)	US Percentage	Taiwan Number (N=224)	Taiwan Percentage
1.	Gender				
	Male	112	52.83%	122	54.46%
	Female	100	47.17%	102	45.54%
		212	100%	224	100%
2.	Age Group				
	17-20	50	23.58%	158	70.54%
	21-25	145	68.40%	62	27.68%
	26-30	13	6.13%	3	1.34%
	31-35	1	0.47%	0	0.00%
	36-40	1	0.47%	0	0.00%
	41-45	2	0.94%	0	0.00%
	Greater than 45	0	0.00%	1	0.44%
		212	100%	224	100%
3.	Web Experience				
	Never used before	4	1.89%	13	5.80%
	A few times before	3	1.42%	11	4.91%
	A few times/month	5	2.36%	5	2.23%
	Every week	25	11.79%	28	12.50%
	Almost every week	175	82.55%	167	74.55%
		212	100%	224	100%

In addition, a pretest was conducted to review the site design and questionnaire for validity (measuring what is intended), completeness (including all relevant variable items), and readability/understandability for the Taiwanese research subjects. This included approximately 100 subjects, and none of these data points is included in this study. In addition, the authors examined language translation to ensure that the interpretation of both sites and the questionnaire had the same linguistic interpretations for all subjects.

In order to ensure that variables comprising each proposed research construct (privacy, trust, and behavioral intention) were internally consistent, a reliability assessment was carried out using Cronbach's alpha for both the US and Taiwan data that was used in this study. A low value of Cronbach's alpha (i.e., close to 0) implies that the variables are not internally related in the manner expected (Churchill, 1979). Table 2 presents the research constructs, measurement, and reliability assessment. As one can see in Table 2, the reliability coefficients for each of the research constructs in this study are all above the 0.70 level, and therefore, the reliability of three research constructs is supported for both US and Taiwanese samples.

Summary of Results

Table 3 summarizes the mean values, standard deviations, and t test results for the High and Low Privacy treatment groups, as well as the US and Taiwanese samples combined.

T-Tests

The t-test statistical technique was applied as a preliminary analysis for viewing differences concerning privacy, trust, and behavioral intentions. A t-test is especially useful for testing differences between two groups. As can be seen in Table 3, the p values given by $Prob > |t|$ is so small (< .0001), we may conclude that there are statistically

Table 2. Research constructs, measurements, and reliability assessment

Research Constructs	Measure Components	Cronbach's alpha		
		US (N=212)	Taiwan (N=224)	Total (N=436)
Privacy	notice, access, choice, security	0.78	0.89	0.85
Trust	level on notice_trust, level on access_trust, level on choice_trust, level on security_trust	0.93	0.95	0.94
Behavioral Intentions	purchase, re-visit, recommend positive comment	0.92	0.95	0.94

significant differences between the two treatment groups for each US and Taiwan sample, as well as the combined US and Taiwanese samples. The t-tests show not only strength, but also the direction (positive or negative) of the differences. The results strongly suggest that there is a difference in perceptions concerning privacy, trust, and behavioral intentions for the High Privacy and Low Privacy treatment groups for the US sample, the Taiwan sample, and even when the samples are combined. Subsequently, we may conclude that there are perceptual differences between the Low Privacy and High Privacy treatment groups. This suggests that two different Web sites used in this study provide the desired perceptions for the two treatment groups. In addition, there appears to be strong support for the research model in Figure 1 and hypotheses concerning privacy, trust, and behavior intention. However, there appears to be no support for cultural differences concerning privacy, trust, and behavioral intention between the US and Taiwanese subjects.

MANOVA

In addition to the t-tests, a MANOVA statistical technique was applied. MANOVA is a dependence technique that measures the differences for two or more metric variables based on a set of qualitative independent variables. It is useful as a multivariate procedure to access group differences across multiple metric-dependent variables simultaneously. Here, the MANOVA test was conducted for two samples from the US and Taiwan for the High Privacy and Low Privacy treatment groups. Figures 2, 3, and 4 summarize the F and p values for the MANOVA tests.

Table 3. Summary of descriptive statistics and t-test results

Research Constructs	US High Privacy Group (N=106)		US Low Privacy Group (N=106)		T Value	Taiwan High Privacy Group (N=112)		Taiwan Low Privacy Group (N=112)		T Value	US & Taiwan High Privacy Group (N=218)		US & Taiwan Low Privacy Group (N=218)		T Value
	Mean	Std	Mean	Std		Mean	Std	Mean	Std		Mean	Std	Mean	Std	
Privacy	**5.12**	**1.06**	**3.89**	**1.02**	**8.70***	**5.31**	**1.23**	**3.82**	**1.54**	**7.94***	**5.22**	**1.15**	**3.89**	**1.32**	**11.26***
(notice,	5.21	1.32	4.21	1.26	5.69 *	5.22	1.28	3.90	1.78	6.38 *	5.22	1.30	4.11	1.55	8.07 *
choice,	5.29	1.30	3.61	1.33	9.27 *	5.30	1.42	3.49	1.79	8.38 *	5.30	1.36	3.58	1.59	12.10 *
security,	5.21	1.23	3.96	1.33	7.09 *	5.34	1.28	3.97	1.81	6.54 *	5.23	1.26	3.99	1.58	9.43 *
access)	4.84	1.83	3.79	1.67	4.35	5.37	1.79	3.95	1.99	5.60 *	5.11	1.83	3.87	1.83	7.04 *
Trust	***4.76***	***1.26***	***3.77***	***1.06***	***5.85****	***5.02***	***1.25***	***3.72***	***1.71***	***6.48****	***4.90***	***1.26***	***3.79***	***1.45***	***8.29****
(level on															
notice_trust,	4.94	1.38	3.87	1.19	6.03 *	5.16	1.56	3.91	1.89	5.62 *	5.05	1.39	3.94	1.60	7.75 *
access_trust	4.62	1.57	3.85	1.29	3.87 *	5.12	1.55	3.86	2.03	5.21 *	4.87	1.58	3.91	1.72	6.12 *
choice_trust,	4.64	1.60	3.64	1.30	4.96 *	4.67	1.63	3.41	1.91	5.30 *	4.65	1.61	3.60	1.66	6.91 *
security_trust	4.89	1.52	3.63	1.33	5.34 *	5.13	1.45	3.64	1.83	6.72 *	5.01	1.43	3.82	1.61	8.19 *
Behavioral Intentions	***5.09***	***1.26***	***4.08***	***1.23***	***5.58****	***4.92***	***1.36***	***3.89***	***1.86***	***4.92****	***5.00***	***1.31***	***4.02***	***1.58***	***7.06****
(re-visit,	5.30	1.43	4.39	1.53	4.45 *	5.14	1.60	4.04	2.02	4.53 *	5.22	1.52	4.24	1.81	6.11 *
recommend,	5.16	1.27	4.28	1.36	4.83 *	5.02	1.55	3.97	2.00	4.36 *	5.09	1.42	4.16	1.74	6.09 *
purchase,	5.07	1.54	3.90	1.52	5.11 *	4.97	1.49	3.78	2.05	4.98 *	5.02	1.51	3.90	1.82	7.01 *
comment)	5.23	1.28	3.53	1.53	5.59 *	5.10	1.60	4.02	1.99	4.48 *	4.85	1.52	3.65	1.69	6.09 *

* *p<0.0001*

As illustrated in Figure 2, the analysis reveals a significant multivariate effect on privacy concerns between the High Privacy and Low Privacy treatment groups (F=126.74, p<0.0001), but the privacy concerns are not significant when we compare the US and Taiwanese samples (F=0.47, p>0.4913). Similar results also were found on the trust and behavioral intention constructs. As can be seen in Figure 3, there is a significant

Figure 2. Privacy concerns test based on regions and types

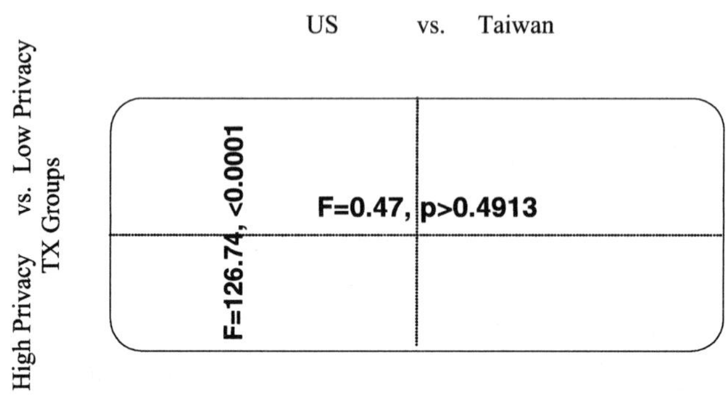

Figure 3. Level of trust test based on regions and types

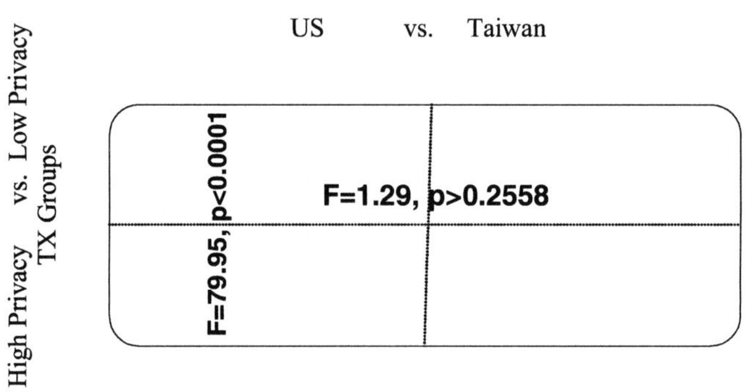

Figure 4. Behavioral intentions test based on regions and types

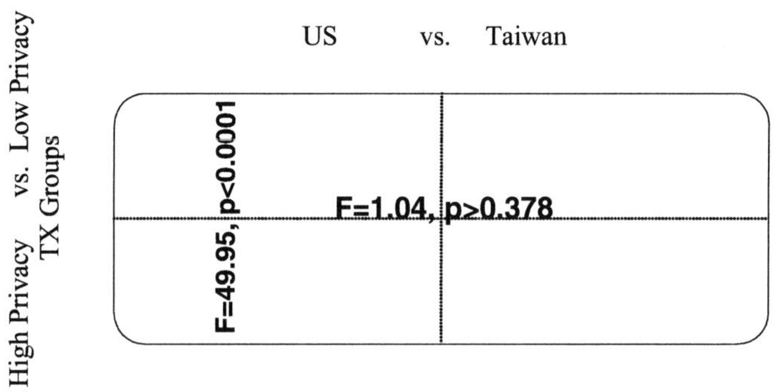

multivariate effect on level of trust for the High Privacy and Low Privacy treatment groups (F=79.95, p<0.0001), but no real significance when we compare the US and Taiwanese samples (F=1.29, p>0.2558). Lastly, Figure 4 illustrates that there is a significant multivariate effect on behavioral intentions when we compare the High Privacy and Low Privacy treatment groups (F=49.95, p<0.0001), but no real significant difference exists between the US and Taiwanese samples (F=1.04, p>0.378). Once again, there appears to be no support to indicate any cultural differences between the US and Taiwanese subjects.

Correlation Analysis

Mean values and a matrix of intercorrelations among the research constructs were also calculated based on the 228 responses of the High Privacy treatment groups (106 from the US and 122 from Taiwan). Dimensions were created by taking the average response for trust with notice, trust with choice, trust with security, and trust with access to develop an overall value for trust. In addition, an average for the purchase, revisit, recommendation, and making of positive comments about the site variables was created as an overall dimension for behavioral intention. If the overall trust value was correlated positively and significantly with the four privacy dimensions (notice, choice, security, and access), then we can conclude that there is a positive relationship between privacy and the degree or level of trust an individual has when making an online transaction. Also, if the behavioral intentions value correlated positively and significantly with the overall trust, then we can conclude that there is a positive relationship between the level or degree of trust an individual has with an online business and the individual's behavioral intentions. The means, standard deviations, and matrix of intercorrelations are presented in Table 4 and Table 5.

As can be seen in Table 4, there is a strong relationship between the level of trust and the notification variable concerning how a business partner will use personal information (r=0.65, p<.0001). A strong relationship also exists between the level of trust and allowing an individual to access his or her personal information (r=0.43, p<.0001). A similar conclusion also can be drawn between the level of trust and an individual's choice

Table 4. Matrix of intercorrelations among privacy and trust constructs (N=218)

Construct	Mean	St.D.	1	2	3	4	5	6
1. Level of Trust	4.90	1.26	1.00 (0.0)					
2. Notice	5.22	1.30	0.65 (.0001)	1.00 (0.0)				
3. Choice	5.30	1.36	0.69 (.0001)	0.55 (.0001)	1.00 (0.0)			
4. Security	5.28	1.26	0.72 (.0001)	0.58 (.0001)	0.72 (.0001)	1.00 (0.0)		
5. Access	5.11	1.83	0.43 (.0001)	0.38 (.0001)	0.47 (.0001)	0.52 (.0001)	1.00 (0.0)	
6. Overall Privacy	5.22	1.15	0.75 (.0001)	0.75 (.0001)	0.78 (.0001)	0.86 (.0001)	0.78 (.0001)	1.00 (0.0)

* *Note: (1) p values are in the (). (2) The measurement scale of mean values is from 1 (completely disagree) to 7 (completely agree).*

Table 5. Matrix of intercorrelations among trust and behavioral intention Constructs (N=218)

Construct	Mean	St.D.	1	2	3	4	5	6
1. Level of Trust	4.90	1.26	1.00 (0.0)					
2. Recommend	5.09	1.42	0.70 (.0001)	1.00 (0.0)				
3. Purchase	5.02	1.51	0.64 (.0001)	0.79 (.0001)	1.00 (0.0)			
4. Visit again	5.22	1.52	0.61 (.0001)	0.76 (.0001)	0.74 (.0001)	1.00 (0.0)		
5. Say positive	4.85	1.28	0.76 (0.001)	0.80 (0.001)	0.77 (0.001)	0.76 (.0001)	1.00 (0.0)	
6. Overall Intention	5.09	1.52	0.75 (.0001)	0.90 (.0001)	0.90 (.0001)	0.88 (.0001)	0.89 (.0001)	1.00 (0.0)

* *Note: (1) p values are in the (). (2) The measurement scale of mean values is from 1 (completely disagree) to 7 (completely agree)*

to opt out or opt in ($r=0.69$, $p<.0001$). In addition, there is a strong relationship between the level of trust and reasonable attempts to keep data secure ($r=0.72$, $p<.0001$). Overall, there is a positive relationship between privacy and the degree or level of trust an individual has when making an online transaction ($r=0.75$, $p<.0001$). These results suggest that the four privacy dimensions have an important positive and significant influence on an individual's degree of trust concerning an e-commerce Web site.

Table 5 provides the results for testing relationships between level of trust and behavioral intentions. As indicated in Table 5, there is a strong relationship between the level of trust and whether an individual would make an online purchase again ($r=0.64$, $p<.0001$). Moreover, a strong relationship also exists between the level of trust and whether an individual would visit the online business's Web site again ($r=0.61$, $p<.0001$). A similar conclusion can also be drawn between the level of trust and whether an individual would recommend the online business's Web site to others ($r=0.70$, $p<.0001$). In addition, there is a strong relationship between the level of trust and whether an individual would make positive comments about the online business's Web site ($r=0.76$, $p<.0001$). Overall, there is a positive relationship between degree or level of trust an individual has and his or her behavioral intentions ($r=0.75$, $p<.0001$). These results suggest that the degree of trust is an important intermediary variable for behavioral intention. More specifically, trust is positively related to whether an individual will purchase again, revisit a site, recommend the site to others, and make positive comments about the site.

CONCLUSIONS AND IMPLICATIONS

The results of this study suggest that strong support exists for the theoretical model. The implication is that the four dimensions of privacy have a strong influence on the level of trust an individual has with an e-commerce business. In turn, it appears that this level of trust will influence a customer's behavioral intentions in terms of whether they would purchase from or visit the site again, whether they will have positive things to say about the business, and whether they would recommend this e-commerce site to others.

However, the results of this study indicate that no significant differences were found between the US and Taiwan samples. It would appear that people from these two different areas of the globe have the same privacy and trust concerns when making a purchase online. It suggests that all privacy protection ingredients should be included when developing a corporate Web site to conduct e-commerce on a global scale.

One reason may be that Internet technology amplifies the power of individuals. For example, the Internet may create a new cultural environment for those who shop online. In this new environment, an individual's desires, aspirations, and perceptions of their rights may be the same regardless of their cultural background. Second, shopping online is an individual activity. An individual may attain an identity from such things as a user ID or a credit card number. As a result, there may be less of a connection with an affiliated organization or a social group. Third, the notion of a global village supported by Internet and telecommunications technology may blur many cultural differences in modern societies. Subsequently, there may be a trend where consumer shopping practices in the Asian culture are becoming more similar to those in the West. Therefore, more research is needed to further test the model proposed in this study and to understand more fully the effects of culture dimensions in a global marketplace.

As the potential number of global customers continues to grow, companies interested in e-commerce will face both opportunities and challenges. Companies can take advantage of these opportunities by tailoring their products and services and by customizing the online buying experience for those customers. Although multilingual versions of a Web site provide an important means for customization, the explicit and implicit collection of personal information provides another important way to distinguish their Web site from their competition.

The data collected about customers is a valuable asset to the company. The company may use this data to learn more about the online and purchasing habits of its customers. Or a company can sell this data to other third parties.

The challenges of global e-commerce include language barriers as well as political, social, cultural, and legal issues. In addition, many individuals are becoming increasingly concerned about the privacy of the personal information that e-commerce companies are collecting, storing, using, and selling. Subsequently, many customers have said that they either do not provide any personal information or that the information they do provide is false.

Both the challenge and opportunity is to better understand individual perceptions concerning the level of trust an individual has with an electronic commerce storefront and their behavioral intentions. Therefore, the findings of this study should be of interest to both academics and practitioners.

For academics, the privacy (trust) behavioral intention model presented and tested in this chapter provides strong support for this important relationship. Although many variables and relationships exist and act interdependently, developing and testing a model that attempts to explain why an individual will or will not purchase from an electronic commerce storefront would be too complex and unrealistic. However, the experiment as a research strategy provides an opportunity to hold constant or at least minimize the effect of many of these variables and relationships while allowing for the manipulation and study of only a few key components. While the model proposed in this chapter focuses only on a small segment, it does provide an important window for understanding a much larger and complex phenomenon. Future studies that expand or adapt the privacy-trust-behavior intention model can provide key insights and increase our understanding.

However, the findings of this study should be of particular interest to practitioners. More specifically, having a privacy policy may lead to more repeat visits and more purchases. It would also appear that a company can increase the level of trust and therefore influence a customer's behavioral intention by integrating the notice, access, choice, and security dimensions into the design of their e-commerce Web site. For example, an electronic commerce storefront could provide notices regarding how any personal information collected will be used and then give the customer access to that information in order know what data is being collected and the option to change it, if necessary. In addition, having security controls in place and reinforcing the idea that the company is taking precautions to protect that data may be reassuring to many individuals and thereby increase their level of trust with that company. By including the four privacy dimensions into electronic storefronts, e-commerce companies may find that customers are more willing to provide personal information. As a result, companies can better customize and tailor their products and services to their customers' needs while improving their competitive position in a global marketplace.

Although strong support for the privacy-trust-behavioral intention model exists, the authors acknowledge several limitations. Most importantly, the samples only focused on US and Taiwanese perceptions. A much more stringent and deeper understanding of culture and how it influences one's perceptions should be included. In addition, the US and Taiwan samples may have included students who are not American or Taiwanese, as no distinction was made to identify international students. However, this would have a minor impact on the results, because these students would be a minority when compared to the entire sample from each country.

The use of the experiment as a research strategy provides a higher degree of internal validity. Unfortunately, the experiment as a research strategy has several weaknesses and generally does not provide a high degree of external validity that would allow for drawing more general conclusions across different populations. However, every research strategy has its own inherent limitations, and future research should include a myriad of empirical research designs to extend the external validity of the model and to improve our understanding of the privacy, trust, and behavioral intention relationships.

REFERENCES

Adler, N.J. (1991). *International dimensions of organizational behavior* (2nd ed.). Boston: PWS-Kent.

Ajzen, I., & Fishbein, M. (1980). *Understanding attitudes and predicting social behavior*. Englewood Cliffs, NJ: Prentice-Hall.

Albarracin, D., Johnson, B.T., Fishbein, M., Muellerleile, P.A., (2001). Theories of reasoned action and planned behavior as models of condom use: A meta-analysis. *Psychological Bulletin, 127*(1), 142-161.

Alderman, E., & Kennedy, C. (2000). The Internet, consumers and privacy. Retrieved November 10, 2001 from *http://www.internetpolicy.org/briefing/alderman_kennedy.html*

Bowlby, J. (1982). *Attachment and loss*. New York: Basic Books.

Cheung, C., & Lee, M. (2000). Trust in Internet shopping: A proposed model and measure instrument. *Proceedings of the 2000 AMCIS Conference*, Long Beach, CA.

Churchill, G.A. Jr. (1979). A paradigm for developing better measures of marketing constructs. *Journal of Marketing Research, 16*(2), 64-73.

CIO Metrics. (2001). Latin American e-commerce to hit $67 billion. Retrieved January 31, 2001 from *http://www2.cio.com/forums/viewmetric_content.cfm?ID=all&METRIC=159*

Clark, R.A. (1988). Information technology and dataveillance. *Communications of the ACM, 31*(5), 498-512.

Culnan, M.J. (2000). Georgetown Internet Privacy Policy Survey. Retrieved May 20, 2000, from *www.msb.georgetown.edu/faculty/culnanm/gippshome.html*

D'Amico, E. (2001, September 26). Global e-commerce. *Chemical Week, 8*.

Dickson, G.W. (1989). A programmatic approach to information systems research: An experimentalist's view. In Benbaset (Ed.), *Information systems research challenge: Experimental research methods* (vol 2., pp. 147-170). Harvard Business School Research Colloquium.

Direct Marketing (2001). Study examines e-commerce/Internet use in Latin American. *63*(12), 9.

Early, P.C. (1993). East meets west meets mideast: Further explorations of collectivistic and individualistic work groups. *Academy of Management Journal, 36*(2), 319-348.

Engler, N. (1999). Global e-commerce: How products and services help sites expand worldwide. *Informationweek*. Retrieved May 20, 2000, from *http://www.informationweek.com/755/global.htm*

Fishbein, M., & Ajzen, I. (1975). *Belief, attitude, intention, and behavior: An introduction to theory and research*. Reading, MA: Addison-Wesley.

Garbarino, E., & Johnson, M.S. (1999). The different roles of satisfaction, trust, and commitment in customer relationships. *Journal of Marketing, 63*(2), 70-87.

Garfield, M.J., & Watson, R.T. (1997). Differences in national information infrastructures: The reflection of national cultures. *Journal of Strategic Information Systems, 6*(4), 313-337.

Gilbert, J. (1999). Research points to the growth of online ads. *Advertising Age, 70*(35), 6, 25.

Goldstein, H. R., Roth, A. E., & Young, T. L. (2001). International personal data safe harbor program launched. *Intellectual Property & Technology Law Journal, 13*(4), 24-25.

Hoffman, D.L., & Novak, T. (1999). Building consumer trust online. *Communications of the ACM, 42*(4), 80-85.

Hofstede, G. (1980). *Culture's consequences: International differences in work-related values*. Beverly Hills, CA: Sage.

Hofstede, G. (1991). *Cultures and organizations*. London: McGraw-Hill.

Jarvenppa, S.L., Knoll, K., & Leidner, D.E. (1998). Is anybody out there? Antecedents of trust in global virtual teams. *Journal of Management Information Systems, 14*(4), 29-64.

Katz, J.P., & Townsend, J.B. (2000). The role of information technology in the "fit" between culture, business strategy, and organizational structure of global firms. *Journal of Global Information Management, 8*(2), 24-35.

Kim, C.S., & Peterson, D.K. (2003). A comparison of the perceived importance of information systems development strategies by developers from the United States and Korea. *Information Resources Management Journal, 16*(2), 1-18.

Lee, H.G. (1998). Do electronic marketplaces lower the price of goods. *Communications of the ACM, 41*(1), 73-80.

Lewis, J.D., & Weigert, A. (1985). Trust as a social reality. *Social Forces, 63*(4), 967-985.

Liu, C., & Arnett, K.P. (2002). An examination of privacy policies in Fortune 500 Web sites, *Mid-American Journal of Business, 17*(1), 13-21.

Lu, L.C., Rose, G.M., & Blodgett, J.G. (1999). The effects of cultural dimensions on ethical decision making in marketing: an exploratory study. *Journal of Business Ethics, 18*(1), 91-105.

Madden, T.J., Ellen, P.S., & Ajzen, I. (1992). A comparison of the theory of planned behavior and the theory of reasoned action. *Personality and Social Psychology Bulletin, 18*(1), 3-9.

Mason, R.O. (1986). Four ethical issues of the information age. *MIS Quarterly, 10*(1), 4-12.

Mayer, R.C., Davis, J.H., & Schoorman, F.D. (1995). An integrative model of organizational trust. *Academy of Management Review, 20*(3), 709-734.

Mehta, A. (1994, May/June). How advertising response modeling (ARM) can increase AD effectiveness. *Journal of Advertising Research*, 62-74.

Milberg, S.J., Burke, S.J., Smith, H.J., & Kallman, E.A. (1995). Values, personal information privacy, and regulatory approaches. *Communications of the ACM, 38*(12), 65-84.

Milne, G.R., & Boze, M. (1999). Trust and concern in consumers' perceptions of marketing information management practices. *Journal of Interactive Marketing, 13*(1), 5-24.

Ricer, W.H. (1971). The nature of trust. In J.T. Tedeschi (Ed.), *Perspectives on social power* (pp. 63-81). Chicago: Aldine Publishing Company.

Sarker, S., Valacich, J.S., & Sarker, S. (2003). Virtual team trust: Instrument development and validation in an IS educational environment. *Information Resources Management Journal, 16*(2), 35-55.

Scott, W.R. (1992). *Organizations: Rational, natural, and open systems*, Englewood Cliffs, NJ: Prentice Hall.

Shapiro, S.P. (1987). The social control of impersonal trust. *American Journal of Sociology*, (93), 623-658.

Triandis, H.C. (1995). *Individualism and Collectivism*. Boulder, CO: Westview Press.

Viega, J., Kohno, T., & Potter, B. (2001). Trust (and mistrust) in secure applications. *Communications of the ACM, 44*(2), 31-36.

Warren, S.D., & Brandeis, L.D. (1890). The right to privacy. *Harvard Law Review, 4*(5), 193-220.

Webster, J. (1998). Desktop videoconferencing: Experiences of complete users, wary users, and non-users. *MIS Quarterly, 22*(3), 257-286.

Whiting, R. (2000). DoubleClick gets double trouble with database plan. *Informationweek*, 776, 24.

Zeuthaml, V.A., Berry, L.B., & Parasuraman, A. (1996). The behavioral consequences of service quality. *Journal of Marketing, 60*(4), 31-46.

APPENDIX 1: ONLINE QUESTIONNAIRE

I was informed about what information the company would collect about me.
Strongly Disagree: ° 1 ° 2 ° 3 ° 4 ° 5 ° 6 ° 7 Strongly Agree

The Husky Virtual Bookstore explained why they were collecting personal information.
Strongly Disagree: ° 1 ° 2 ° 3 ° 4 ° 5 ° 6 ° 7 Strongly Agree

The Husky Visual Bookstore explained how they would use the information collected about me.
Strongly Disagree: ° 1 ° 2 ° 3 ° 4 ° 5 ° 6 ° 7 Strongly Agree

I feel that the Husky Virtual Bookstore is making an effort to keep my personal information and credit card information out of the hands of unauthorized individuals.
Strongly Disagree: ° 1 ° 2 ° 3 ° 4 ° 5 ° 6 ° 7 Strongly Agree

I feel that the Husky Virtual Bookstore will not release personal information about me without my express permission.
Strongly Disagree: ° 1 ° 2 ° 3 ° 4 ° 5 ° 6 ° 7 Strongly Agree

I feel that the Husky Virtual Bookstore made a reasonable effort to ensure that the information collected about me was accurate.
Strongly Disagree: ° 1 ° 2 ° 3 ° 4 ° 5 ° 6 ° 7 Strongly Agree

The Husky Virtual Bookstore gave me a clear choice before disclosing personal information about me to third parties.
Strongly Disagree: ° 1 ° 2 ° 3 ° 4 ° 5 ° 6 ° 7 Strongly Agree

The Husky Virtual Bookstore has a mechanism to review and change incorrect personal information.
Strongly Disagree: ° 1 ° 2 ° 3 ° 4 ° 5 ° 6 ° 7 Strongly Agree

The Husky Virtual Bookstore's policy on how it would use any personal information about me makes me feel that the company is trustworthy.
Strongly Disagree: ° 1 ° 2 ° 3 ° 4 ° 5 ° 6 ° 7 Strongly Agree

The Husky Virtual Bookstore's policy with respect to how they will share my personal information with third parties makes me feel the company is trustworthy.
Strongly Disagree: ° 1 ° 2 ° 3 ° 4 ° 5 ° 6 ° 7 Strongly Agree

The ability to access my personal information to ensure that it is accurate and complete makes me feel that Husky Virtual Bookstore is trustworthy.
Strongly Disagree: ° 1 ° 2 ° 3 ° 4 ° 5 ° 6 ° 7 Strongly Agree

The Husky Virtual Bookstore's security policy makes me feel that the company is trustworthy.
Strongly Disagree: ° 1 ° 2 ° 3 ° 4 ° 5 ° 6 ° 7 Strongly Agree

The Husky Virtual Bookstore's level of encryption and other security measures make me feel that the company is trustworthy.

Strongly Disagree: ° 1 ° 2 ° 3 ° 4 ° 5 ° 6 ° 7 Strongly Agree

The Husky Virtual Bookstore's privacy policy concerning the notice of personal information collection makes me feel this company is trustworthy.

Strongly Disagree: ° 1 ° 2 ° 3 ° 4 ° 5 ° 6 ° 7 Strongly Agree

After visiting the Husky Virtual Bookstore, I would be willing to provide my personal information to this site.

Strongly Disagree: ° 1 ° 2 ° 3 ° 4 ° 5 ° 6 ° 7 Strongly Agree

I would be willing to recommend the Husky Virtual Bookstore site to others interested in buying textbooks.

Strongly Disagree: ° 1 ° 2 ° 3 ° 4 ° 5 ° 6 ° 7 Strongly Agree

I would be willing to purchase from the Husky Virtual Bookstore again if I needed additional books.

Strongly Disagree: ° 1 ° 2 ° 3 ° 4 ° 5 ° 6 ° 7 Strongly Agree

I would be willing to visit the Husky Virtual Bookstore again.

Strongly Disagree: ° 1 ° 2 ° 3 ° 4 ° 5 ° 6 ° 7 Strongly Agree

I have positive things to say about the Husky Virtual Bookstore.

Strongly Disagree: ° 1 ° 2 ° 3 ° 4 ° 5 ° 6 ° 7 Strongly Agree

Gender: ° Male ° Female

To what age group do you belong?

° 17-20 ° 21-25 ° 26-30 ° 31-35 ° 36-40 ° 41-45 ° 46-50 ° over 50

Which statement best describe your level of experience of using World Wide Web?

° I have never used the Web before this survey
° I have used the Web a few times before this survey
° I use the Web a few times a month
° I use the Web every week
° I use the Web almost every day

Aside from participating in this study, I have purchased online:

° never
° less than 5 times
° greater than or equal to 5, but less than 10 times
° greater than or equal to 10, but less than 15 times
° greater than or equal to 16, but less than 20 times
° greater than or equal to 20 times

I have heard about "Privacy Policy" on Internet before this experiment:

o Yes o No

I am aware that many companies in Taiwan have a privacy policies on their Web sites:

o Yes o No

I have clear understanding of what "Privacy Policy" is:

o Yes o No

When is your first time to learn about "Privacy Policies"?

o this time when I participate this experiment
o when I purchased from other Web sites before
o classroom
o from magazine or newspapers
o from Web site newsgroup
o my friend
o e-newspapers

When I visit a Web site, I do read the "Privacy Policy" very carefully

o Yes, I do read very carefully
o No, I only glance at it
o No, I don't read at all.

About the Authors

M. Gordon Hunter is currently an associate professor of information systems in the Faculty of Management, University of Lethbridge. Dr. Hunter has previously held academic positions at universities in Canada, Hong Kong, and Singapore. He has held visiting positions at universities in Australia, Monaco, Germany, the US, and New Zealand. He has a Bachelor of Commerce from the University of Saskatchewan in Saskatoon, Saskatchewan, Canada, and a doctorate from Strathclyde Business School, University of Strathclyde in Glasgow, Scotland. He has also obtained a Certified Management Accountant (CMA) designation from the Society of Management Accountants of Canada. He is a member of the British Computer Society and the Canadian Information Processing Society (CIPS), where he has obtained an Information Systems Professional (ISP) designation. Dr. Hunter chairs the Executive Board of the Information Institute, an information policy research organization. He has extensive experience as a systems analyst and manager in industry and government organizations in Canada. He is an associate editor of the *Journal of Global Information Management*. He is the Canadian World Representative for the Information Resource Management Association. He serves on the editorial board of the *Journal of Global Information Technology Management*, and the *Journal of Information Technology Cases and Application*. Dr. Hunter has published articles in *MIS Quarterly, Information Systems Research, Journal of Strategic Information Systems, Journal of Global Information Management, Information Systems Journal*, and *Information, Technology and People*. He has conducted seminar presentations in Canada, the US, Asia, New Zealand, Australia, and Europe. His current research interests relate to the productivity of systems analysts with emphasis on the personnel component, including cross-cultural aspects, the use of information systems by small business, and the effective development of information systems.

Felix B. Tan is professor of information systems and head of the School of Computer and Information Sciences at Auckland University of Technology, New Zealand. He serves as editor-in-chief of the *Journal of Global Information Management* and the director of the Centre for Research on Information Systems Management. He is also the vice president of Research for the Information Resources Management Association, the editor of the *ISWorld Net's EndNote Resources* page, and is the Region 3 (Asia Pacific)

Council for the Association for Information Systems. He has held visiting positions with the Department of Information Systems, National University of Singapore, and Richard Ivey School of Business, University of Western Ontario. Dr. Tan is internationally known for his work in the global IT field and is a frequent guest speaker at international conferences. His current research interests are in electronic commerce, global information management, business-IT alignment, and the management of IT. He is actively using cognitive mapping and narrative inquiry methods in his research. Dr. Tan has published in *MIS Quarterly*, *Information & Management*, *Journal of Information Technology*, and other journals, and has refereed conference proceedings. He has also published six books in the global IT field and is editor of the book series *Advanced Topics in Global Information Management*. He has more than 20 years experience in information systems management and consulting with large multinationals, as well as university teaching and research in Singapore, Canada, and New Zealand.

* * *

Irwin Brown is currently a senior lecturer in information systems at the University of Cape Town, South Africa. He graduated from the University of Zimbabwe with a BSc Engineering (Honors) (Electrical) degree, and from Curtin University of Technology, Australia, with a Master of Information Systems. His research interests are in the areas of technology adoption, systems success, and strategic information systems planning. He has published in several international journals, including the *International Journal of Information Management*, *Information Resources Management Journal*, *Journal of Global Information Management*, *Journal of Global Information Technology Management*, and *Electronic Journal of Information Systems in Developing Countries*.

Steven Tung-lung Chang is currently a professor of marketing and international business at the College of Management, Long Island University – C.W. Post (USA). He has centered his research on the global expansion strategy of multinational corporations. His research has been published by the *Journal of World Business*, *International Marketing Review*, *Journal of Global Marketing*, *Competitiveness Review*, and *International Finance Review*, among others. Dr. Chang is the recipient of a Fulbright Grant, Keller Grants, and a United Nations Development Program Grant. He formerly served as a member of the New Jersey State Export Finance Company Advisory Council.

Ja-Shen Chen is currently chairman and associate professor of the Business Administration Department at Yuan Ze University, Taiwan. He received his PhD in decision sciences from Rensselaer Polytechnic Institute, New York. His current research focus is the investigation of empirical studies, as well as modeling techniques on e-commerce and e-business management, especially in the areas of customer relationship management and supply chain management. He has published in a number of journals and conference papers in decision sciences and information management areas.

Russell K.H. Ching is a professor of MIS and chair of the Management Information Science Department at California State University, Sacramento, California (USA). Dr. Ching received his PhD from the University of Arkansas, Fayetteville. He is a member of the Association for Information Systems, INFORMS, Asia-Pacific Decision Sciences

Institute, and the International Consortium of Electronic Business. His publications have appeared in various journals and numerous conference proceedings.

Wing S. Chow is an associate professor of MIS in the School of Business, HK Baptist University, Hong Kong. He received his master's and doctoral degrees from the University of Manitoba. Dr. Chow published many papers in the journals and conference proceedings. His latest edited book is titled *Multimedia Information Systems in Practice* (Springer, 1999). His research interests are in the areas of manufacturing, supply chain information management, and business and electronic information systems.

Jan Damsgaard is professor in the Department of Informatics, Copenhagen Business School, Denmark. His research focuses on the diffusion and implementation of networked and standard-based technologies such as intranet, extranet, Internet portals, EDI, mobile technologies (GSM, GPRS, UMTS, WAP, 802.11x) and corporate information infrastructures. He presents his work at international conferences (ICIS, ECIS, PACIS, HICSS, IFIP 8.2, and 8.6) and in international journals (ISJ, JSIS, JGIM, JOCEC, ITP, CAIS, and JAIS). He serves on the advisory board of *Scandinavian Journal of Information Systems*, as an associate editor of *European Journal of Information Systems* and on the editorial board of *Information and Organization*.

Roberto Evaristo is on the faculty of the Information and Decision Sciences Department at the University of Illinois, Chicago (USA). He is currently involved in several projects related to the management of distributed projects, with work done in Japan, the US, and Europe. He has published in outlets such as *Communications of the ACM*, *International Journal of Project Management*, *Database*, *Journal of Global Information Management*, *European Management Journal*, *Human Systems Management*, *Journal of Organizational Computing and Electronic Commerce*, and others. He also serves on the editorial board of the *Journal of Global Information Management* and the *Journal of Global Information Technology Management*.

Jai Ganesh is a junior research associate with Infosys SET Labs, India. His research is on emerging areas such as Web services. He holds a PhD in information systems from the Indian Institute of Management Bangalore (IIMB). He is a graduate of Nagpur University where he completed his MBA in marketing. He earned a BSc in physics before that. His papers have been selected for leading peer-reviewed international conferences such as ICIS, ICEC, ICEB, and others. His papers have been published by IIMB Management Review, World Markets Research Centre, and others. He has received a research grant from NS Raghavan – GIV Centre for Entrepreneurial Learning. He is a reviewer for *Journal of Electronic Commerce in Organizations*, *International Journal of Electronic Business*, and conferences such as AMCIS and ICIS. He has also consulted for software firms in Bangalore.

Douglas Havelka is an assistant professor of management information systems in the Richard T. Farmer School of Business Administration at Miami University, USA. His education includes an AB in chemistry and an MBA, with a concentration in finance from Miami University, and an MSBA and a PhD in management information systems from

Texas Tech University. Dr. Havelka is a CPA, and prior to joining Miami in Fall 2000 was a project manager for industry standards at AT&T.

Rudi Hoppe graduated with a Bachelor of Business Science degree in information systems (Honors) from the University of Cape Town, South Africa (2001).

Xiaorui Hu is an assistant professor of management information systems at the John Cook School of Business, Saint Louis University. She received her PhD in economics from the University of Texas at Austin. Her research focuses on trust-related issues in electronic commerce, B2B markets, telecommunication market evolution, and culture impact on international business.

J.H. Huang is associate professor, deputy chair of Department of Management Science and Engineering, School of Economics and Management, Tsinghua University, China. She has conducted her research at Sloan School of MIT, University of Waterloo as a visiting scholar. In 1997, she spent one year at the University of Illinois as a Fulbright Scholar. Her main research interests include electronic business, decision support systems, and information systems analysis. Since 1996, more than 20 papers has been published in some journals and important international conferences, such as Computers & Industrial Engineering, IEEE International Conference on Systems, Man and Cybernetics, IEEE International Conference on Web Intelligence, World Multiconference on Systemics, Cybernetics, and Informatics. She has published seven books in China (Mainland and Taiwan).

Wayne Huang is an associate professor in the MIS Department, College of Business, Ohio University, USA. He has worked as a faculty member in universities in Australia, Singapore, and Hong Kong. His main research interests include group support systems (GSS), electronic commerce, e-learning, knowledge management systems, and software engineering. He has published papers in leading IS journals such as the *Journal of Management Information Systems* (*JMIS*); *IEEE Transactions on Systems, Man, and Cybernetics*; *Information & Management* (*I&M*); *IEEE Transactions on Professional Communication*; *Decision Support Systems* (*DSS*), and others. He is on the editorial board in *I & M, JGIM,* and *Journal of Data Management* (*JDM*).

P.D. Jose is an assistant professor of corporate strategy & policy at the Indian Institute of Management Bangalore, India. He completed his PhD at IIM Ahmedabad and his post-graduate work in forestry management (PGDFM) from the Indian Institute of Forest Management (IIFM), Bhopal. He received a BSc in physics prior to that. Jose is a recipient of the prestigious Fulbright Fellowship for 1999-2000 at the Massachusetts Institute of Technology and Kenan Flagler Business School. He has been a faculty member at the Administrative Staff College of India. He has written several academic papers and case studies in journals such as *Business Strategy and the Environment*, and the *Journal of Long Range Planning*. He has consulted for several public sector organizations and governments. He has also successfully completed projects for institutions such as the World Bank, among others.

Michelle L. Kaarst-Brown is an assistant professor at the School of Information Studies, Syracuse University (USA). Drawing upon two decades of business and information technology (IT) management, she now specializes in how social, cultural, and knowledge factors impact IT strategy and IT management. Her current research focuses on cross-cultural issues in global electronic commerce and firm/industry transitions in the e-business environment. Dr. Kaarst-Brown has published in a number of top academic and business journals including *MIS Quarterly*, *Information Technology and People*, *Journal of Strategic Information Systems*, *Journal of Organizational Change Management*, and *CIO Canada*.

Jan Karlsbjerg is an assistant professor in the Department of Computer Science, Aalborg University, Denmark. His research deals with the diffusion and implementation of networked and standard-based information technologies. The subject of his doctoral thesis is organizational information infrastructures and their emergence under the influence of network effects, market competition, and de facto standards. Jan has presented his work at local conferences (IRIS) as well as international conferences (ECIS, HICSS, and IFIP 8.4).

Chung Kim is a professor in the Department of Computer Information Systems at Southwest Missouri State University (USA). She received her PhD in MIS from Texas Tech University. Her current research interests include system development methodologies, business reengineering, and cultural issues in system development practices. She has published in numerous journals including the *Journal of Global Information Management*, *Information & Management*, *Journal of Global Information Technology Management*, and *Journal of Computer Information Systems*.

Seokha Koh is a professor of the Department of Management Information Systems at Chungbuk National University, South Korea. He received a PhD and Master of Science in Management Science from the Korea Advanced Institute of Science and Technology. Professor Koh is active in research and has published several articles that have appeared in *Information & Management*, *Operations Research Letters*, *Journal of Optimization Theory & Applications*, and *Journal of Computer Information Systems, Telecommunication Systems, and Industrial Management & Data Systems* among others. He was also a visiting scholar of Miami University, Ohio.

Catherine Ku, DBA, is an associate professor and chair of accounting at National Dong Hwa University of Taiwan. She received her Doctor of Business Administration from Mississippi State University (1994). Dr. Ku's research interests are in accounting information systems and electronic commerce.

Sooun Lee is a professor of the Department of Decision Sciences and Management Information Systems, Richard T. Farmer School of Business Administration at Miami University (USA). He received an MBA from the University of Colorado, Boulder, and a PhD in MIS from the University of Nebraska, Lincoln. Professor Lee has been teaching in Miami University for 17 years. His research interests are IS education; Web database, analysis, and design; networking; and project management. He has published a number of articles in several academic and practitioners' journals.

Hans Lehmann is a management professional with 25 years of business experience with international information technology. His work experience spans continental Europe, Africa, the UK, North America, and Australasia. He worked for 11 years as an international management consultant with Deloitte's, specializing in the development and implementation of international information systems for multinational companies in the financial and manufacturing sectors. In 1991, he changed careers and joined the University of Auckland, New Zealand, where his research focuses on the strategic management of international information technology, especially the emergence of transnational electronic business and commerce applications.

Peter Ping Li is a professor of management at California State University at Stanislaus (USA). He received his PhD in international business from the George Washington University. His primary academic interest lies in the development of holistic, dynamic, and paradoxical theories with regard to the impact of culture and change on strategic management of multinational corporations. In particular, his recent research focuses on reexamining the extant theories from the perspective of China and East Asia. Dr. Li has published in various academic journals, including *Organization Studies*, *Asia Pacific Journal of Management*, *Journal of Organizational Change Management*, *International Executive*, *Competitiveness Review*, *American Asian Review*, *International Financial Review*, *Advances in International Comparative Management*, *Advances in Chinese Industrial Studies*, and *Advances in Pacific Basin Business, Economics and Finance*.

John Lim is associate professor in the School of Computing at the National University of Singapore. Concurrently, he heads the Information Systems Research Lab. Dr. Lim graduated with First Class Honors in electrical engineering, an MSc in MIS from the National University of Singapore, and a PhD from the University of British Columbia. His current research interests include collaborative technology, negotiation support, IT and education, and IS implementation. He has published in MIS and related journals, including *Journal of Management Information Systems*, *Journal of Global Information Management*, *Decision Support Systems*, *International Journal of Human Computer Studies*, *Organizational Behavior and Human Decision Processes*, *Behavior and Information Technology*, *Journal of Database Management*, and *Small Group Research*. A major part of his current research plan addresses the multilingual and cross-cultural aspects of negotiation support.

Z.X. Lin is an assistant professor of information systems in the Rawls College of Business Administration at Texas Tech University (USA). He received an MEng in computer science from Tsinghua University in 1982, an MS in economics in 1996, and a PhD in information systems in 1999 from the University of Texas at Austin. His research interests include electronic commerce, network traffic management, information economics, and IT in China. He is currently working on several projects: online risk relief services, IT strategies for China, network resource allocation with the economic approach, and distance learning.

Chang Liu, DBA, is an associate professor of management information systems at Northern Illinois University (USA). He received his doctor of business administration from Mississippi State University in 1997. His research interests are electronic commerce, Internet computing, and telecommunications. His research works have published in *Information & Management*, *International Journal of Electronic Commerce and Business Media*, *Internet Research*, *Journal of Computer Information Systems*, *Mid-American Journal of Business*, *Communications of the International Information Management Association*, and *Journal Informatics Education Research*. Dr. Liu teaches database and electronic commerce courses.

T.R. Madanmohan is an associate professor of technology and operations management at the Indian Institute of Management Bangalore (IIMB), India. He also holds an adjunct research professor position at Eric Sprott School of Business, Carleton University, Canada. He received a BE in civil engineering from Gulbarga University in 1985, and a PhD in management studies from the Indian Institute of Science, Bangalore, in 1992. He subsequently worked at the Institute of Rural Management, Anand, and Management Development Institute, Gurgaon, as assistant professor. His teaching, training, and research have been in the areas of e-business, technology management, and service operations.

Jack T. Marchewka, PhD, is an associate professor and the Barsema professor of management information systems in the Department of Operations Management and Information Systems (OMIS) at Northern Illinois University, USA. In addition, he is the director of the Business Information Technology Transfer Center (BITTC). His current research interests include IT project management, electronic commerce, and knowledge management. His articles have appeared in the *Information Resources Management Journal*, *Information Technology and People*, *Journal Informatics Education Research*, *Communications of the International Information Management Association*, *Mid-American Journal of Business*, and *Journal of International Information Management*.

Robert M. Mason is professor of management information systems and Sprint professor of business administration at Florida State University. His undergraduate and master's degrees are in electrical engineering (MIT), and his PhD is in industrial and systems engineering (Georgia Tech). He has spent time in industry and managed his own consulting firm before returning full time to academia. He is a former president of the International Association for the Management of Technology (IAMOT) and serves as an associate editor for *Technovation*. His research interests include technology management, especially knowledge management and learning, and the cultural issues associated with information technology.

Pauline Mugera graduated from the University of Cape Town, South Africa, with a Bachelor of Commerce in Information Systems (Honors) (2001).

Paul Newman graduated with a Bachelor of Business Science in Information Systems (Honors) from the University of Cape Town, South Africa (2001).

David J. Pauleen is a senior lecturer and director of the MCA/Honors program in information systems at the School of Information Management at Victoria University of

Wellington, New Zealand. Current research interests include virtual teams and related issues, particularly information and communication technologies; intercultural communication and cultural biases in the development and use of communications technologies; team leadership and facilitation; and knowledge management of team processes and outcomes. His work has appeared in *Leadership and Organizational Development Journal* as well as *Journal of Information Technology and Internet Research - Electronic Networking Applications and Policy*. He is also the editor of the forthcoming book *Virtual Temas: Projects, Protocols and Processes* (Idea Group Inc.)

Norman E. Pence is professor of management science at Metropolitan State College of Denver, Colorado. He received his PhD from the Colorado School of Mines in Mineral Economics (operational research). He spent 10 years working in industry using quantitative and qualitative techniques to help solve business problems before pursuing a teaching career. For the past 26 years, he has taught management science and business statistics at MSCD. Some of the organizations he has had the opportunity to work for or consult for are as follows: A. E. Staley Manufacturing Company, McDonnell Automation Company, the Great Western Sugar Company, Great Western United Corporation, Great Western Cities, Shakey's Pizza Corporation, Coors Porcelain Company, United States Geological Survey, Air Force Accounting and Finance Center, and Time Design Software.

Dane K. Peterson is a professor of quantitative business analysis at Southwest Missouri State University. He has a PhD in quantitative and applied psychology from Southern Illinois University. His research interests include issues related to global information management and business and computer ethics. He has published in numerous journals including the *Journal of Applied Psychology*, *Journal of Global Information Management*, *Organizational Behavior and Human Decision Processes*, *International Journal of Information Management*, and *Journal of Business Ethics*.

László A. Poók is professor of MIS at Metropolitan State College of Denver, Colorado. He received his DBA and PhD from the University of Colorado in the field of information systems and management science. He has more than 33 years of experience in information systems and international business. A native of Central Europe, he has lived, worked, and taught in Europe, South America, and the US. He was an invited lecturer in Bolivia in 1992 and in South Africa, Hungary, and Romania in 2002-2003; a Soros Foundation award recipient in 1993 and 1994 to Hungary, Poland, and Slovakia; and a Fulbright Scholar in Hungary in 1995. He has presented his research findings at international conferences and has published in several international academic journals.

Rens Scheepers is a senior lecturer in the Department of Information Systems, University of Melbourne (Australia). His research focuses on the implementation and use of information technologies such as intranets, extranets, enterprise information portals, and mobile information technologies in organizations. He is also involved in research that examines the role of information technology in fostering knowledge exchange within communities of practice. He has presented his work at international conferences (ECIS, PACIS, IFIP 8.2) and in international journals (*ISJ*, *ITP*, *JAIS*, and *JIT*).

Sudhi Seshadri is a professor of marketing at the Indian Institute of Management Bangalore (IIMB). He holds a PhD in business administration from the Pennsylvania State University and a BTech in electrical engineering from the Indian Institute of Technology. For over a decade, he has worked in India and abroad in the areas of marketing management and marketing research. He has taught and conducted research in three graduate business schools for more than 10 years in the US and at IIMB. He has published more than a dozen refereed papers in leading international management journals including *Management Science*, *Marketing Science*, *Managerial & Decision Economics*, and *Industrial Marketing Management*. He is on the editorial board of the *Review of Marketing Science*. Dr. Seshadri has consulted for the public sector and has technical marketing and sales experience with major multidivisional engineering firms. He was also associated with the Indian School of Business, Hyderabad, as Associate Dean, Executive Programmes.

Nitish Singh is an associate professor of marketing at California State University, Chico (USA). He obtained his PhD from Saint Louis University. His research interests include international e-business strategy and cross-cultural consumer research. His recent and forthcoming publications include *Journal of International Business Studies*, *Journal of Business Research*, *Journal of E-Business*, *Journal of Web Site Promotion*, and several conference proceedings.

Adrie Stander is currently a senior lecturer in information systems at the University of Cape Town, South Africa. He obtained BSc degrees in information systems and psychology from the University of South Africa, and an MTech from Cape Technikon, South Africa. He has more than 20 years experience in the South African IT industry. He has a keen interest in the human side of IT and has done research in many related areas such as mobile usability, cultural aspects of user interfaces, and gender issues in the IT industry. He has spoken at conferences in Africa, Europe, and the US.

Chingning Wang is a doctoral student at the School of Information Studies at Syracuse University (USA). Having a background in communications and experience with various forms of media, her research interests incorporate information system issues at both the micro and macro levels. Her interests include cognitive-affective aspects of human-computer interaction, visual persuasion, issues of user behaviors and organizational behaviors, managing information technology-enabled change, and the impact of new media on marketing strategy. She has presented on human-computer interaction (HCI) studies at the Americas Conference on Information Systems and published in the *2002 Proceedings*.

David C. Yen is a professor of MIS and chair of the Department of Decision Sciences and Management Information Systems at Miami University (USA). He received a PhD in MIS and Master of Sciences in computer science from the University of Nebraska. He is active in research and has published two books and over 100 articles that have appeared in *Communications of the ACM*, *Decision Support Systems*, *Information & Management*, *International Journal of Information Management*, *Information Sciences*, *Journal of Computer Information Systems*, *Interfaces*, *Telematics and Informatics*, *Computer Standards and Interfaces*, *Information Society*, and *Internet Research*, among others.

He was also one of the co-recipients of a number of grants such as Cleveland Foundation (1987-1988), GE Foundation (1989), and Microsoft Foundation (1996-1997).

Hongxin Zhao is an associate professor of international business at Boeing Institute, Saint Louis University. He obtained his PhD from George Washington University and taught previously at the National University of Singapore. He is currently consulting for the Boeing Aircraft Company on global joint ventures. Dr. Zhao's research interests are in FDI and technology management. His studies have appeared in *International Business Review, Management International Review, Journal of International Markets, Asia Pacific Journal of Management, Thunderbird International Business Review*, and *Journal of Development Studies*.

Index